The Collected Courses of the Academy of European Law
Series Editors: Professor Marise Cremona
Professor Bruno de Witte, and
Professor Francesco Francioni,
European University Institute, Florence
Assistant Editor: Anny Bremner, *European University Institute, Florence*

VOLUME XVII/2

New Technologies and Human Rights

The Collected Courses of the Academy of European Law
Edited by Professor Marise Cremona,
Professor Bruno de Witte, and Professor Francesco Francioni
Assistant Editor: Anny Bremner

This series brings together the Collected Courses of the Academy of European Law in Florence. The Academy's mission is to produce scholarly analyses which are at the cutting edge of the two fields in which it works: European Union law and human rights law. A 'general course' is given each year in each field, by a distinguished scholar and/or practitioner, who either examines the field as a whole through a particular thematic, conceptual, or philosophical lens, or who looks at a particular theme in the context of the overall body of law in the field. The Academy also publishes each year a volume of collected essays with a specific theme in each of the two fields.

New Technologies and Human Rights

Edited by

THÉRÈSE MURPHY

OXFORD
UNIVERSITY PRESS

OXFORD
UNIVERSITY PRESS

Great Clarendon Street, Oxford OX2 6DP

Oxford University Press is a department of the University of Oxford.
It furthers the University's objective of excellence in research, scholarship,
and education by publishing worldwide in

Oxford New York

Auckland Cape Town Dar es Salaam Hong Kong Karachi
Kuala Lumpur Madrid Melbourne Mexico City Nairobi
New Delhi Shanghai Taipei Toronto

With offices in

Argentina Austria Brazil Chile Czech Republic France Greece
Guatemala Hungary Italy Japan South Korea Poland Portugal
Singapore Switzerland Thailand Turkey Ukraine Vietnam

Oxford is a registered trade mark of Oxford University Press
in the UK and in certain other countries

Published in the United States
by Oxford University Press Inc., New York

Reprinted 2010

ISBN 978-0-19-956257-2

Printed in the United Kingdom by
Lightning Source UK Ltd., Milton Keynes

Acknowledgements

The project *New Technologies and Human Rights* was conceived by the Directors of the Academy of European Law, in particular Professor Francesco Francioni. I am very grateful to them for inviting me to participate in it. I would also like to say thank you to Anny Bremner at the European University Institute, and Alex Flach and Lucy Page at Oxford University Press, for their assistance and good humour throughout the production stages of this book.

Thérèse Murphy,
Nottingham

Summary Contents

Table of Cases xiii
Table of Legislation xvii
List of Abbreviations xxv
List of Contributors xxvii

1. **Repetition, Revolution, and Resonance: An Introduction to *New Technologies and Human Rights*** 1
Thérèse Murphy
2. **Human Dignity, Ethical Pluralism, and the Regulation of Modern Biotechnologies** 19
Roger Brownsword
3. **Regulating Human Genetics in a Neo-Eugenic Era** 85
Han Somsen
4. **Constitutional Patriotism and the Right to Privacy: A Comparison of the European Court of Justice and the European Court of Human Rights** 128
Francesca Bignami
5. **New Technologies, the Precautionary Principle, and Public Participation** 161
Laurence Boisson de Chazournes
6. **The Texture of Reproductive Choice: Law, Ethnography, and Reproductive Technologies** 195
Thérèse Murphy
7. **The International Law of Genetic Discrimination: The Power of 'Never Again'** 222
Iulia Voina Motoc
8. **Individual Human Rights in Genetic Research: Blurring the Line between Collective and Individual Interests** 246
Hélène Boussard

Index 273

Contents

Table of Cases xiii
Table of Legislation xvii
List of Abbreviations xxv
List of Contributors xxvii

1. Repetition, Revolution, and Resonance: An Introduction to *New Technologies and Human Rights* 1
Thérèse Murphy

1. Repetition 7
2. Revolution 10
3. The Resonance of Human Rights 15

2. Human Dignity, Ethical Pluralism, and the Regulation of Modern Biotechnologies 19
Roger Brownsword

1. Ethical Pluralism, Human Dignity, and the UNESCO Regulatory Axiom 21
   A. The Nature of the Ethical Plurality and the Bioethical Triangle 22
   B. Human Dignity 26
   C. Taking Stock 32
2. Bioethical 'Floaters': Harm, Consent, and Precaution 33
   A. The Harm Principle 33
   B. Consent 35
   C. Precaution 37
   D. Taking Stock 39
3. Ethics as an External Benchmark 40
   A. Therapeutic Cloning and Human Embryonic Stem Cell Research 41
   B. PGD and PTT 50
4. Ethical Pluralism and Internal Regulatory Standards 60
   A. Ethics as an Explicit Regulatory Standard 61
   B. Ethics as an Implicit Regulatory Standard: The *Pro-Life Alliance* Case 71
5. The Revolutionary Sting 76
   A. The Chaplain's Concern 78
   B. Panopticon Technologies 79
   C. Exclusionary Technologies 81
6. Conclusion 83

3. Regulating Human Genetics in a Neo-Eugenic Era 85
Han Somsen

1. Human Genetics: The Mountain Comes to Mohammed 85
2. PGD and 'Neo-Eugenics' 88
   A. PGD: Applications, Clinical Risks, and Ethical Dilemmas 88
   B. PGD: A Eugenic Project? 91
   C. Interim Summary 93
3. Regulation of Technology, Deliberative Democracy, and Cosmopolitan Legitimacy 94
   A. International Judicial Review and Domestic Deliberative Democracies: A Legitimacy Paradox 94
   B. Cosmopolitan Legitimacy of Risk Regulation: An Alternative Perspective to the Dilemma 95
   C. Lessons for Regulating PGD 102
4. The Bioethical Triangle 103
   A. Utilitarianism, Human Rights, and Human Dignity 103
   B. Human Rights as Meta-Norm for the Regulation of PGD 108
5. The Minister's Opinion Reviewed 116
   A. Brief History 116
   B. The Importance of Transparency 116
   C. The Ethical Plurality Employed by the Minister 117
   D. Answers to the Questions Regarding Availability of PGD 119
   E. The Opinion Critically Reviewed 125
6. Conclusions and Final Considerations 126

4. Constitutional Patriotism and the Right to Privacy: A Comparison of the European Court of Justice and the European Court of Human Rights 128
Francesca Bignami

1. Constitutional Patriotism 131
2. The Right to Privacy 134
   A. Council of Europe 135
   B. The European Union 139
3. The European Constitutional Culture of Privacy 143
   A. *Criminal Proceedings against Bodil Lindqvist* 144
   B. *Von Hannover v Germany (Princess of Monaco)* 151
4. Tolerant and Intolerant Constitutional Patriotism 155
5. Conclusion 159

5. New Technologies, the Precautionary Principle, and Public Participation 161
Laurence Boisson de Chazournes

1. Risks and International Law: The Virtues of Precaution 163
   A. Precaution and International Law 163
   B. Precaution: A Means for Law to Address New Concerns 166
   C. The Notion of Precaution: Identifying the Criteria of Definition 168
2. The Precautionary Principle and Public Participation 178
   A. Access to Information, Public Participation, and the Notion of Risk in the Context of the Cartagena Protocol on Biosafety and of the *Codex Alimentarius* 179
   B. The Aarhus Convention and Genetically Modified Organisms (GMOs) 184
   C. Practice in Litigation Fora 187
   D. The Issue of Democratic Governance of a Society Facing New Technologies and their Use 190
3. Concluding Remarks: Understanding Precaution through its Societal Dimension 191

6. The Texture of Reproductive Choice: Law, Ethnography, and Reproductive Technologies 195
Thérèse Murphy

1. A Bad Argument? 196
   A. Choked by Controversy 197
   B. Reproductive Choice Does Not Seem to Fit the 'Facts' of Assisted Reproductive Technologies 202
2. Securing the Speakability of Reproductive Choice 213
   A. Reproductive Choice 'in Action' 213
   B. Is PGD Different? 216
   C. The Challenge of 'Speakability' 218
3. Conclusion 220

7. The International Law of Genetic Discrimination: The Power of 'Never Again' 222
Iulia Voina Motoc

1. The Definition of Genetic Discrimination 224
   A. International Legal Instruments 225
2. The Challenge of Privacy in the Genetic Context 231

3. From International Law to Domestic Implementation: Insurance, Employment, and Criminal Investigation 233
A. Insurance and Employment 233
B. Criminal Investigation 237
4. Vulnerable Groups 241
5. Conclusion 244

8. Individual Human Rights in Genetic Research: Blurring the Line between Collective and Individual Interests 246
Hélène Boussard
1. Candidate Human Rights in Genetics 248
A. The Protection of Genetic Identity as New Rationale for Human Rights 249
B. Protection of the Individual *Per Se*: The Right to Genetic Privacy 252
C. Protection of Genetic Diversity: Collectivization of the Human Genome 256
2. Legal Constructions to Protect the Interests of Genetic Relatives 259
A. Disclosure of the Results of Genetic Tests: Protection of Family Members 261
B. The Collectivization of Entitlements in Epidemiological Research: Protection of the Group 265
3. Conclusion 270

Index 273

Table of Cases

COURT OF FIRST INSTANCE OF THE EUROPEAN COMMUNITIES

Pfizer Animal Health SA v Council of the European Union, Case T-13/99 [2002] ECR II–3305 . . . . 38, 95, 101, 171, 172, 191

EUROPEAN COURT OF JUSTICE

Cassis de Dijon (Rewe-Zentral AG v Bundesmonopolverwaltung für Branntwein), Case 120/78 [1979] ECR 649 96
Criminal Proceedings against Bodil Lindqvist, Case C-101/01 [2003] ECR I–12971 . . . . 130, 144, 149, 150, 155, 156, 158, 159, 160
Criminal Proceedings against Keck & Mithoard, Cases C-247/91 and C-268/91 [1993] ECR I–6097 148
European Parliament v Council and Commission, Cases C-317/04 and C-318/04 [2006] ECR I–4721 136
Monsanto Agricultura Italia v Presidenza del Consiglio dei Ministri, Case C-236/01 [2003] ECR I–8105 101
Netherlands v Council, Case C–377/98 [2001] ECR I-7079 . . . . 62, 105
Omega Spielhallen-und Automatenaufstellungs-GmbH v Oberbürgermeisterin der Bundesstadt Bonn, Case C–36/02 [2004] ECR I–9069 . . . . 31, 37, 67, 148, 149, 150, 151
Ordre des barreaux francophone et germanophone and Others, Case C-305/05 [2007] ECR I–5305 139
Portugal v Commission, Case C–149/96 [1999] ECR I-8395 101
Stoke-on-Trent v B & Q plc, Case C–169/91 [1991] ECR I-6635 148
Torfaen BC v B & Q plc, Case C–145/88 [1989] ECR 3851 . . . . 148, 149

EUROPEAN COURT OF HUMAN RIGHTS

Amann v Switzerland, 27798/95 [2000] ECHR 88 (16 February 2000) 136
Evans v United Kingdom, 6339/05 [2006] ECHR 200 (7 March 2006); [2007] ECHR 264 (10 April 2007) [GC] 11, 24, 207, 208
Guerra and Others v Italy, 14967/89 [1998] ECHR 7 (19 February 1998) 190
Handyside v United Kingdom, 5493/72 [1976] ECHR 5 (7 December 1976) 152
Lopez Ostra v Spain, 16798/90 [1994] ECHR 46 (9 December 1994) 190
Odievre v France, 42326/98 [2003] ECHR 86 (13 February 2003) 253
Segerstedt-Wiberg and Others v Sweden, 62332/00 [2006] ECHR 597 (6 June 2006) 136
Vo v France, 53924/00 [2004] ECHR 326 (8 July 2004) 24, 105, 115
Von Hannover v Germany (Princess of Monaco), 59320/00 [2004] ECHR 294 (24 June 2004) 130, 131, 136, 152, 154, 155, 156, 159, 160

EUROPEAN PATENT OFFICE

Harvard Onco-Mouse, OJ EPO 10/1992, 590 . . . . 23, 44, 63
HOWARD FLOREY/Relaxin [1995] EPOR 541 64
LELAND STANFORD/Modified Animal [2002] EPOR 2 63

INTERNATIONAL COURT OF JUSTICE

Gabčíkovo-Nagymaros Project (Hungary/Slovakia), Judgment, ICJ Rep (1997) 7 162
Legal Consequences for States of the Continued Presence of South Africa in Namibia (South West Africa) notwithstanding Security Council Resolution 276, Advisory Opinion, ICJ Rep (1971) 16 268

WORLD TRADE ORGANIZATION

Australia – Measures Affecting the Importation of Salmon, WT/DS18/AB/R, 20 October 1998 177
European Communities – Measures Affecting Asbestos and Asbestos-containing Products, Report of the Appellate Body, WT/DS135/R, 18 September 2000 and WT/DS135/AB/R, 12 March 2001 . . . . 177, 181
European Communities – Measures Affecting the Approval and Marketing of Biotech Products, WT/DS291/R, WT/DS292/R, WT/DS293/R, 29 September 2006 . . . . 71, 93, 101, 102, 171, 176, 177, 192, 194, 198
European Communities – Measures Concerning Meat and Meat Products (Hormones), WT/DS26/AB/R, WT/DS48/AB/R, 16 January 1998 . . . . 69, 100, 169, 176, 177, 178, 180
Japan – Measures Affecting the Importation of Apples, WT/DS245/AB/R, 26 November 2003 . . . . 102, 178
United States – Measures Affecting the Cross-Border Supply of Gambling and Betting Services, WT/DS285/AB/R, 7 April 2005 66

OTHER INTERNATIONAL COURTS AND TRIBUNALS

ICSID, Técnicas Medioamnientales Tecmed, SA v United Mexican States, Case No ARB (AF)/00/2 . . . . 189, 190
ITLOS, The MOX Plant Case (Ireland v United Kingdom), Case No 10, Request for Provisional Measures, Order of 3 December 2001 187
Permanent Court of Arbitration, The MOX Plant Case (Ireland v United Kingdom), Order No 3, 24 June 2003 187
Permanent Court of Arbitration, Dispute Concerning Access to Information under Article 9 of the OSPAR Convention (Ireland v United Kingdom), Final Award, 2 July 2003 187

NATIONAL COURTS

Costa Rica

Ana Victoria Sanchez, Villalobos and Others v Costa Rica, Decision No 2000-02306 (Constitutional Supreme Court) 199

France

Conseil d'Etat (27 October 1995) req no 136–727 (Commune de Morsang-sur-Orge) 31
Conseil d'Etat (27 October 1995) req no 143-578 (Ville d'Aix-en-Provence) 31

United Kingdom

Airedale NHS Trust v Bland [1993] 1 All ER 821 23
Leeds Teaching Hospitals NHS Trust v A [2003] 1 FLR 1091 58
McFarlane v Tayside Health Board [2000] 2 AC 59 72

Naomi Campbell v MGN Limited [2004] UKHL 22 156
Natallie Evans v Amicus Healthcare Ltd and Others [2003] EWHC 2161 (Fam); [2004] EWCA Civ 727 208
R v Chief Constable of South Yorkshire, ex parte LS and Marper [2002] EWCA Civ 1275; [2004] UKHL 39 77
R v Human Fertilisation and Embryology Authority, ex parte Blood [1997] 2 All ER 687 . . . . 201, 206
R v Kelly [1998] 3 All ER 741 49
R (Quintavalle on behalf of Comment on Reproductive Ethics) v Human Fertilisation and Embryology Authority [2002] EWHC 2785 (Admin); [2003] EWCA Civ 667; [2005] UKHL 28 . . . . 51, 55, 59, 60, 61, 71, 211
R v Secretary of State for Health, ex parte Quintavalle (on behalf of Pro-Life Alliance) [2001] EWHC 918 (Admin); [2002] EWCA Civ 29; [2003] UKHL 13 . . . . . . 72, 73, 74, 75, 76
Royal College of Nursing of the United Kingdom v Department of Health and Social Security [1981] AC 800 75

United States

Gonzales v Carhart 550 US 124 (2007) 218
Safer v Pack 677 A 2d 1188 264
Webster v Reproductive Health Services 492 US 490 (1989) 203

Table of Legislation

INTERNATIONAL LEGISLATION AND INSTRUMENTS

Aarhus Convention on Access to Information, Public Participation in Decision-Making and Access to Justice in Environmental Matters 1998 . . . . 184, 185, 186, 187
Art 1 185
Art 2, para 3 185
Art 3, para 5 186
Art 4 185
para 1 185
para 4 185
Art 6 185
para 7 185
para 11 186
Art 6*bis* 186
para 1 186
Annex I 185
Annex I*bis* 186
African Charter on Human and Peoples' Rights 1981 268
Agreement on Sanitary and Phytosanitary Measures 1994 . . . . 96, 97, 98, 99, 101, 109, 171, 176
Art 2.2 . . . . 97, 98, 177
Art 5 171
Art 5.1 . . . . 98, 99, 101, 102, 169, 171, 176, 180, 181
Art 5.2 . . . . 171, 180
Art 5.3 171
Art 5.7 . . . . 98, 101, 102, 171, 176, 177, 178
Annex A(1) 97
Annex A(4) 99, 101, 102
Annex C(1)(a) 99
American Convention on Human Rights 1969 199
American Declaration of the Rights and Duties of Man 1948 198
Convention for the Protection of the Marine Environment and the Coastal Region of the Mediterranean 1976 175
Convention for the Protection of the Marine Environment of the North-East Atlantic 1992 175
Convention on Biological Diversity 1992 164
Preamble 164, 166
Cartagena Protocol on Biosafety 2000 . . . . 165, 169, 170, 171, 175, 181, 182, 186
Preamble 165
Art 20 182
Art 21, para 1 182
Art 23 182
Art 26 182
para 2 183

Annex III 169, 170, 182
para 3, 4 169
para 5, 6 170
Convention on the Protection and Use of Transboundary Watercourses and International Lakes 1992 174
Convention on the Rights of the Child 1989 199
Convention on the Settlement of Investment Disputes between States and Nationals of other States 1965 189
Council of Europe Convention for the Protection of Individuals with Regard to Automatic Processing of Personal Data 1980 . . . . . 135, 136, 137, 140, 142, 159
Art 5a 137
Art 5b 137, 138
Art 5c 137, 138
Art 5d 137, 138
Art 5e 137, 138
Art 6 137
Art 7 137
Art 9 139
Additional Protocol regarding supervisory authorities and transborder data flows 2001 140
Council of Europe Convention on Human Rights and Biomedicine 1997 . . . . . 28, 48, 52, 107, 227, 228, 246, 247, 249
Art 1 252
Art 2 257
Art 10 . . . . . 120, 254
Art 11 227
Art 12 227
Art 14 242
Art 18 48
(2) 43, 44
Additional Protocol on the Prohibition of Cloning Human Beings 1998 247
Art 1 257
Additional Protocol on Transplantation 2002 247
Additional Protocol concerning Biomedical Research 2005 247
European Convention on Human Rights 1950 . . . . . 24, 62, 117, 135, 139, 151, 152, 154, 159, 227
Art 2 . . . . . 24, 208
Art 8 . . . . . 135, 136, 151, 154, 190, 207, 208, 253
(1), (2) 52
Art 10 . . . . . 151, 190
Art 12 208
Art 14 . . . . . 208, 209
General Agreement on Tariffs and Trade (GATT) 1994
Art XX 177
(a) 65, 66, 98
General Agreement on Trade in Services (GATS) 1994
Art XIV 67
(a) 65, 66, 67, 98
Art XVI 66

ILO Convention No 169 concerning Indigenous and Tribal Peoples in Independent Countries 1989 266
International Covenant on Civil and Political Rights 1966 . . . . . 26, 199, 223
Preamble 26
Art 1 268
(1) 268
Art 7 . . . . . 226, 249
International Covenant on Economic, Social and Cultural Rights 1966 . . . . . 26, 223
Preamble 26
Art 1 268
(1) 268
Art 7 248
Art 15 . . . . . 223, 248
(1), (2) 216
International Declaration on Human Genetic Data 2003 . . . . . 1, 21, 28, 224, 228, 232, 247
Art 1 106
(c) 228, 232, 241
Art 2(i) 228
Art 3 . . . . . 229, 250, 257
Art 4(ii) 261
Art 5 228
Art 6 245
Art 7 229
Arts 8–12 241
Art 8 232
Art 9 254
Art 10 120
Art 12 241
Arts 13–15 241
Art 14 233
(a), (b), (c), (d), (e) 233
Art 15 233
Arts 16–19 241
Arts 20–22 241
Art 21(b), (c) 241
Art 23 224
Rio Declaration on Environment and Development 1992 . . . . . 163, 175
Principle 10 179
Principle 15 . . . . . 163, 174
UNCITRAL Arbitration Rules 1976 188
UN Framework Convention on Climate Change 1992 . . . . . 164, 174
Preamble 164
UNESCO Recommendation on the Status of Scientific Researchers 1974
Art 8 248
UNGA Declaration on the Use of Scientific and Technological Progress in the Interests of Peace and for the Benefit of Mankind 1974 248
United Nations Convention on the Law of the Sea 1982
Art 123 187

Art 287 187
Annex VII 187
United Nations Convention on the Rights of Persons with Disabilities 2006 243
Art 17 243
United Nations Declaration on Human Cloning 2005 . . . . 2, 207
United Nations Declaration on the Rights of Disabled Persons 1975 . . . . 242, 243
Universal Declaration on Bioethics and Human Rights 2005 . . . . 1, 21, 22, 28, 229, 242, 247
Art 2(d) 2, 21
Art 3
(1) 21
(2) 23
Art 4 23
Art 6 35
(3) 270
Art 7 242
Art 12 21
Art 16 38
Art 20 38
Universal Declaration of Human Rights 1948 . . . . 26, 247
Preamble 26, 223, 226
Art 1 . . . . 27, 106, 223
Art 5 225
Art 27 248
Universal Declaration on the Human Genome and Human Rights 1997 . . . . 1, 21, 28, 107, 223, 224, 226, 228, 246, 247, 261
Preamble 226, 227
Art 1 . . . . 226, 252
Art 2 . . . . 223, 227, 252
Art 2b 252
Art 5 . . . . 120, 227, 252, 256
Art 5a 249
Art 5b 255
Art 5c 252, 253
Art 6 . . . . 227, 255, 258
Art 7 . . . . 227, 252, 255, 256
Art 9 261
Art 10 . . . . 223, 226, 227, 261
Art 11 . . . . 223, 227, 256, 257
Art 12 227
Arts 17–19 223
Art 17 227
Art 18 227
Art 19 227
Vienna Convention for the Protection of the Ozone Layer 1985 164
Preamble 164
WTO Dispute Settlement Understanding 1994
Art 23 97
(1) 97

EUROPEAN UNION

Charter of Fundamental Rights of the European Union
2000 . . . . 139, 159
Art 3 232
Art 3.2 91
Art 7 139
Art 8 139
Directive 90/220/EEC on the deliberate release into the environment of genetically modified organisms, OJ L 117/15 99
Directive 95/46/EC on the protection of individuals with regard to the processing of personal data, OJ 1995 L 281 . . . . 140, 141, 142, 144, 145, 146, 147, 159
Art 6 140
Art 7 140
(f) 140
Art 8 140
Art 9 141
Art 10 141
Art 11 141
Art 12 . . . . 140, 141
Art 13 141
Art 15 141
Art 17 140
Arts 18–120 141
Art 25 141
Art 26 141
Art 28 141
Directive 98/44/EC on the legal protection of biotechnological inventions, OJ 1998 L 213/13 . . . . 28, 61, 62, 64
Recital 26 . . . . 62, 105
Art 6(2) 107
(a), (c) 63
Directive 2001/18/EC on the deliberate release into the environment of genetically modified organisms, OJ 2001 L 106/1 99
Art 2(2) 87
EC Treaty (EEC Treaty/Treaty of Rome) 1957 . . . . 97, 148, 149
Art 28 148
Art 130r(2) 173
Art 155 95
European Patent Convention 1973 63
Art 53(a) 23, 61, 63
Europol Convention 1995 142
Regulation 258/97/EC concerning novel foods and novel food ingredients, OJ 1997 L 43/1 99
Regulation 515/97/EC on mutual assistance between the administrative authorities of the Member States and cooperation between the latter and the Commission, OJ 1997 L 82/1 142

Regulation 2821/98/EC amending, as regards withdrawal of the authorization of certain antibiotics, Directive 70/524/EEC concerning additives in feedingstuffs, OJ 1998 L 351/4 . . . 172
Regulation 1829/2003/EC on genetically modified food and feed, OJ 2003 L 268/1 . . . 99
Art 2(1) . . . 97
Art 7(1) . . . 97
Regulation 1830/2003/EC concerning the traceability and labelling of genetically modified organisms and the traceability of food and feed products produced from genetically modified organisms and amending Directive 2001/18/EC, OJ 2003 L 268/24 . . . 99
Treaty on European Union (Maastricht Treaty) 1992
Title V . . . 145
Title VI . . . 145

NATIONAL LEGISLATION AND REGULATIONS

France

Acts on bioethics 1994 . . . 28
Civil Code . . . 235
Art 16-10 . . . 235
Code of Public Health
Art 1131 . . . 264
Law on respect for the human body 1994 . . . 235
Basic Law . . . 131, 132

Netherlands

Embryo Act . . . 124
Statute on Medical Examinations 1997
Art 3 . . . 234

United Kingdom

Abortion Act 1967 . . . 199
Human Fertilisation and Embryology Act 1990 . . . 48, 51, 56, 60, 72, 75, 202, 207, 208, 209
s 1(1) . . . 72, 73, 74
(a) . . . 73
s 2(1) . . . 56
s 3(3)
(c) . . . 59
(d) . . . 72, 73, 74, 76
s 13(5) . . . 202
Sch 2
para 1 . . . 56
para 3(2), (6) . . . 43
Human Fertilisation and Embryology Authority (Disclosure of Donor Information) Regulations 2004 . . . 205
Human Fertilisation and Embryology (Deceased Fathers) Act 2003 . . . 205
Human Fertilisation and Embryology (Research Purposes) Regulations 2001 . . . 43
Human Reproductive Cloning Act 2001 . . . 48, 73
Human Rights Act 1998 . . . 52, 207

United States

Americans with Disabilities Act 1990 236
Clean Air Act 1990 31
Clean Water Act 1977 . . . . 31, 188
Genetic Information Nondiscrimination Act 2008 235
Health Insurance Portability and Accountability Act 1996 . . . . 231, 235, 237
Illegal Gambling Business Act 1970 66
Interstate Horseracing Act 1978 67
Travel Act 1961 66
Violent Crime Control and Law Enforcement Act 1994 . . . . 238, 239
Wire Act 1961 66

List of Abbreviations

AB	Appellate Body
AIA	Advance Informed Agreement
ART	assisted reproductive technology
CFI	Court of First Instance
CIOMS	Council for International Organizations of Medical Science
CNR	cell-nuclear replacement
ECHR	European Convention on Human Rights
ECJ	European Court of Justice
ECtHR	European Court of Human Rights
EPO	European Patent Office
EU	European Union
GMO	genetically modified organism
HFEA	Human Fertilisation and Embryology Authority
HFEA 1990	Human Fertilisation and Embryology Act 1990
HIPAA	Health Insurance Portability and Accountability Act 1996 (US)
ICCPR	International Covenant on Civil and Political Rights
ICESCR	International Covenant on Economic, Social and Cultural Rights
ICSI	intracytoplasmic sperm injection
IDHGD	International Declaration on Human Genetic Data
IVF	*in vitro* fertilization
MOX	mixed oxide
MTBE	methyl tertiary butyl ether
PGD	pre-implantation genetic diagnosis
PGS	pre-implantation genetic screening
PTT	pre-implantation tissue-typing
UDBHR	Universal Declaration on Bioethics and Human Rights
UDHGHR	Universal Declaration on the Human Genome and Human Rights
UDHR	Universal Declaration of Human Rights
UNESCO	United Nations Educational, Scientific and Cultural Organization
WMA	World Medical Assembly
WTO	World Trade Organization

List of Contributors

Francesca Bignami is Professor at the George Washington University Law School. Her areas of expertise include comparative privacy law and European administrative law. She is the author of numerous articles on comparative privacy law, comparative administrative law, and rights and accountability in European governance. In 2006–2007, Professor Bignami was a visiting professor at Harvard Law School. She has also taught at the European University Institute in Florence (Academy of European Law), at the Boston College Law School, in the Masters Program on Public Administration at the University of Rome *'La Sapienza'* and at the *Libera Università Internazionale degli Studi Sociali* in Rome, Italy. She has been a Visiting Fellow at the Center for European Studies, Harvard University, and the Jean Monnet Center at the NYU School of Law. Her research has been supported by, among others, the German Marshall Fund and the Fulbright Program.

Laurence Boisson de Chazournes is Professor of international law and Head of the Department of Public International Law and International Organization at the Faculty of Law of the University of Geneva (Switzerland). Between 1995 and 1999, she was Senior Counsel with the Legal Department of the World Bank. A consultant and a member of groups of experts with various international organizations, including the World Bank, WHO, UNDP, and ILO, she is the author of a large number of publications. Recent publications include: *Protection internationale de l'environnement* (in collaboration with R. Desgagne, M. Mbengue, and C. Romano) (Paris: Pedone, 2005); *Freshwater and International Economic Law* (with E. Brown Weiss and N. Bernasconi-Osterwalder (eds)) (Oxford: OUP, 2005); *Une société internationale en mutation: quels acteurs pour une nouvelle gouvernance?* (with R. Mehdi (eds)) (Bruxelles: Bruylant, 2005).

Hélène Boussard was appointed as Scientific Officer in the Bamako 2008 Secretariat, in February 2008. She is in charge of coordination between the WHO, UNESCO, the World Bank, the Council on Health Research for Development, and the Global Forum for Health Research for the organization of the 2008 Global Forum on Research for Health (Bamako, 2008). She received her doctorate in law from the European University Institute in 2008. She is the author of various articles on bioethics and international human rights law in the field of human genetic research.

Roger Brownsword, who is a graduate of the London School of Economics, is Professor of Law at King's College London where he is Director of TELOS (the KCL centre for the study of technology, ethics, and law in society); he is also an Honorary Professor in Law at the University of Sheffield. Professor Brownsword has written on a wide range of topics in law and ethics, including co-authoring *Human Dignity in Bioethics and Biolaw* (2001) and *Consent in the Law* (2007) (both with Deryck Beyleveld). His most recent book, *Rights, Regulation, and the Technological Revolution*, was published by OUP in Spring 2008.

Iulia Motoc studied law and philosophy. She has an LLB from the University of Bucharest; a DEA and a Doctorate in Law from the University Aix-Marseille III (1996) '*L'interprétation des exceptions de l'article 2 & 4 de la Charte ONU par le Conseil de* *sécurité*'; an Habilitation à diriger des recherches (1998); and a Doctorate in Philosophy (1999) 'The theory of international relations: the sources of political and moral philosophy'. She is currently Professor of International Law at the University of Bucharest. Professor Motoc holds visiting academic positions at NYU School of Law and Yale University School of Law. She is the author of three books and co-author of a textbook on international law, as well as articles in the fields of theory of international law, the use of force, human rights, and the ethics of international relations. Before joining academia, she was a judge at the Bucharest Court. From 2001–2004, Professor Motoc was UN Special Rapporteur on the human rights situation in the Democratic Republic of Congo. She has also been a member and former President of the Sub-commission on Human Rights, and a member of the Council of Europe's Advisory Committee on National Minorities. She is currently UN Special Rapporteur on Human Rights and the Human Genome. She is also a member of the UN Human Rights Committee.

Thérèse Murphy is Professor of Law & Critical Theory at the University of Nottingham. She is the author of *Civil Liberties Law: The Human Rights Act Era* (OUP, 2001) (with Stephen Livingstone and Noel Whitty) and *Health and Human Rights* (forthcoming, Hart Publishing). Her research in the field of health and human rights has been supported by, among others, the Fulbright Programme, the Jean Monnet Fellowship scheme at the European University Institute, and the UK's Arts and Humanities Research Council. She is on the editorial board of the Human Rights Law Review, and is a member of the expert team that provides thematic and annual reports on the UK for the EU's Fundamental Rights Agency.

Han Somsen is professor and director at Tilburg Institute for Law, Technology and Society (TILT), a multi-disciplinary research centre that focuses on the regulation of new technologies. He is founder and editor of *The Yearbook of European Environmental Law* (Oxford University Press). His current research concentrates on medical and agricultural biotechnology. Recent publications include 'Cloning Trojan Horses' in R. Brownsword and K. Yeung, *Regulating Technologies* (2008) and *The Regulatory Challenge of Biotechnology* (2006).

1

Repetition, Revolution, and Resonance:

An Introduction to *New Technologies and Human Rights*

Thérèse Murphy

'And'. It's such a little word but, in international legal circles where linkage is *de rigueur*, it is almost ridiculously popular. The 'trade and' (or 'trade linkage') literature is possibly the premier example of this,[1] but it does not stand alone. There is also, for instance, a growing literature on regulation and rights,[2] health and rights,[3] security and rights,[4] and risk and rights.[5] Is it the case, then, that 'new technologies and human rights', the focus of this book, is merely an attempt to tag along or be in fashion by invoking the apparent power of that one little word, 'and'? An even more cynical view might suggest that the linkage is an attempt by either the champions of new technologies or human rights advocates, or indeed by both of these groups, to engage the appeal of its counterpart and thereby help staunch, or at least deflect from, its own alleged shortcomings.

A more neutral view on the linkage is also available, however. This view would point to the flurry of international instruments on aspects of science and technology, many of which have been championed by the United Nations Educational, Scientific and Cultural Organization (UNESCO), citing them as obvious sites of study for human rights lawyers. These instruments include the Universal Declaration on the Human Genome and Human Rights (1997), the International Declaration on Human Genetic Data (2003), and the Universal Declaration on

[1] For a review of this literature and commentary on its discursive structures, see Lang, 'Reflecting on "Linkage": Cognitive and Institutional Change in the International Trading System' (2007) 70 MLR 523.

[2] See eg B. Morgan (ed), *The Intersection of Rights and Regulation: New Directions in Sociolegal Scholarship* (2007).

[3] See eg R.J. Cook and C.G. Ngwena (eds), *Health and Human Rights* (2007); S. Gruskin *et al* (eds) *Perspectives on Health and Human Rights* (2005); T. Murphy, *Health and Human Rights* (forthcoming).

[4] See eg B.J. Goold and L. Lazarus (eds), *Security and Human Rights* (2007); I. Loader and N. Walker, *Civilizing Security* (2007).

[5] See eg Murphy and Whitty, 'Risk and Rights in UK Prison Governance' (2007) 47 British Journal of Criminology 798.

Bioethics and Human Rights (2005). All three provide clear evidence of the importance of human rights as a jumping-off point in the evolving body of international law and practice on the regulation of new technologies.

These instruments also provide human rights advocates with a second, far more compelling reason for exploring the linkages between new technologies and human rights. That reason, which is explained in detail in Roger Brownsword's essay in this book, is as follows: human rights may be a point of departure in the emerging international regulatory framework, but it does not stand alone—rather, it is one part of an ethical plurality, sharing the regulatory terrain with both a dignitarian ethic and a utilitarian one. The three points of this plurality, or 'bioethical triangle', are readily apparent from even the most cursory examination of UNESCO's clutch of biotechnology instruments. Take, for example, the Declaration on Bioethics and Human Rights, wherein Article 2(d) states that one of the aims of the Declaration is 'to recognize the importance of freedom of scientific research and the benefits derived from scientific and technological developments, while stressing the need for such research and development to occur within the framework of ethical principles set out in this Declaration and to respect human, dignity, human rights and fundamental freedoms'. What is not so apparent, however, is the problem that lurks within this regulatory framework: put bluntly, the three ethics can lead to entirely different outcomes, which would seem to make the plurality rather less than viable over the longer term. Occasionally, of course, as appears to be the case at present with human reproductive cloning, the plurality will be able to converge around a consensus.[6] More often however, as Han Somsen notes in his essay in this book, 'choices will have to be made.'[7] And, as he goes on to note, the crucial point for human rights advocates is that '[u]sing the human rights perspective is not a matter of course: there are alternative meta-regimes out there, and each has its advocates.'[8]

If further grounds are needed to justify the technology/rights linkage that drives this book, we can turn to the individual 'cultural lives'[9] of human rights and new technologies. Here we find another reason for human rights advocates to engage with the relationship between new technologies and human rights: namely, a set of shared characteristics. To my mind, hope stands out as amongst the most prominent of these characteristics. Consider, first, that technologies such as *in vitro* fertilization (IVF) and stem cell science are portrayed—and also widely seen— as sources of great, unprecedented hope. Even the US President's Council

[6] It is important, though, to emphasize that the consensus concerns human reproductive cloning, not human cloning in general. Not surprisingly, then, the UN Declaration on Human Cloning (2005), which prohibits all forms of human cloning inasmuch as they are incompatible with human dignity and the protection of human life, achieved only 84 votes when it secured acceptance at the UN General Assembly; 34 members were against it and there were 37 abstentions.

[7] Somsen at 114. [8] *Ibid*, at 103.

[9] The term is taken from Loader, 'The Cultural Lives of Security and Rights', in Goold and Lazarus (eds), *supra* note 4, at 27.

on Bioethics, a body widely associated with a 'red light' or prohibition-perspective on certain biotechnology matters, speaks of entering upon a 'golden age for biology, medicine, and biotechnology.'[10] The evidence that human rights link to hope is also plentiful. Indeed, faith in rights—specifically, faith in rights as trumps over the everyday ups and downs of politics, with an ability to generate genuine social change—seems to have an astonishing ability to survive even the hardest knocks. Neither the self-evident limits of rights, nor the repeated empirical debunking of hope in rights by socio-legal scholars, nor indeed the determined claims of a range of rights sceptics, have put much of a dent in this faith.

Time and again, rights have been linked to hope: our faith in them has been widespread, deep-rooted and hugely resilient. Whilst making an argument *against* faith in rights, Duncan Kennedy, a CLS scholar, takes us some way towards understanding why this has been the case. He notes that rights talk has benefited greatly from the fact that it is connected to daring, specifically 'to daring to claim things on a basis that might previously have been disqualifying, to claiming things *for* blacks, women, gays, or Hispanics, when the feeling before might have been that *because* one was one of these things one was disentitled to make claims.'[11] Kennedy also points out that rights talk has profited from the widespread feeling that 'if "we" lose our faith in rights rhetoric but "they" don't, then they will gain an advantage over us.' And, as he acknowledges, this feeling is entirely understandable given the 'sense of righteousness, of mediation, that rights have historically provided':

> 'Giving up' rights would be like a professional athlete giving up steroids when all his or her competitors are still wedded to them.[12]

Hope is a complex feeling, however. In particular, in practice it can be hard to separate it from hype, and fear is generally also close at hand. This leads me directly to the second characteristic common to new technologies and human rights: namely that, in addition to their links to hope, both human rights and new technologies have strong associations with hype and fear.

Looking first at technology, human reproductive cloning seems to stand out as that which has been most widely feared, that which raises most clearly the spectre of the 'endangered human'.[13] As regards hype and technology, it's almost a case of 'take your pick'. In part, this is because hype often acts as a kind of engine of technological development. By this I mean that science often develops from a speculation or projection about what might be achievable following further

[10] President's Council on Bioethics, *Beyond Therapy* (2003), at 5.

[11] Kennedy, 'The Critique of Rights in Critical Legal Studies', in W. Brown and J. Halley (eds), *Left Legalism/Left Critique* (2002) 178, at 214.

[12] *Ibid*, at 217.

[13] Annas, Andrews, and Isasi, 'Protecting the Endangered Human: Toward an International Treaty Prohibiting Cloning and Inheritable Alterations', in S. Gruskin *et al* (eds), *supra* note 3, at 135.

research and investment—we can call this 'fiction science'[14]—to the realization of that possibility by capitalizing on hype in order to attract the venture capital that can make this happen and then benefit financially from it. Hype can also help to turn fiction science into real science by making the technology acceptable to the public, or at least a sector thereof, by emphasizing the benefits it would bring.

Moving on to human rights, as noted earlier, the hype or 'myth' of rights is a claim that is familiar from empirical socio-legal work on rights. And human rights as threatening, or redolent with risk, has been a perennial theme in rights-sceptical challenges. Moreover, lately, new voices have been added on the latter point, including what Roger Brownsword calls the 'dignitarian alliance', which depicts human rights as a dangerous libertarian approach to the regulation of new technologies that could, it says, place human dignity and humanity itself in peril. A second new voice can be traced to the post 9/11 conversation about life in a 'time of crisis'. Its position, in crude terms, is that exceptional times mandate exceptional measures. It has been exploring extra-legality in a serious, detailed manner and, typically, it conjures human rights as a potentially dangerous artefact of times past.

An association with revolution is the third characteristic common to new technologies and human rights. I say more about this later in Part 2; here, however, we should note a couple of preliminary points. Looking first at human rights, it seems that the familiar term 'rights revolution' has recently acquired at least two meanings. On the one hand, it evokes the astonishing ascent of human rights in the popular imagination and in national, international, legal, and political orders. We have witnessed, as Neil Walker points out, a widespread 'institutionalisation and internationalisation of the idea of law as a system of rights which frames and contains government power.'[15] The potential for rights to constrain private power also seems to have gained momentum. Evidence of these trends is everywhere. What is most striking, however, is the way in which rights have become *the* language of claims-making. And they have become this language not just for activists, advocates, and victims who seek to engage the power of law but also for the 'everyday': in short, as Duncan Kennedy has pointed out, rights seem to have seeped into 'just about every milieu where people argue about who should do what.'[16]

On the other hand, however, and as hinted at earlier, the term 'rights revolution' seems increasingly to capture both a new defensiveness in human rights advocacy and, its correlate, a new anti-rights sentiment which queries the legitimacy of rights in a time of alleged crisis or exception. The widespread and high-level engagement with the question 'is it sometimes acceptable to use torture?' has been astonishing. More generally, as Liora Lazarus and Benjamin Goold have

[14] See P. Baldi, *The Shattered Self: The End of Natural Evolution* (2001).

[15] Walker, 'Review of M. Loughlin, Sword & Scales: An Examination of the Relationship Between Law & Politics' [2001] PL 644, at 646.

[16] *Supra* note 11, at 180.

pointed out, 'the idea that certain human rights can be "turned off" when necessary' has acquired remarkable power; indeed, it is now widely regarded as a 'thoroughly reasonable reaction to the dangers allegedly faced by democratic societies.'[17] And as they go on to emphasize:

> In effect, the exceptionalism argument has become pivotal, so much so that liberals and human rights organisations must either rebut claims that our conditions are unique, or respond to these supposedly exceptional conditions by adjusting our institutions, practices, procedures and laws.[18]

The status of the exceptionalism argument is not the only reason for the new defensiveness of human rights. It also seems to have been hit hard by the now-undeniable promiscuity of human rights claims[19] and, relatedly, by the burgeoning awareness amongst liberal rights enthusiasts that 'their' rights initiatives have no 'always and everywhere'[20] destiny. In addition, as Roger Brownsword and Han Somsen explain in this book, in the evolving law of biotechnology, 'respect for human dignity not only sets limits on pushing ahead with the science, it is also open to an interpretation that puts it in tension with respect for human rights.'[21] Moreover, as Brownsword demonstrates, the dignitarian alliance has been pushing the latter interpretation very strongly indeed.

Talk of revolution has not been confined to rights, however. It has also been all-pervasive in the cultural life of new technologies, but (as is the case with the rights revolution) pinning down the essence of the technological revolution is genuinely difficult. Take, for example, the so-called 'reproductive revolution'. If we take the UK as our country study, did the reproductive revolution happen with the arrival of the female contraceptive pill? Or with the introduction of IVF and the birth of Louise Brown, the world's first test-tube baby? Perhaps the true revolution was the birth of Dolly the sheep in 1996 at Edinburgh's Roslin Institute: Dolly's birth confirmed the viability of somatic cell nuclear replacement, and provoked an outpouring of hope, hype, and fear in relation to what Ian Wilmut, one of her creators, describes as our new capacity for 'biological control'.[22] Alternatively, we might want to classify the

[17] Lazarus and Goold, 'Security and Human Rights: The Search for a Language of Reconciliation', in B.J. Goold and L. Lazarus (eds), *supra* note 4, 1, at 4.

[18] *Ibid.*

[19] That is to say, the ways in which rights have become the language of choice for a radically diverse, often oppositional and expanding range of social movements: see eg Goldberg Hiller, '"Subjectivity is a Citizen": Representation, Recognition and the Deconstruction of Civil Rights' (2003) 28 Studies in Law, Politics, and Society 139, at 173, describing the rise of an 'equal rights, not *special* rights' idiom amongst US groups opposed to gay marriage, and noting that arguments such as this 'invert the nature of harm and redefine majorities as victims of anarchic demands and as authentic rights subjects.'

[20] D. Kennedy, *The Dark Sides of Virtue: Reassessing International Humanitarianism* (2004), at 32–3.

[21] Brownsword at 21–22.

[22] On Dolly, see generally S. Franklin, *Dolly Mixtures: The Remaking of Genealogy* (2007).

'IVF-Stem cell interface'[23]—in particular, the heavy reliance of human embryonic stem cell research on a supply of embryos donated from assisted reproduction clinics—as the revolutionary revolution. A rather different perspective on the reproductive revolution might be offered, however, by a women's human rights campaigner, who would surely argue that a reduction in maternal mortality is the revolution that is yet to be, and also the one on which we need to focus our minds. Yet another perspective might be offered by campaigners for fathers' rights, for whom the question of reproductive choice for *men* is a live matter.[24] In the US, for instance, a number of men's groups have campaigned against sperm 'banditry' (the term they use to describe women who trick men into unprotected sexual intercourse with the intention of trying to become pregnant) and the imposition of financial support obligations on the genetic father in circumstances where a woman, who has the opportunity to have a safe, legal abortion, decides to bring a pregnancy to term.[25]

Of course, the problems of pinning down 'what' and 'when' are in no way peculiar to the reproductive revolution; the genetic revolution, for instance, would be just as difficult to pin down. At one point not so long ago, it was claimed that the sequencing of the human genome would be the breakthrough of breakthroughs. A draft sequence of the genome was published in early 2001, but the much-anticipated breakthrough did not materialize in the way that was expected. As the sociologist Nikolas Rose points out, '[t]he first map of the Human Genome did not meet expectations. It did not deliver the language of life; "the digital instructions" that make us what we are.'[26] Moreover, he adds, from today's vantage point, '[t]he brute reductionism of much of the genomics of the closing decades of the 20th century already seems old-fashioned.'[27]

Perhaps then we should seek the revolution elsewhere. We could, for example, look to the relationship between science and democracy and, in particular, to the remarkable rise in new forms of citizen participation in science over the past decade. The championing of participation has been especially notable in the United Kingdom, where crises such as BSE/vCJD and GM have coalesced with other factors to generate a shift from expert risk management, and a begrudging or patronizing 'public understanding of science' approach, to an officially endorsed

[23] See Franklin, 'Embryonic Economies: The Double Reproductive Value of Stem Cells' (2006) 1 BioSocieties 71.

[24] See, generally, R. Collier and S. Sheldon, *Fragmenting Fatherhood: A Socio-Legal Study* (2008); Sheldon, 'Reproductive Choice: Men's Freedom and Women's Responsibility?', in J.R. Spencer and A. Du Bois-Pedain (eds), *Freedom and Responsibility in Reproductive Choice* (2006) 175. Sheldon speculates that 'as a rhetoric of more involved fathering gains dominance in discussions of parenting practices, a similar trend may be seen in understandings of reproductive responsibilities' (at 195).

[25] See, respectively, Sheldon, '"Sperm Bandits", Birth Control Fraud and the Battle of the Sexes', (2001) 21 LS 460; Sheldon, 'Unwilling Fathers and Abortion: Terminating Men's Child Support Obligations?' (2003) 66 MLR 175.

[26] Rose, 'The Politics of Life Itself' (2001) 18 Theory, Culture & Society 1, at 14.

[27] *Ibid.*

and funded engagement with citizen participation as a vital component of policy-making in the biosciences and biomedicine.

I say more about revolution later. First, however, I want to examine its opposite, repetition. Specifically, I want to use the essays in this book to draw out ways in which the relationship between new technologies and human rights could be said to be no more than 'more of the same'. Putting that another way, I query the newness of the 'new'. Thereafter, in Part 2, I revert to the theme of revolution. I examine this theme by using the book's essays to pinpoint a range of ways in which new technologies do seem to present deep challenges for human rights. In Part 3, the final section of the chapter, I do some 'tooling-up'; that is, I propose a range of resources on which human rights could draw in order to bolster their role and usefulness in the regulation of new technologies. Overall, the chapter is designed as a preliminary or ground-clearing exercise. It is of course also intended as an introduction to the essays that follow. I have, however, decided against offering essay-by-essay summaries: these summaries have been provided by the authors themselves in the introductions and conclusions to their individual chapters, and there is no good reason to reword and repeat that material here. I have opted instead to focus on drawing out what I see as the most salient points from across the chapters, using the frames 'repetition', 'revolution', and 'resonance'.

1. Repetition

Any discussion of new technologies and human rights is likely to face a number of basic definitional problems. To begin with, 'human rights' has at least two different meanings: one refers to a body of law, the other to movements or groupings of lawyers, non-governmental organizations, and others who seek to secure and defend a particular right, group of rights, or human rights in general. The slipperiness of human rights seems a minor matter, however, when compared with the problems that arise in defining new technologies. Consider the following question: when, or why, is it that a technology counts as a 'new technology'? To answer, I could refer you to the technologies discussed in this book. These include: agricultural technologies, such as genetically modified organisms (GMOs) for human consumption; genetic reproductive technologies, such as pre-implantation genetic diagnosis (PGD); and the digital technologies which, as Francesca Bignami points out in her essay in this book, have made spying 'easy for everyone'.[28] That approach, though, would not take us towards a plausible working definition of new technologies. Most obviously, it would be too static, excluding other extant

28 Bignami at 130: 'A standard, pocket-size camera can take photographs, record conversations, and make video clips. A common flash drive can store vast data sets containing many gigabytes worth of electronic information. With a simple Internet connection, these digital files can be shared with hundreds of thousands of people throughout the world. These digital files can be combined with other files, taken from other sources, to create more complete and intrusive personal profiles.'

technologies, such as nanotechnology,[29] that surely deserve to be included. It might also be difficult to get from the particular technologies discussed in this book (or any other select grouping of technologies) to a set of general characteristics, or criteria, which capture the essence of new technology.

In her chapter, Laurence Boisson de Chazournes emphasizes that there is no widely accepted definition of 'new technologies' but notes that the definition most commonly found in legal writings treats it as 'a synonym of "high technology activities", encompassing technologies in the domains of biology, space, communications, computer, and nuclear activities'.[30] As she goes on to explain, '[t]his definition relies on the novelty of the technological means at stake, meaning that a "new technology" of today will no longer be "new" in the future, and that, more generally, all technologies are "new" when they are first introduced into society.'[31] But, as Thérèse Murphy's essay explains, it is also entirely possible that novelty will come and go in relation to an individual technology. Taking the example of IVF in the UK as her focus, Murphy reminds us that the first IVF baby, Louise Brown, was born in Oldham in the UK in 1978 and that, at that time, Brown and other IVF children were commonly described as test-tube babies. IVF itself tended to be described as an 'artificial reproductive technology' or, indeed, a 'new reproductive technology'. Subsequently, the descriptors shifted: the birth of more IVF babies, and the developing understanding of nature itself as in need of a 'helping hand', helped to popularize IVF as a form of 'assisted', rather than 'artificial', conception.[32] More recently, however, as IVF has become a platform for genetic technologies, such as pre-implantation genetic diagnosis (PGD), and surplus IVF embryos are in demand for controversial stem cell research, this technology seems to be acquiring a new streak of 'artificiality' or, indeed, 'newness'. To sum up, then, a short 'life history' of IVF in the UK might go something like this: it was new in 1978,[33] then for a time it was no longer 'new' and, arguably, it came to be seen as less technological and more akin to assisted nature, but more recently its use as a platform technology has rendered it newly controversial. One indicator of the emergent controversy is the arrival of a distinction between 'good' IVF and 'bad' IVF.

Repetition is also a feature of law's engagement with new technologies. The legal scholar Margot Brazier has noted that in medical law we seem destined to

[29] For discussion, see eg G. Hodge *et al* (eds), *New Global Frontiers in Regulation: The Age of Nanotechnology* (2007).

[30] Boisson de Chazournes at 161. [31] *Ibid.*

[32] See, in particular, M. Strathern, *Reproducing the Future: Essays on Anthropology, Kinship and the New Reproductive Technologies* (1992) for the argument that reproductive technologies have the capacity to remake allegedly natural facts and assumptions about sex and reproduction, and about kinship.

[33] More precisely, IVF practised with human gametes and embryos was new in 1978; IVF as an agricultural technique was certainly not new. See the special issue, 'Between the Farm and the Clinic: Agriculture and Reproductive Technology in the Twentieth Century' (2007) 38 Studies in History and Philosophy of Biological and Biomedical Sciences, 303 ff.

address the same problems again and again.[34] Moreover, several of the essays in this book suggest that, in at least some respects, the challenges of new technologies are old challenges—that is to say, familiar ones—for human rights law and practice. Thérèse Murphy's essay, for example, documents the ways in which public framings of reproductive and genetic reproductive technologies pitch arguments for and against reproductive choice that seem familiar from abortion politics. In Hélène Boussard's essay, it is emphasized that the 'geneticization' of rights (that is, the increasing references to the 'right to genetic identity', the 'right to genetic privacy', and the 'right to genetic confidentiality') that is taking place in the law and also in both scholarly and popular commentary does not refer to any new right: rather, it highlights the new context in which the 'old rights' (that is, rights already enshrined in human rights treaty or customary international law) are to be protected. Boussard and Iulia Motoc discuss both the relationship between the individual and the family, communities and the human species, and the relationship between informed consent and the protection of human rights.[35] These two themes have already been the subject of extensive debate in human rights law. Similarly, the references to eugenics in the essays by Motoc and Han Somsen cannot but draw us back to the events that helped to forge international human rights law.

In the essays by Francesca Bignami, Roger Brownsword, and Han Somsen, one especially thorny point of lasting dispute is brought into focus: namely, how to secure the legitimacy of the review function of courts and other reviewing bodies? This question is difficult in the context of domestic courts; but, as all three essays emphasize and explore, it is typically a great deal more difficult when one is dealing with supra- or international courts or review bodies. All three essays also address a related issue: namely, to what extent does, and should, human rights offer flexibility and room for local variation? The universality of human rights was seen traditionally as a core strength, but in recent decades it has become a point of deep contention, amidst claims that international human rights law is no more than the universalization of the particular. Relatedly, as Bignami explores in the context of the right to privacy, although the interaction of the European Court of Human Rights and also the European Court of Justice with domestic courts in the Council of Europe and the European Union, respectively, may demonstrate very clearly that it would be a mistake to think of, say, domestic and European human rights law by means of the classical domestic/international dichotomy, what is far less clear is how we should think of or define this relationship. So, for example, as we think about the interactions, or judicial conversations, between the different courts in Europe, should we deploy a

34 Brazier, 'Regulating the Reproduction Business' (1999) 7 Med L Rev 166.

35 See, more generally, D. Beyleveld and R. Brownsword, *Consent in the Law* (2007); Jacob, 'Form-Made Persons: Consent Forms as Consent's Blind Spot' (2007) 30 PoLAR: Political and Legal Anthropology Rev 249; N.C. Manson and O. O'Neill, *Rethinking Informed Consent in Bioethics* (2007).

constitutionalist frame or a pluralist one? And, relatedly, how should we conceive of tools such as the margin of appreciation doctrine and the evolutive approach to interpretation, both of which have been deployed repeatedly by the European Court of Human Rights? Those who seek certainty and coherence are likely to deplore the margin of appreciation as a source, in practice, of sophistry or fudging. And those who want the European legal order to be a full and conventionally constitutionalist regime may see the doctrine as an unwelcome political element in a system that ought to be focused on a common or unified set of norms. As Bignami points out however, constitutionalists might also want to consider whether 'in the European search for solidarity', strong constitutional patriotism might have a dark side: namely, if diversity among the national communities is suppressed in the interests of greater unity, isn't there a risk that patriotism 'might very well betray its normative aspirations of tolerance and inclusiveness'?[36] Bignami's own view is that, for the European Court of Human Rights and the European Court of Justice, deference to national courts on difficult questions of constitutional law 'is an essential instrument of tolerant—and patriotic—constitutionalism'.[37] Finally on this point, all of us—whether constitutionalists, pluralists, or something else—surely face additional challenges as we think about the doctrine of the margin of appreciation, and more generally the question of local variation, in the specific context of new technologies where, quite clearly, disagreement is deep and entirely reasonable.

2. Revolution

I imagine that by this point it will be clear that, in my view, talking about new technologies and human rights by reference to the language of revolution invites intense questioning and, possibly, defeat. Yet, as hinted at earlier and as a number of the essays in this book confirm, entirely to exclude revolution would be to risk missing salient properties of the relationship between human rights and new technologies. In an attempt to address this latter risk, this part of the chapter singles out aspects or features of our contemporary circumstances that seem to be viable contenders for the label 'revolutionary'.

The embryo is my first nominee. It is being imbued with new meaning[38] amidst battles over human embryonic stem cell research and the disposition of surplus embryos resulting from assisted reproductive technologies (ARTs). It is

[36] Bignami at 129. [37] *Ibid*, at 160.

[38] The embryo as a political actor or agent is not new, however. See Morgan, 'Embryo Tales', in S. Franklin and M. Lock (eds), *Remaking Life & Death: Toward an Anthropology of the Biosciences* (2003) 261: '[in the US] human embryos were invoked as political actors and agents as early as the 1910s and 1920s . . . Yet the images that scientists produced during that time did not immediately result in personification of the embryo, nor did they much influence attitudes towards abortion . . . Embryos were hitched to an entirely different set of social problems, namely, evolution, the "race problem," and the relationship of humans to nonhuman animals.'

also being 'brought to life' by the enactment of legislation that seeks to prohibit the creation of certain kinds of genetically manipulated embryos: indeed, as Isabel Karpin points out:

> It is not the discourse of science alone, or even primarily, that gives rise to these new possibilities. Instead, it is through the enactment of prohibitory legislation that (legislative) life is given to entities that are yet to be made. In so doing, the law gives reality to the fantasy of the very beings that it seeks to deny. Law through both regulation and prohibition carries us forward in the imaginary leap that is necessary to take us from the embryonic being to the *post-human* being.[39]

Moreover, the case of *Evans v UK*,[40] decided by the Grand Chamber of the European Court of Human Rights in the spring of 2007, suggests that the ART embryo also has the capacity to make us ask: 'reproductive choice for whom?' In *Evans*, the Court had to grapple with the question 'what if a woman and a man store frozen embryos but at a later point disagree about their use?' Facts such as these demonstrate very clearly that ARTs have the capacity to challenge long-standing assumptions about reproductive choice. Perhaps most crucially they raise the question of whether arguments grounded in reproductive choice require 'a commitment to gender neutrality which is difficult to achieve in all cases and may be undesirable in some.'[41]

The second potentially revolutionary feature can be brought into view by asking 'who counts as a rogue?'[42] Because regulators need to know whom and how they ought to regulate, this is a relatively standard question—albeit one that tends to be framed in a less dramatic manner than I have used here. So, looking specifically at human rights, how might we answer this question? Rogue states will surely be the most prominent candidates. Recently, however, in disputes about access to essential medicines, the World Trade Organization (WTO), 'Big Pharma', and trade more generally have been depicted by some as rogues. There has also been concern about rogue scientists; recall, for example, the reports following the revelation that apparently ground-breaking stem cell research in South Korea by Dr Woo-Suk Hwang had in fact involved both the fabrication of data and the unethical recruitment of egg donors.[43]

However, none of those examples provides the revolutionary twist: for this, we need to focus on individuals, individual families, and communities. It might be objected that 'rogues' is surely a misdescription here, and also potentially

[39] Karpin, 'The Uncanny Embryos: Legal Limits to the Human and Reproduction Without Women' (2006) 28 Sydney Law Rev 599, at 602.

[40] ECtHR, Judgment of 10 April 2007, *Evans v The United Kingdom*, Case No 6339/05.

[41] Morgan, 'Review of Jackson, *Regulating Reproduction*' (2003) 11 Med L Rev 262, at 263.

[42] See, further, Brownsword, 'Red Lights and Rogues: Regulating Human Genetics', in H. Somsen (ed), *The Regulatory Challenge of Biotechnology: Human Genetics, Food, and Patents* (2007), 39.

[43] See Jackson, 'Fraudulent Stem Cell Research and Respect for the Embryo' (2006) 1 BioSocieties 349.

offensive: that, although there have been concerns about practices such as 'reproductive tourism', individuals, families, and groups are better described as reproductive citizens, 'genetic citizens',[44] or, more generally, 'biocitizens'.[45] Indeed, focusing on reproductive technology, it might even be claimed that the best descriptor would be 'moral pioneers'.[46] What all of these terms try to capture is the ways in which new technologies, including those that are still fiction science, are forging new forms of subjectivity and biosociality.[47] Put crudely, new technologies provoke challenges and opportunities for the ways in which we think about 'me', 'us', and 'them', and about rights, obligations, and responsibilities. In this book, these points are brought out very clearly in a range of essays, including those by Hélène Boussard, Iulia Motoc, and Thérèse Murphy.[48] Boussard and Motoc demonstrate that genetic testing, although designed primarily to accomplish medical ends by assessing the future risk of disease, has effects that reach far beyond these ends, creating new forms of biosociality and subjectivity. Thérèse Murphy's chapter summarizes ethnographies of assisted reproductive technologies which challenge contemporary anxieties about 'designer' babies, 'designing' parents, and the alleged onset and impact of mere, sheer choice. Moreover, one point brought out by all three chapters is that not everyone has equal citizenship in our new biocitizenship age.

The precautionary principle is the third contender for the tag 'revolutionary'. At its core, this principle—which was developed in the field of environmental protection—permits the imposition of restrictions on otherwise legitimate activities, if there is a risk of environmental damage, even if that risk has not yet been fully demonstrated by science. In this book, Laurence Boisson de Chazournes focuses on the precautionary principle; there is also discussion of the principle in the essays by Brownsword and Somsen. Boisson de Chazournes emphasizes that international law's traditional reactive approach to risk is not fit-for-purpose with respect to risks deriving from new technologies: 'a new approach based on *uncertainty* and *future events*',[49] and new legal tools are thus needed. Put shortly, '[i]nternational law cannot ignore precaution';[50] moreover,

[44] See eg Rapp, 'Cell Life and Death, Child Life and Death: Genomic Horizons, Genetic Diseases, Family Stories', in S. Franklin and M. Lock (eds), *supra* note 38, at 129.

[45] See, in particular, Rose and Novas, 'Biological Citizenship', in A. Ong and S.J. Collier (eds), *Global Assemblages: Technology, Politics, and Ethics as Anthropological Problems* (2005) 439; N. Rose, *The Politics of Life Itself* (2007). For an account of the differences between the biocitizenship of the contemporary biological age and the racialized and nationalized forms of biocitizenship of the so-called eugenic age, see Rose and Novas, *ibid.*

[46] R. Rapp, *Testing Women, Testing the Fetus: The Social Impact of Amniocentesis in America* (2000).

[47] The latter term was coined by Paul Rabinow: see eg P. Rabinow, *Essays on the Anthropology of Reason* (1996).

[48] Intellectual property is another area where biotechnology has been forging new ways of thinking about 'me', 'us', and 'them': see eg Hayden, 'Taking as Giving: Bioscience, Exchange, and the Politics of Benefit Sharing' (2007) 37 Social Studies of Science 729.

[49] Boisson de Chazournes at 162.

[50] *Ibid*, at 164.

precaution will bring changes to international law. Boisson de Chazournes identifies precaution as the bearer of four specific disturbances in the law. First, a remodelling of the interaction between law and science. Secondly, a remodelling of the interaction between law and economics, in relation to which she argues that the principle must be seen as a substantive element of sustainable development such that it operates not to destroy economic activity, but rather to calibrate economic benefits alongside social equity and ecological sustainability. Thirdly, a remodelling of the interaction between law and effectiveness, given that courts and tribunals will need to objectivize the elements of the precautionary principle in ways that 'arbitrate between the contradictory interests of society's different agents' in order to implement and underwrite this new body of law, 'whose essence is "soft".'[51] Finally, the principle illustrates the presence of overlapping systems of governance, highlighting both the datedness of the classical international/domestic dichotomy and the increasing importance of being in tune with public opinion. Boisson de Chazournes provides the following example of this point:

> The crises of contaminated blood and mad cow disease which erupted in European countries proved the precautionary principle's applicability in the international sphere, and required greater cooperation between states. But they also demonstrated its applicability in the domestic sphere, and confronted decision-makers with public opinion, which is increasingly exacting as to the measures that should be taken to manage the threats and risks of today.[52]

As noted earlier, the precautionary principle is also discussed in the chapters by Roger Brownsword and Han Somsen. For Brownsword, it is one amongst a number of key principles which feature in different ways at the three different points—utilitarianism, human rights, and dignitarianism—in the bioethical triangle that is incorporated in the emerging international law of biotechnology. Working with a commitment to human rights, he points out that 'when we adopt a human rights perspective, it is precaution for the sake of human rights that becomes focal.' There is thus a 'precautionary form of argument available, and waiting to be fully articulated, within human rights thinking.'[53] Somsen, like Brownsword, is committed to the articulation of an ethic grounded in human rights as the most promising means of meeting the challenges and opportunities of new technologies. He differs from Brownsword, however, on the role of the precautionary principle: for Somsen, precaution is a revolution that could be taken too far. He is concerned, in particular, that in a regulatory framework for human genetics, the precautionary principle could turn out to be an enemy of the

[51] *Ibid*, at 193. [52] *Ibid*, at 194.

[53] Brownsword at 38–39. See relatedly Murphy and Whitty, 'Is Human Rights Prepared? Risk, Rights and Public Health Emergencies' (forthcoming), urging human rights lawyers to build a human-rights based approach to risk and rights, using *inter alia* the frames 'risk within rights' and 'rights as risk'.

human-rights based version of dignity—that is, dignity as empowerment, as opposed to dignity as red light constraint. Precaution, he says, 'leads to arbitrariness and provides governments with a blank cheque for limiting individual liberties.'[54]

Human dignity—specifically, the rise of what Brownsword describes as human dignity as constraint and its primary proponent, the dignitarian alliance—represents the fourth revolutionary development. The revolution lies not in the presence of human dignity *per se*, but rather in the severing of it from human rights and the emergence of two conflicting notions of dignity: namely, a human rights version of human dignity as empowerment, and a dignitarian version of it as a red light constraint. For human rights advocates, what this means is that the battle (for dignity) needs urgently to be engaged.

Human rights advocates might also wish to consider whether the sluggishness of human rights in addressing the tensions within its own deployments of human dignity is in part responsible for the situation in which it finds itself today. At all events, the fight for dignity is here and its existence brings to mind another related revolution: namely, the 'rights revolution' mentioned earlier in this chapter, or at least that version of it which has put human rights on the defensive. To recap: of late, amidst talk of crisis and exceptionalism, human rights advocates seem to be facing new or heightened aversions towards rights. In particular, rights are being attached to minorities, allegedly dangerous minorities; increasingly, they are presented as not being for, or indeed needed by, the majority. Moreover, as the criminologist Ian Loader and others have pointed out, a new right[55]—the right to be safe—seems to be clambering to the top of the popular wish list, and other rights alleged to threaten this new right seem to be under threat. All in all, it seems that rights today are seen as for 'them', not for 'us', and as a 'conspiracy against common sense and practical wisdom, against the things that "we" all know to be true.'[56]

To close this section, I want to make reference to one final revolution, which is discussed at length in Roger Brownsword's essay. For me, Brownsword's essay cuts through all the to-ing and fro-ing about the newness of the new and how one ought to define revolution or, indeed, where we should be looking for these revolutions. In simple, neutral terms, Brownsword's message is that human rights lawyers, and lawyers more generally, ought to bear in mind that new technologies represent regulatory tools, not just regulatory targets.[57] Stated in a slightly less neutral manner, the message is that as regulators focus their attention on the

[54] Somsen at 115.

[55] A more precise description would frame this not as a new right but as a specific version thereof, thereby recognizing the other claims that circulate under the heading of the 'right to security': see further Goold and Lazarus (eds), *supra* note 4.

[56] Loader, *supra* note 9, at 39.

[57] On technology as both a regulatory tool and a regulatory target, see R. Brownsword, *Rights, Regulation, and the Technological Revolution* (2008).

development of frameworks to control new technologies, technology itself is increasingly likely to be deployed in order to bolster, and possibly even to replace, legal regulation. But it is the most strongly worded version of the message that makes us stop in our tracks. Setting the scene for what he calls the 'revolutionary sting', Brownsword asks the following question: 'if ethical pluralism makes it difficult for regulators to set agreed limits to the development and exploitation of modern biotechnology, how are regulators themselves to be limited as they turn (as they surely will) to the development and exploitation of modern technologies as regulatory tools?'[58] The sting, then, is that:

> [I]f regulation by legal prescription, by rules issued by legislative assemblies and courts, gives way to regulation by technological design-in or design-out, it matters little whether we agree that this amounts to technological revolution – for the displacement of law by technology, the Rule of Technology rather than the Rule of Law, is a sea-change however we view it.[59]

To this mix we might also add that technology seems increasingly adept at survival and adaptation: indeed, as the sociologist Sarah Franklin has noted, ethical concerns are increasingly being 'built into' new life forms,[60] as for example in the development of a stem cell extraction procedure which removes one cell from the blastocyst without impinging on its ability for further development. It is hard to know how to respond to the rise of this sort of 'technological fix'. It could be a sign of an encouraging new engagement between science, markets, and the public. At the same time, however, I cannot avoid the sense that the technological fix as a way around deep social dispute is both insufficient and a potentially dangerous step on the path towards the Rule of Technology.

3. The Resonance of Human Rights

Having been pulled up short by the potential for technology to displace law, by the prospect of 'the Rule of Technology rather than the Rule of Law',[61] the question is: where do we go from here? Somewhat perversely I propose that we consider 'what works'. Specifically, I propose that the essays in this book show

58 Brownsword at 20.

59 Brownsword, *supra* note 57, at 6. He closes his essay in this collection by emphasizing: 'As agents who value our autonomy, we need to be aware that the cumulative effect of seemingly benign individual choices could represent a threat to the conditions that make those choices meaningful. And, as agents who aspire to moral community, we need to be particularly mindful of the corrosive impact of a (bio)technological approach to social control.'

60 Franklin, 'Ethical Biocapital', in S. Franklin and M. Lock (eds), *supra* note 38, at 97. For the argument that biotechnology has challenged and eroded the conceptual form of law, see Pottage, 'The Socio-Legal Implications of the New Biotechnologies' (2007) 3 Annu Rev Law Soc Sci 321.

61 Brownsword, *supra* note 59.

that a human rights approach 'works'—or, more precisely, that there are good reasons to make it work and that, with some adjustment, it could work.

The essays provide a range of jumping-off points for the project I am calling 'human rights works'. In what follows, I single out just a few of these. For starters, as Laurence Boisson de Chazournes reminds us in her study of the precautionary principle, the principle 'does not stand alone, but is interwoven with other international law norms and principles.'[62] Crucially, one of the principles with which it interacts is public participation:

> The state cannot be the only agent responsible for the evaluation of whether precaution should be applied in a particular situation. Scientists, companies, NGOs, local populations, and other concerned agents must participate in the decision-making process. In addition, adequate information as a corollary to participation guarantees a transparent decision-making process.[63]

This leads Boisson de Chazournes to endorse the view of Marie-Angèle Hermitte who has described the precautionary principle as the bearer of 'a subversive capacity for the whole of the legal order and leads the way to democratic renewal.'[64] And this in turn should lead us, as human rights advocates, to engage more deeply with the apparent take-up of citizen participation in the regulation, and increasingly the design, of new technologies. Is citizen participation, as it is being practised today, the path away from top-down, technocratic, or expert approaches to regulation? Is it, in addition, the path towards increased trust in policy-making by ensuring that even those who disagree fundamentally with any individual decision or policy are content with the procedure that led to it? In the UK, citizen participation is currently a site of great hope. However, a number of social scientists have expressed concerns about the apparent assumption that participation leads automatically to increased trust. Others have emphasized that if we are all experts now, there is surely a need to calibrate and configure different types of expertise. To my mind, and as mentioned above, these developments suggest that human rights advocates need urgently to engage with citizen participation and public participation more generally. It is no longer sufficient (if it ever was) simply to promote participation as a good in itself.

It could be said that the theme of citizen participation is also taken up in Thérèse Murphy's essay, albeit in a less conventional manner. In a bid to safeguard what she describes as the 'speakability' of reproductive choice in the area of ARTs, Murphy invites us to incorporate a new dimension into the way that we, as human rights lawyers, theorize and investigate reproductive choice. That new dimension is ethnography, and the hope is that this move will help to flesh out the human rights argument for reproductive choice by interrupting the popular dominion of bare 'for or against' positions on choice. More broadly, of course, the hope is that ethnographic work might help human rights advocates to

[62] Boisson de Chazournes at 178. [63] *Ibid*, at 179. [64] *Ibid.*

fight back against the claim that we are willing to play fast and loose with human life and, more generally, with human dignity.

Voices, or stories, are also central to the argument made in Iulia Motoc's chapter on genetic discrimination. Motoc invokes the power of 'never again', encouraging us to look to history—specifically, World War II—as we unravel the complexities of the relationship between new technologies and the protection of human rights, and develop an international law of genetic discrimination. She argues that 'we must consciously and deliberately respond to eugenics not only so that we avoid repeating the mistakes of the past, but also so that unwarranted fear and guilt do not render us incapable of properly appropriating and incorporating genetic information into our social structure, our policies, and our jurisprudence.'[65]

In a way, Motoc's focus on 'never again' is complemented by Han Somsen's essay, which cautions against the assumption that the regulatory challenge posed by human genetics is of an entirely new kind. The latter position might usefully be classified as 'never before'; that is, it assumes that the challenges of the present day are of a kind that lacks precedent or parallel. Somsen attacks this assumption by drawing on literature and case law from the World Trade Organization (WTO) on risk regulation in the context of GMOs as part of his analysis of how we might regulate the genetic reproductive technology of pre-implantation genetic diagnosis (PGD). He asks: 'How can we make use of these lessons in internal and cosmopolitan legitimacy for the development of a regulatory framework for PGD?'[66] He is fully aware that many assert that the differences between GMOs and human embryos are 'far too profound to enable us to apply insights obtained from the regulation of GM corn to the regulation of PGD.'[67] But he is not convinced by such assertions:

> First of all, it is simply not logical to place higher demands on the legitimacy of a policy on GM corn than on a policy on embryos. On the contrary, given the ethical implications of PGD, we should strive for a debate on PGD that is as clear and transparent as possible, in a manner befitting modern deliberative democracies. WTO and EU case law regarding GMOs offers a number of useful guidelines for such a transparent and clear debate. Insofar as this debate concerns the *clinical risks* associated with PGD, the lessons from the WTO and EU jurisprudence are in any case of direct relevance.... In addition, the precautionary principle has started to surface in the debate about human genetics, and the insights derived from WTO case law are instructive here as well.[68]

Somsen fully accepts that reviewing ethical decisions is far more complex and contentious than review of risk regulation of GMOs. Complexity and contention arise, in part, because multiple, overlapping courts and review bodies are

[65] Motoc at 223. On our attitude to history, see Judt, 'What Have We Learned, If Anything?', 1 May 2008, *The New York Review of Books* 16. He notes *inter alia* that: 'We have become stridently insistent that the past has little of interest to teach us. Ours, we assert, is a new world: its risk and opportunities are without precedent.'

[66] Somsen at 102. [67] *Ibid.* [68] *Ibid.*

involved. Another complicating factor, at least in the short term, is that there has been far more biocitizenship around GMOs, and agricultural or green biotechnology more generally, than around red biotechnology focused on interventions in human biology. Hence, as Sheila Jasanoff has emphasized, we need to ask 'Could it be that formal ethical discourse has been too quick and too successful in colonising [deliberation around red biotechnology]?' If this is the case, then, 'we very much need to open up the whole notion of bioethics to permit a more inclusive, and in some ways more contentious, debate than we've promoted on many of these issues thus far.'[69] A third problem brings us full circle, that is to say, it brings us back to the reasons for linking new technologies and human rights. In reviewing risk regulation on GMOs, the WTO was able to call upon the SPS rule which dictates that decisions have to be based on a scientific assessment: this rule helped to legitimate the WTO's review function and overall decision-making. But as we have seen, red biotechnology currently lacks the support of a 'cosmopolitan meta-norm'. What Somsen and others are arguing is not only that such a norm is needed but that a human rights perspective is the best candidate.

To conclude, I want to be a little less upbeat, a little less focused on jumping-off points within human rights and rather more focused on the fact that, for human rights advocates, formidable challenges are already in play, with many more yet to be. I want to echo the arguments of Boussard, Brownsword, Somsen, and others that it is simply not enough to assert rights as an always and everywhere, no-explanation-needed good in themselves. Human rights advocates need to engage, not evangelize. Our stance needs to be clear and fully developed (which isn't to say that new issues won't arise or that shifts in position are ruled out, but rather that we should not repeat the mistake of ducking the issue that we made with human dignity). Our stance also needs to be explained. And it needs, where necessary, to understand and acknowledge the limits of rights. Finally, to borrow from the legal scholar Conor Gearty, for human rights advocates to ignore these challenges would be to risk something worse than controversy and conflict:

> [W]ithout a reworking of what the term 'human rights' means today, designed to give it contemporary intellectual confidence, some theoretical zest, then the time might come when firing the human rights argument will be greeted neither with warmth nor dismay but rather with blank indifference, or (which is worse) mute incomprehension: whatever can that term mean?[70]

[69] Jasanoff, 'Experiments Without Borders: Biology in the Labs of Life', BIOS Annual Lecture, LSE 15 June 2006, available at <http://www.lse.ac.uk/bios> (accessed 31 March 2008).

[70] Gearty, *Can Human Rights Survive?* (2006) 20.

2

Human Dignity, Ethical Pluralism, and the Regulation of Modern Biotechnologies

Roger Brownsword

According to the President's Council on Bioethics:

> By all accounts we have entered upon a golden age for biology, medicine, and biotechnology. With the completion of (the DNA sequencing phase of) the Human Genome Project and the emergence of stem cell research, we can look forward to major insights into human development, normal and abnormal, as well as novel and more precisely selected treatments for human disease . . . In myriad ways, the discoveries of biologists and the inventions of biotechnologists are steadily increasing our power ever more precisely to intervene into the workings of our bodies and minds and to alter them by rational design.[1]

When we add developments in information and communication technology into this mix, we might think that this golden age is also an age of revolutionary proportions. Pierre Baldi is one who takes such a view:

> We are witnessing a revolution brought about by scientific and technological advances, one in which change occurs at an accelerated pace. By 1945 we had the ability to destroy life on a large scale. By 2045, only a hundred years later, we ought to be able to create life from scratch, both in real space and in cyberspace.[2]

As Baldi sees it, the rapid and interacting developments in information technology and biotechnology signify 'the end of our evolutionary odyssey'.[3] But, of course, such revolutionary claims are easier made than vindicated. One difficulty is the particular way that we conceive of a 'revolution'. What does it take before a particular development, disruption, or acceleration qualifies as 'revolutionary'?[4] Even if our conception of a revolution is accepted, another difficulty is that we might be looking for the relevant development, disruption, or acceleration in

[1] President's Council on Bioethics, *Beyond Therapy* (2003), at 5–6.
[2] P. Baldi, *The Shattered Self* (2002), at 163. [3] *Ibid.*
[4] Compare M. Klang, *Disruptive Technology* (2006); and A. Burgess, *Cellular Phones, Public Fears, and a Culture of Precaution* (2004) at 59.

rather different places (in the underlying science, in society, in our culture, and so on).[5] Let us grant, then, that claims that announce a revolutionary change need to be treated with caution.[6] Nevertheless, as a raft of modern technologies (including nano and neurotechnologies) becomes mutually enabling and interactive, this surely signals developments that are significant both for their technological innovation and for their social and cultural impact. Indeed, the sting at the end of this chapter is that the development of these technologies is highly significant for the future of law and the way that we regulate social life.

But, first things first: although technological revolution might be in the background, the foreground issues in this paper concern ethical pluralism, human dignity, and regulatory legitimacy. Accordingly, our first task is to sketch what I see as a three-way ethical plurality—captured in the form of a 'bioethical triangle'—in which the idea of human dignity is located and contested (in Part 1). Secondly, we need to clarify where the ideas of 'harm', 'consent', and 'precaution' stand in relation to the principal constituencies that make up the bioethical triangle (in Part 2). In the absence of such clarification, these ideas (which are pervasive in debates about the legitimate regulation of modern biotechnologies) lack explicit ethical anchoring and they can operate as if they were, so to speak, 'bioethical floaters'. Thirdly, we can illustrate the general import of the plurality as an *external* benchmark for the regulation of biotechnology, sometimes actively shaping the standards set by regulators and always relevant to any assessment of the legitimacy of the regulatory position (Part 3). Here, two hard case illustrations are presented, namely: (i) therapeutic cloning and stem cell research; and (ii) the use of pre-implantation diagnostic and pre-implantation tissue-typing techniques for human reproductive purposes. Fourthly, we deal with the embedding of ethical pluralism as an *internal* regulatory standard (Part 4): sometimes the grounding is *explicit*, as is the case with express morality clauses of the kind that we meet in (i) patent and (ii) trade regimes; sometimes, the embedding is *implicit*, as when particular regulatory provisions call for interpretation and application. Finally, we return to regulation and the revolutionary sting (Part 5): if ethical pluralism makes it difficult for regulators to set agreed limits to the development and exploitation of modern biotechnology, how are regulators themselves to be limited as they turn (as they surely will) to the development and exploitation of modern technologies as regulatory tools? Or, to put this more darkly, what are the prospects for law once regulators take up their (bio)technological options?

[5] For an interesting discussion of the relationship between technological development and cultural change, see Somerville, 'Searching for Ethics in a Secular Society', in UNESCO, *Ethics of Science and Technology: Explorations of the Frontiers of Science and Ethics* (2006) 17–39.

[6] Compare Price, 'The Newness of New Technology' (2001) 22 Cardozo L Rev 1885. On the extent to which the new information society may be regarded as revolutionary, see Fried, 'Perfect Freedom, Perfect Control?' (2000) 114 Harvard L Rev 606.

1. Ethical Pluralism, Human Dignity, and the UNESCO Regulatory Axiom

On 19 October 2005, the General Conference of UNESCO adopted the Universal Declaration on Bioethics and Human Rights. Having already adopted the Universal Declaration on the Human Genome and Human Rights in 1997, and the International Declaration on Human Genetic Data in 2003, this was the third, and arguably the most ambitious of UNESCO's attempts to 'globalize' ethics in the face of increasingly globalized science and technology.[7]

The key feature of the latest Declaration is the continuing commitment to ethically clean science coupled with a willingness to engage with cultural diversity. First, and wholly characteristically, UNESCO promotes the idea that, while science and technology have the capacity to function as forces for the good (making people better as well as making for a better environment for human flourishing), scientific research and development and the application of science must be compatible with respect for human rights and human dignity (making better people, or making a better kind of person, might not be so compatible).[8] This is the position consistently pressed by UNESCO and, once again, it is reflected in this most recent Declaration.[9] Secondly, while UNESCO elaborates these cosmopolitan principles, it also supports the idea of cultural diversity.[10] Thus, to the extent that pluralism and cultural diversity are consistent with the agreed cosmopolitan principles, they are welcome.

However, the drafting of the Declaration glosses over two major difficulties. First, the line and the relationship between the cosmopolitan and the local is unclear (although, to some degree, these are problems that can be managed by adopting flexible notions such as 'the margin of appreciation', or the like). Secondly, a conflictual ethical pluralism is already written into the UNESCO cosmopolitan principles. Crucially, respect for human dignity not only sets limits

[7] See, Brownsword, 'What the World Needs Now: Techno-Regulation, Human Rights and Human Dignity', in R. Brownsword (ed), *Human Rights* (Volume IV of *Global Governance and the Quest for Justice*) (2004) 203.

[8] See Brownsword, 'Making People Better and Making Better People: Bioethics and the Regulation of Stem Cell Research' (2005) 1 J of Academic Legal Studies 3 (online).

[9] See, eg Art 2(d), in which one of the aims of the Declaration is said to be 'to recognize the importance of freedom of scientific research and the benefits derived from scientific and technological developments, while stressing the need for such research and development to occur within the framework of ethical principles set out in this Declaration and to respect human dignity, human rights and fundamental freedoms'; and Art 3(1), which provides as the first principle of the Declaration, that '[h]uman dignity, human rights and fundamental freedoms are to be fully respected.'

[10] See, eg Art 12: 'The importance of cultural diversity and pluralism should be given due regard. However, such considerations are not to be invoked to infringe upon human dignity, human rights and fundamental freedoms, nor upon the principles set out in this Declaration, nor to limit their scope.'

on pushing ahead with the science, it is also open to an interpretation that puts it in tension with respect for human rights.

This part of the chapter is in two sections. First, I map the leading bioethical positions (such as those compressed within the UNESCO Declaration) that comprise the background ethical plurality. For this purpose, I sketch what I have elsewhere termed the 'bioethical triangle'.[11] Secondly, I comment on the increasingly important idea of human dignity, an idea which features strongly at two points of the triangle.

A. The Nature of the Ethical Plurality and the Bioethical Triangle

If we were to try to tease out the nature of the cosmopolitan commitments from the UNESCO Declaration, we would find three competing ethics—the three ethical views that make up the bioethical triangle. These are: the utilitarian view, advocating the pursuit of human welfare (human health, wealth, and happiness); the human rights view; and the view of a constituency that I have previously termed 'the dignitarian alliance'.[12] Each of these constituencies has its own distinctive ethical perspective. Occasionally, the perspectives converge to invite regulators to act on a consensus (as is the case at present with human reproductive cloning); sometimes, a synthesis of utilitarian and human rights thinking will emerge;[13] but, too often for regulatory comfort, these perspectives generate competing and conflicting views (as is the case, for instance, with therapeutic cloning and human embryonic stem cell research).[14]

Utilitarians count utility (encompassing individual pleasure and preference satisfaction, and the like, together with convenience and economy), the sum of utilities being aggregated in the credit column; and they count disutility (encompassing individual pain, suffering, and distress, and the frustration of

[11] See Brownsword, 'Three Bioethical Approaches: A Triangle to be Squared', paper presented at international conference on the patentability of biotechnology organised by the Sasakawa Peace Foundation, Tokyo, September 2004 (on file with the author) and 'Stem Cells and Cloning: Where the Regulatory Consensus Fails' (2005) 39 New England L Rev 535.

[12] See Brownsword, 'Bioethics Today, Bioethics Tomorrow: Stem Cell Research and the "Dignitarian Alliance"'(2003) 17 Notre Dame J of Law, Ethics and Public Policy 15.

[13] See Brownsword, 'Regulating Human Genetics: New Dilemmas for a New Millennium' (2004) 12 Med L Rev 14.

[14] All 191 members of the United Nations support a prohibition on human reproductive cloning. However, after four years of deliberation, during which time efforts were made to achieve a consensus covering the regulation of all uses of cloning technology (reproductive and therapeutic) in humans, the nations remained divided. On 18 February 2005, the Legal Committee voted 71 in favour, 35 against, with 43 abstentions, to recommend to the General Assembly that members should be called on 'to prohibit all forms of human cloning inasmuch as they are incompatible with human dignity and the protection of human life' (UN press release GA/L/3271: <http://www.un.org/News/Press/docs/2005/gal3271.doc.htm>). On 8 March 2005, the General Assembly accepted this recommendation, 84 members voting in favour of the (non-binding) UN Declaration on Human Cloning, with 34 against and 37 abstentions (UN press release GA/10333:<http://www.un.org/News/Press/docs/2005/ga10333.doc.htm>).

preferences and the like, together with cost, inconvenience, and the general expenditure of resources), such disutilities being totalled in the debit column. For utilitarians, the maximization of utility and the minimization of disutility is all that counts; to do the right thing is to identify and adopt whichever option seems most likely to maximize utility or minimize disutility. Broadly speaking, Article 4 of the UNESCO Declaration—providing that '[i]n applying and advancing scientific knowledge, medical practice and associated technologies, direct and indirect benefits to patients, research participants and other affected individuals should be maximized and any possible harm to such individuals should be minimized'—speaks this kind of language.

Although utilitarians count the benefits/harms in relation to all those who are affected by an action, they restrict the calculation to those who are judged capable of experiencing pain or pleasure, having preferences, and so on. Insofar as a human embryo is a pre-sentient life form, utilitarianism is not directly engaged—pre-sentient life fails to show interests in the relevant sense, and the logic of utilitarian thinking is that the same applies to post-sentient humans—as Lord Mustill famously remarked in the *Bland* case,[15] to talk about the 'best interests' of a human who is in a persistent or permanent vegetative state is to ignore the 'distressing truth' that they no longer have any interests in a material sense.[16] On the other hand, utilitarianism is directly engaged by sentient non-human life. Accordingly, in the *Harvard Onco-Mouse* case,[17] the utilitarian-minded examiners thought it perfectly sensible to weigh in the moral balance (that they took to be required by Article 53(a) of the European Patent Convention) the distress occasioned to a genetically engineered mouse.[18]

For human rights theorists, the key is not positive or negative consequences so much as respect for individual human rights. Much of the history of bioethics hinges on ethical opposition to the idea that the interests of individuals may be legitimately sacrificed for the greater good. While Article 3(2) of the UNESCO Declaration emphatically confirms this opposition, insisting that 'The interests and welfare of the individual should have priority over the sole interest of science or society', which is to say that humans have rights and that they must be taken seriously, it remains moot who precisely qualifies for such protection. Like the utilitarians, human rights theorists do not usually recognize pre-sentient life as a bearer of rights.[19] Indeed, in the human rights tradition, an even more restrictive view is taken—even sentience does not suffice. Thus, in two recent decisions at Strasbourg, the European Court of Human Rights has declined to affirm that

[15] *Airedale NHS Trust v Bland* [1993] 1 All ER 821.

[16] *Ibid*, at 894.

[17] OJ EPO 10/1992, 590.

[18] See further, *infra* at 63.

[19] Compare the position presupposed by Carl Cohen in C. Cohen and T. Regan, *The Animal Rights Debate* (2001) at 35: 'To be a moral agent is to be able to grasp the generality of moral restrictions on our will. Humans understand that some acts may be in our interest and yet must not be willed because they are simply wrong. This capacity for moral judgment does not arise in the animal world; rats can neither exercise nor respond to moral claims.'

either a human embryo or a six-month foetus is protected by the Convention right to life.[20] If the embryo has no *direct* protective entitlement, the only caveats that will be imposed by this ethical approach either relate to the need for free and informed consent by relevant rights-holders (such as women who donate eggs or couples who donate embryos), or they reflect any *indirect* or precautionary considerations.

It should be said, of course, that the caveats relating to free and informed consent are by no means trivial, unproblematic, or easily satisfied. Indeed, under pressure from proponents of human rights, these requirements have tended to become ever more demanding (particularly so in relation to the information to be supplied). It should also be emphasized that the indirect or precautionary arguments carry some weight. Given that the human rights ethic is fundamentally committed to human freedom, the claim that certain acts should not be permitted lest they *indirectly* impact negatively on rights-holders, or that precaution should be exercised before excluding those life forms that might (for all we know) have the required characteristics, will not be accepted just like that. Nevertheless, any reflective proponent of the human rights ethic will recognize that claims of this kind do need to be taken seriously.[21]

By now, we are well-versed in the tension between utilitarian promotion of the general good (where little or no attention is paid to the *distribution* of utility or disutility) and the constraints imposed if individual rights are to be respected.[22] Where human rights has made its mark, it is axiomatic, as we have just said, that best practice demands careful attention to free and informed consent, that the capacity for autonomous decision-making should be respected, that privacy and confidentiality should be protected, and so on. However, if bioethics was once a two-way contest between utilitarians and human rights theorists, this is no longer the case.

Where a technology impacts on the human body, as is particularly the case with the human genetics applications of biotechnology, this is widely seen as raising concerns about human dignity. Even in the case of information and communication technology which, issues of (Internet) content regulation apart, largely raises questions within and between utilitarian and human rights thinking, we find concerns about human dignity being expressed once a bio-application is proposed—witness, for example, the discussion by the European Group on Ethics in Science and New Technologies with regard to ICT implants in the

[20] *Vo v France* (2005) 40 EHRR 12 (a six-month foetus); *Evans v UK* [2006] ECHR 200 (7 March 2006) (a human embryo). In *Evans*, on further appeal to the Grand Chamber, the Court held unanimously that there had been no violation of Art 2 (transcript, 10 April 2007).

[21] See, further, Beyleveld and Brownsword, 'Principle, Proceduralism and Precaution in a Community of Rights' (2006) 19 Ratio Juris 141.

[22] For some vivid examples of researchers putting their projects before their research participants, see A. Plomer, *The Law and Ethics of Medical Research—International Bioethics and Human Rights* (2005), Ch 2 (US human radiation experiments), Ch 3 (UK Porton Down experiments), and Ch 6 (research trials in developing countries).

human body.[23] At all events, what we now have is a third ethical constituency, an alliance of dignitarian views, making up the plurality.[24]

The dignitarian perspective condemns any practice, process, or product—human reproductive cloning, therapeutic cloning, and stem cell research using human embryos being prime examples[25]—which it judges to compromise human dignity. Such condemnation (by reference to human dignity) operates as a 'conversation stopper';[26] but the dignitarians are not troubled—to say that something violates human dignity is the ultimate condemnation. The emergence of the new dignitarian view creates a genuinely triangular contest, the dignitarians disagreeing as much with the utilitarians as they do with the human rights constituency—with the former because they do not think that consequences, even entirely 'beneficial' consequences (that is, 'beneficial' relative to a utilitarian standard), are determinative; and with the latter because they do not think that informed consent cures the compromising of human dignity.

Insofar as the dignitarian view expresses a religious or a Kantian approach, the duty to respect human dignity (or not to compromise human dignity) will be treated as cosmopolitan; and, because the sharp end of dignitarian ethics is to identify which practices should be prohibited, it invites being regarded as imperialistic.[27] Insofar as it is communitarian, the dignitarian view might be of more limited range—for example, insisting on embryo protection at home but without arguing that all nations should do the same thing. Amongst its arguable cosmopolitan principles, dignitarianism decrees that human life should be protected and respected (from the point of conception onwards), that life

[23] Opinion No 20 of the European Group on Ethics in Science and New Technologies to the European Commission, *Ethical Aspects of ICT Implants in the Human Body* (adopted 16:03:2005).

[24] Compare Brownsword, 'Biotechnology and Rights: Where are We Coming from and Where are We Going?', in M. Klang and A. Murray (eds), *Human Rights in the Digital Age* (2005) 219; and Millns, 'Consolidating Bio-rights in Europe', in F. Francioni (ed), *Biotechnologies and International Human Rights* (2007) 71.

[25] Similarly, dignitarians have objected to the recently proposed use of hybrid cytoplasmic embryos as research tools. For discussion, see: House of Commons Science and Technology Select Committee, *Government Proposals for the Regulation of Hybrid and Chimera Embryos* (Fifth Report of Session 2006–07) (HC 272-I, 5 April 2007); and the Academy of Medical Sciences, *Inter-Species Embryos* (2007) at 28–30 (especially, para 8.3.1, for consideration of the objection that 'the creation of human embryos containing animal material is unacceptable because it subverts the animal-human species distinction and undermines human dignity and human rights').

[26] See Birnbacher, 'Do Modern Reproductive Technologies Violate Human Dignity?', in E. Hildt and D. Mieth (eds), *In Vitro Fertilisation in the 1990s* (1998) 325. Compare Nuffield Council on Bioethics, *Genetically Modified Crops: the Ethical and Social Issues* (London, 1999), at 96 (those who contend that genetic modification is intrinsically wrong or unnatural present views that 'have something of an "unarguable" quality, inasmuch as no amount of information, explanation or rationalisation would move a person with such views from their position').

[27] Compare Brownsword, 'Cloning, Zoning, and the Harm Principle', in S.A.M. McLean (ed), *First Do No Harm* (2006) 525.

should be recognized as having no price (hence, it should not be commercialized), and that life should not be instrumentalized (hence, it should not be commodified).[28]

Finally, it should be understood that, while the third point of the bioethical triangle brings together a range of duty-driven approaches, each taking it as axiomatic that human dignity must not be compromised, the dignitarian alliance does not have a monopoly on the idea that human dignity should be respected. For, human rights thinking also subscribes (albeit in a rights-led way) to the fundamental importance of human dignity. It is, thus, to the elusive concept of human dignity that we must now turn.

B. Human Dignity

In debates concerning the ethics and regulation of biotechnology, human dignity is an idea that appears in two very different roles, in the one case acting in support of individual autonomy (human dignity as empowerment) and, in the other case, acting as a constraint on autonomy (human dignity as constraint).[29]

We can try to disentangle and clarify these two approaches to human dignity by taking one step at a time. First, we can identify the salient features of human dignity as empowerment; secondly, we can do the same with human dignity as constraint; thirdly, we can point to the contrast between human dignity, on the one hand, as a distinctive characteristic of individual persons that speaks to their moral entitlement and obligation and, on the other hand, as a collective good to be respected by a community; fourthly, we can anticipate where practical tensions are likely to arise between these approaches; and finally, we can consider the possible extension of these rival conceptions of human dignity to matters of the environment (and, thus, to possible applications of green biotechnology).

1. Human Dignity as Empowerment

The conception of human dignity as empowerment is very closely linked with modern human rights thinking. In particular, human dignity is explicitly declared to be one of the foundational ideas in the Universal Declaration of Human Rights (1948), and its partner Covenants on Economic, Social and Cultural Rights (1966), and on Civil and Political Rights (1966). Thus, the Preamble to each of these instruments provides that 'recognition of the inherent dignity and of the equal and inalienable rights of all members of the human

[28] Compare Caulfield and Brownsword, 'Human Dignity: A Guide to Policy Making in the Biotechnology Era' (2006) 7 Nature Reviews Genetics 72.

[29] The terminology here comes from D. Beyleveld and R. Brownsword, *Human Dignity in Bioethics and Biolaw* (2001). Compare R. Dworkin, *Life's Dominion* (1993) especially at 233–7.

family is the foundation of freedom, justice and peace in the world'; and Article 1 of the Universal Declaration famously proclaims that 'All human beings are born free and equal in dignity and rights.' That is to say, each and every human being has inherent dignity; it is this *inherent* dignity that grounds (or accounts for) the possession of *inalienable* human rights; and, because all humans have dignity, they hold rights equally. So understood, the injunction to respect human dignity is much more than a demand that we commonly make in contexts where we detect demeaning or degrading treatment, or where we are trying to give weight to an interest in privacy—it is perceived to be the infrastructure on which the entire superstructure of human rights is constructed.

If the inherent dignity of humans is the justifying reason for the possession of human rights, what is it precisely that dignity connotes? To say that humans have dignity, meaning that humans have a value simply by virtue of being members of the human species, is not altogether convincing—certainly not to those who are *not* members of the human species and not even to fellow humans. For, any attempt to privilege the members of a particular species, including the members of the human species, merely by virtue of their species-membership will attract the charge of 'speciesism'. Such a response is arbitrary and it plainly will not do. If humans have a value, it must be for a better reason than this.

Humans, let us concede, are different from rocks and stones; humans have the capacity to value their existence; they have preferences; and they have the capacity to make free and informed choices—or, at any rate, such is the phenomenology of human experience. When we are throwing pebbles on a beach, we need not worry whether we are acting against the pebbles' preferences or interfering with their purposes, because quite simply they have none. Similarly, although computers are getting ever smarter, I do not yet pause to think that I might be acting unfairly in using my laptop more than my desktop computer. When, however, we discriminate between humans or throw them around, it is a very different matter. So, dignity, in the sense of the capacity to make one's own choices, or to value one's own preferences, and so on, distinguishes humans from inanimate objects and, quite probably, from most other non-human living things. However, we are still some way short of seeing how human dignity justifies human rights in anything more than a limited way.

From the mere fact that humans have the capacity to make choices or to value their preferences, it takes a leap to maintain that humans have a right to make choices and to act on them. Nevertheless, if we accept that human dignity (*qua* the capacity to make choices) is worthy of respect, then the following triple demand is invited: that one's capacity for making one's own choices should be recognized; that, other things being equal, the choices one freely makes should be respected; and that the need for a supportive context for autonomous decision-making should be appreciated and acted upon. Such demands may then be expressed as entitlements that are due to each human as of right.

2. *Human Dignity as Constraint*

Philippe Séguin, President of the National Assembly of the French Republic, remarked in the mid-1990s that the trend towards the enactment of bioethics laws (such as the three French Acts on bioethics of July 1994):

> illustrates a growing awareness around the world that legislators must, despite the difficulties, act to ensure that *science develops with respect for human dignity and fundamental human rights*, and in line with national democratic traditions.[30]

This trend is further illustrated in many different instruments. From the Council of Europe's Convention on Human Rights and Biomedicine[31] through to the EC Directive on the Legal Protection of Biotechnological Inventions[32] we find ringing declarations in favour of human dignity; and, as we have said, respect for human dignity is treated as axiomatic in the three UNESCO Declarations. Insofar as these provisions simply reinforce the demand that human rights should be respected, they say little that is new. However, it is in these most recent appeals to human dignity that we find the new turn in the rhetoric of bioethics, specifically by reliance on human dignity as constraint.

In modern European bioethics, human dignity is articulated in a way that appeals to a coalition of Kantians, Catholics, and communitarians.[33] In fact, both interpretations—human dignity as empowerment and human dignity as constraint—can claim to be rooted in the seminal writing of Immanuel Kant. For, in Kant's work, we find not only the idea that humans have intrinsic dignity (which suggests a conception of human dignity as empowerment, albeit duty-driven rather than rights-driven)[34] but also that human dignity has no price and that humans owe themselves a duty of self-esteem (which might suggest a conception of human dignity as constraint). In *The Metaphysics of Morals*, Kant collects together the strands of his thinking as follows:

> Every human being has a legitimate claim to respect from his fellow human beings and is *in turn* bound to respect every other. Humanity itself is a dignity; for a human being cannot be used merely as a means by any human being . . . but must always be used at the same time as an end. It is just in this that his dignity (personality) consists, by which he

[30] Philippe Séguin, 'Speech by Mr. Philippe Séguin', in *Proceedings of the Third Session of the International Bioethics Committee of UNESCO, September 1995*, vol i (Paris: International Bioethics Committee of UNESCO, 1995) 119, 120 (emphasis added).

[31] This Convention is sometimes referred to as 'the Bioethics Convention'. Its full title is Convention for the Protection of Human Rights and Dignity of the Human Being with Regard to the Application of Biology and Medicine: Convention on Human Rights and Biomedicine, 1996.

[32] Directive 98/44/EC; OJ L 213/13, 30.7.98.

[33] Generally, see J. Rendtorff and P. Kemp, *Basic Ethical Principles in European Bioethics and Biolaw* (vol I: Autonomy, Dignity, Integrity and Vulnerability) (1999).

[34] For the significance of this distinction, see Beyleveld and Brownsword, 'Human Dignity, Human Rights, and Human Genetics', in R. Brownsword, W.R. Cornish, and M. Llewelyn (eds), *Human Genetics and the Law: Regulating a Revolution* (1998) 69; and Beyleveld and Brownsword, *supra* note 29.

raises himself above all other beings in the world that are not human beings and yet can be used, and so over all *things*. But just as he cannot give himself away for any price (this would conflict with his duty of self-esteem), so neither can he act contrary to the equally necessary self-esteem of others, as human beings, that is, he is under obligation to acknowledge, in a practical way, the dignity of humanity in every other human being. Hence there rests on him a duty regarding the respect that must be shown to every other human being.[35]

In these much-quoted remarks, modern writers can (and do purport to) find support for a variety of supposed applications of Kantian morality, not just in practical matters generally but specifically within the fields of bioscience and biocommerce.[36] For, Kant's remarks, if taken literally, are an open invitation to claim that commercialization of the human body is an affront to dignity (by putting a price on something that is beyond price).

Thus, commodification and commercialization of the human body—whether in the form of trade in human organs or tissue, prostitution, surrogacy for profit, or patenting human genes—is just one of a number of practices that are regularly cited as instances of human dignity being compromised. Typically, human dignity as constraint also condemns sex selection and positive (eugenic) gene selection, germ-line gene therapy, embryo research and abortion, euthanasia and assisted suicide, genetic discrimination, and (perhaps top of its current list) human reproductive cloning. The list, though, is hardly closed; and there surely will be additions as technology opens up new bio-options and opportunities—for example, even if it were feasible (say, by the use of pre-implantation diagnostic biopsy techniques) to derive stem cells without harming the developmental prospects of the embryo, dignitarians might still object on the grounds of commodification (treating the embryo as a research tool) and, quite possibly, commercialization of human life.[37]

3. *Two Reference Points for Human Dignity*

When we order our thinking about human dignity, we can fix on two different reference points. One reference point is the idea that human dignity speaks to what is special or specific about humans, that is to say, what is intrinsically and

[35] I. Kant, *The Metaphysics of Morals* (translated and edited by Mary Gregor) (1996) (first published 1797) 209.

[36] See eg Wolbert, 'The Kantian Formula of Human Dignity and its Implications for Bioethics' (1998) 4 Human Reproduction & Genetic Ethics 18.

[37] Although commercialization and commodification are closely related to one another, they are not quite identical. Where an object or resource is exploited for its commercial value, it is necessarily commodified; it is treated as a mere thing. Hence, commercialization presupposes commodification. However, this does not hold in reverse; for, we might commodify a resource without commercializing it. For example, if we have a market for the sale of human babies, babies are both commercialized and commodified. But, if we permit social sex selection of babies, it is arguable that babies are commodified (treated as mere products, like consumables coming off a production line) but they are not yet commercialized.

universally distinctive about humans. As Francis Fukuyama has put it, the demand made in the name of human dignity is one for equal recognition which implies 'that when we strip all of a person's contingent and accidental characteristics away, there remains some essential human quality underneath that is worthy of a certain minimum level of respect.'[38] This reference point is to be contrasted with the idea that human dignity speaks less to what is special about humans *qua* humans and more to what is special about a particular community's idea of civilized life and the concomitant commitments of its members. Here, appeals to human dignity draw on what is distinctively valued concerning human social existence in a particular community—indeed, on the values and vision that distinguish the community as the particular community that it is and relative to which the community's members take their collective and individual identity. In principle, each starting point has the scope to support human dignity as empowerment as much as human dignity as constraint. In practice, though, whereas the former tends to be closely associated with human rights movements aimed at giving individuals the opportunity to flourish as self-determining authors of their own destinies, the latter (as expressed by the dignitarian alliance) combines a (Kantian) view of what is distinctive about humans (their dignity) with views about what defines life as civilized (and, thus, respectful of human dignity) in a particular community.

4. Practical Tension Between the Rival Approaches

Once put into practical application, there are two particular hot spots where we can anticipate a tension between human dignity as empowerment and human dignity as constraint.

First, because human dignity as constraint takes a more inclusive protective approach than human dignity as empowerment, there are likely to be tensions in relation to human life that lies beyond the direct protection of the latter. For example, in the cases already mentioned at the European Court of Human Rights,[39] it was human dignity as constraint that offered the more direct protective shield. Similarly, we will see that, wherever the question of using human embryos for research comes up, it is the dignitarians (relying on human dignity as constraint) who argue against permissive regulation. And, again, if it is proposed to extend reproductive choice by appealing to human dignity as empowerment, it is human dignity as constraint that responds in defence of embryos which are subjected to screening procedures (for positive or negative selection).

[38] F. Fukuyama, *Our Posthuman Future* (2002), at 149. According to Fukuyama, although 'many would list human reason and human moral choice as the most important unique human characteristics that give our species dignity, I would argue that possession of the full human emotional gamut is at least as important, if not more so' (p 169).

[39] *Supra* note 20.

Secondly, whereas human dignity as empowerment pushes for the extension of individual autonomy, human dignity as constraint takes a more limited view.[40] Whilst the former is protective of choices that are made on a free and informed basis, the latter limits autonomy to those choices that do not compromise human dignity. This kind of tension is perfectly illustrated by the famous French dwarf-throwing ('*lancer de nain*') case,[41] where the Conseil d'Etat, upholding the restrictions on dwarf-throwing in clubs, presupposed human dignity as constraint in insisting that the dwarfs compromised their own dignity. Similarly, on at least one reading of the more recent *Omega* case,[42] the Bonn authorities closed the shooting arcade for reasons reflecting a conception of human dignity as constraint. In both cases, the (unsuccessful) argument in support of the autonomy of the dwarfs and games-players presupposed human dignity as empowerment.

5. *Human Dignity and the Environment*

In his book, *Crimes Against Nature*,[43] Robert Kennedy Jr launches a blistering attack on the Bush administration's reversal of established environmental protection policy—the principle that the polluter pays has, it seems, been turned on its head; the Environmental Protection Agency, Kennedy claims, has become more like a country club for America's polluters than a regulatory agency, and so on. At one point, Kennedy despairs of the weak enforcement against the meat barons:

> In violating our federal antipollution laws like the Clean Air and Clean Water Acts, the meat barons are stealing something that belongs to the American public—the purity of our land and waterways, healthy air, abundant fisheries. They have lulled people into accepting unhealthy, unsavoury meat and have gained a stranglehold on US commodity production that poses a genuine threat to our food independence and national security.

[40] Compare Feldman, 'Human Dignity as a Legal Value: Part I' (1999) PL 682, at 685, where it is rightly observed that human dignity can cut both ways: '[W]e must not assume that the idea of dignity is inextricably linked to a liberal-individualist view of human beings as people whose life-choices deserve respect. If the state takes a particular view on what is required for people to live dignified lives, it may introduce regulations to restrict the freedom which people have to make choices which, in the state's view, interfere with the dignity of the individual, a social group or the human race as a whole... The quest for human dignity may subvert rather than enhance choice... Once it becomes a tool in the hands of lawmakers and judges, the concept of human dignity is a two-edged sword.'

[41] Conseil d'Etat (27 October 1995) req nos 136–727 (Commune de Morsang-sur-Orge) and 143–578 (Ville d'Aix-en-Provence). See Roualt's note on the two decisions, *Les petites affiches* (24 January 1996, No 11) 30, at 32. And, for reflections on the case, with dignity being interpreted as the essence of humanity, see Edelman, 'La dignité de la personne humaine, un concept nouveau', Recueil Dalloz 1997, 23e Cahier, Chronique, 185, 187–8.

[42] Case C-36/02 *Omega Spielhallen-und Automatenaufstellungs-GmbH v Oberbürgermeisterin der Bundesstadt Bonn* (14 October 2004); OJ C 300/3, 04.12.2004. For comment, see Ackermann 42 CMLR (2005) 1107. WT/DS285/AB/R, 7 April 2005.

[43] R.F. Kennedy Jr, *Crimes Against Nature* (2005).

But our battle is more than a fight to save American farming and fishing communities. It is a fight to preserve human dignity.[44]

What should we make of this rhetoric of human dignity that emphasizes the importance of respecting the environment and eschewing practices that compromise environmental integrity? Is this the rhetoric of empowerment or of constraint? In fact, it seems to be a rhetoric that might be in line with either the constraint or the empowerment conception of human dignity—even though the gist of the critique is for restraint in both cases.

From the perspective of human dignity as empowerment, the way that we treat the environment is important insofar as it bears on the prospects of flourishing human autonomy.[45] If the sustainability of the environment is connected to the sustainability of autonomous human life, if a healthy environment is a precondition for healthy life, then the empowerment conception will argue for restraint. Hence, if agricultural biotechnology might have indirect negative effects from a human rights perspective, then at the very least it needs to be monitored. But there is also something akin to the voice of human dignity as constraint, possibly captured by a larger ethic of veneration.[46] At all events, where secular communitarianism is in play, a community might well identify itself by its culture, including its attitude to non-human life and the natural environment. On one reading, at least, this seems to be an important element of Kennedy's script.

C. Taking Stock

Disputes concerning the ethics of modern technologies are often simply two-dimensional: utilitarian cost/benefit calculations are set against human rights considerations. In general, unless there are major safety concerns, while utilitarians will assert the 'green light' ethics of proceeding, a case of promotion (of the technology) modulated by a degree of precaution,[47] human rights theorists will take an 'amber light' approach insisting that the technological traffic pauses

[44] *Ibid*, at 75.

[45] Compare the powerful presentation of this line of thinking in Westra, 'Environmental Rights and Human Rights: The Final Enclosure Movement', in R. Brownsword (ed), *Human Rights* (2004) 107.

[46] See Galloux *et al*, 'The Institutions of Bioethics', in M.W. Bauer and G. Gaskell (eds), *Biotechnology—The Making of a Global Controversy* (2002), where three ethical frames are identified—one organized around the principle of utility, a second around the principle of democracy, and a third around the principle of veneration. See, too, Nuffield Council on Bioethics, *Genetically Modified Crops: The Ethical and Social Issues* (London, May 1999) paras 1.32–1.50 (discussing the natural/unnatural boundary, taboos, and moral conservatism).

[47] Compare, Department of Biotechnology, Ministry of Science and Technology, Government of India, *National Biotechnology Development Strategy* (2006) at 23: 'A precautionary, yet promotional approach should be adopted in employing transgenic R & D activities based on technological feasibility, socio-economic considerations and promotion of trade.'

(to ensure rights clearance) before proceeding. By contrast, the ethical plurality in the context of which modern biotechnology is debated is three-dimensional, the dignitarian alliance taking issue with both utilitarians and human rights advocates. Of the three ethical perspectives, it is only the dignitarian view that is genuinely 'red light'. Somewhat confusingly, the idea of human dignity underlies both the human rights and the dignitarian view. However, once this double-take is identified, it is easier to see which version of human dignity is being contended for or presupposed.

2. Bioethical 'Floaters': Harm, Consent, and Precaution

With the bioethical triangle as our frame of reference, our next step is to locate three pervasive principles that, unless clearly anchored, have a tendency to be treated as neutral. For this reason, we can call the principles in question 'bioethical floaters'. They are: (i) that we should not harm others; (ii) that clinicians and researchers should act only with the free and informed consent of their patients/participants; and (iii) the so-called precautionary principle.[48]

A. The Harm Principle

The harm principle is very much at home in the context of utilitarian thinking. Famously, this is the setting for the Millian harm principle.[49] Equally famously, though, the principle that we should define the private sphere by reference to whether our acts occasion any 'harm' to 'others' is problematic. However, the problem is not one that is simply unique to utilitarianism; the problem is that the concepts of 'harm' and 'others' are contested at all points of the bioethical triangle.[50]

1. Utilitarianism

Following a utilitarian approach, 'harm' is largely equated with the stuff of disutility: that is to say, pain and suffering, distress, anxiety, the frustration of plans and non-satisfaction of preferences, and the like. As for 'others', quite simply, they are those beings who are capable of experiencing just this kind of pain and suffering.

[48] Although the precautionary principle is concerned with managing risk and avoiding harm, it is not just another version of the harm principle. Whereas the latter has been invoked across the board to argue against regulatory prohibition (and penal threat) unless there is a clear risk of harm to others, the history of the precautionary principle lies in the sphere of environmental protection where advocates argue in favour of regulatory prohibition notwithstanding the lack of compelling scientific evidence of harm. See further text *infra* at 37.

[49] Mill, 'On Liberty', in J.S. Mill, *Utilitarianism* (1962) (edited by Mary Warnock).

[50] See, further, Brownsword, *supra* note 27.

So, for instance, if the harm principle, so understood, is applied to reproductive human cloning, utilitarians will support prohibition so long as the procedure is unsafe and, thus, capable of generating various types of disutility. Should reproductive cloning become a safe and reliable procedure, and should it prove to be less disruptive of familial relations than some fear, utilitarians might find that they are running out of reasons to maintain their opposition. Of course, those who (for non-utilitarian reasons) are distressed by reproductive human cloning will pose some awkward questions for utilitarian-minded regulators—in response to which, they might simply disregard such distress as irrational or, following the spirit of Mill, judge that the price of freedom is that the preferences of some will have to be overridden.

Applying the same approach to therapeutic cloning, utilitarians will see no objection to terminating the development of a 100-cell human embryo. At that stage of embryonic development, the prevailing wisdom is that there is no possibility of pain and suffering being experienced. The embryo, to be sure, is a potential other; but, at the relevant time, it is not yet in that category (it is not yet an other) and so it is not capable of being harmed.

2. *Human Rights*

In its most robust and distinctive form, human rights theory interprets 'harm' by reference to the set of human rights, including both negative and positive rights within this set, and viewing such rights under a will or choice theory.[51] While consent can figure in other types of ethical thinking, it is the presupposition of will theory that gives informed consent its *integrated* role in human rights thinking, the rights-holder being able to authorize actions that would, consent apart, involve a violation of rights. It follows that, given this approach, it is human 'rights-holders' who are the relevant 'others'.

From such a perspective, human reproductive cloning (to persist with cloning as the illustrative case) is to be opposed where it threatens the well-being of rights-holders (as it might if cloning procedures issue in the birth of damaged children) or where the autonomy of rights-holders is compromised—hence the intense concerns about cloning for characteristics (particular skills and talents) and the closing off of an open future.[52]

As for therapeutic cloning, while human rights theorists might disagree about the stage at which foetal or neonatal life attains rights-holding status (if at all), no will theorist would argue that a 100-cell human embryo is a *paradigmatic* rights-holder. For, there is, at this stage, a very long way to go before the embryo is

[51] On the contrast between will (or choice) and interest theories of rights, see eg Hart, 'Bentham on Legal Rights', in A.W.B. Simpson (ed), *Oxford Essays in Jurisprudence* (Second Series) (1973) 171; and MacCormick, 'Rights in Legislation', in P.M.S. Hacker and J. Raz (eds), *Law, Morality, and Society* (1977) 189.

[52] See Part 3 of this chapter.

capable of actively participating in a community of rights. Accordingly, in the absence of extreme precaution or arguments that are indicative of indirect risk to (unproblematic) rights-holders, the harm principle offers no reason to prohibit therapeutic cloning.

3. The Dignitarians

For dignitarians, 'harm' is understood as any act that compromises human dignity; and 'others' are understood individually, collectively, and inclusively as members of the community. So understood, the harm principle demands that human reproductive cloning should be prohibited because, even if a clonee is not harmed in any sense that would be recognized by utilitarian or human rights standards, this directly compromises human dignity.

As for therapeutic cloning, the embryo is definitely an 'other', it is instrumentalized, possibly commodified, and this is a procedure that is harmful—human dignity is violated.

B. Consent

It is axiomatic in bioethics that those who are subjects of medical treatment or research should participate on the basis of free and informed consent. This is spelled out in Article 6 of the UNESCO Declaration, which provides:

> Any preventive, diagnostic and therapeutic medical intervention is only to be carried out with the prior, free and informed consent of the person concerned, based on adequate information. The consent should, where appropriate, be express and may be withdrawn by the person concerned at any time and for any reason without disadvantage or prejudice.

But, of course, this is very much a version of consent that is driven by the human rights perspective. If, by contrast, we take a utilitarian or a dignitarian view, consent will look much less important.

1. Utilitarianism

Utilitarians, to repeat, count utility and disutility; and, for utilitarians, utility and disutility is all that counts. As such, there is nothing special about consent or the lack of it. In general, it is easy to see the negatives in relation to consent collection. Obtaining consent might not always be practicable; where it is, it nevertheless incurs transaction costs; and, on some occasions, it might be downright distressful. Waiting for consents to be cleared might involve opportunity costs. Moreover, policies might be frustrated if, instead of saying 'yes', those who are asked to consent say 'no'. On the other hand, dealing on the basis of consent might ease matters *ex ante*, it might allay concern and weaken opposition, and it might be a convenient justificatory response *ex post*. Thus, there is no golden rule requiring that the consent of those upon whom an action or decision impacts should be obtained. For example, requiring researchers or doctors to deal on an informed consent basis with research participants or others is not necessarily an improvement on compulsion, ignorance, or paternalism. The calculation always depends on context, convenience, contingency, and circumstance. Having said this, in a culture where preferences strongly favour the currency of consent, even if there is no golden rule requiring consent, utilitarians might well accept the sense of a general rule to this effect.

2. *Human Rights*

Against the utilitarians, human rights theorists hold that what counts is respect for individual autonomy, entailing recognition of the right of individuals to make their own choices, to exercise control over their own person, property, and privacy, and to say 'yes' or to say 'no'. Taking individuals seriously, taking rights seriously, means taking consents and refusals seriously.

Viewed from a human rights perspective, consent by A might signal either a change of position or the creation of a new relationship. There is a tendency to focus more on the former than the latter, that is, on consent signalling A's willingness to modify his or her position in relation to the particular background scheme of rights and duties, permissions and immunities, and the like, that regulates his or her relationship with B (the recipient of the consent). As George Fletcher puts it:

> When individuals consent to undergo medical operations, to engage in sexual intercourse, to open their homes to police searches, or to testify against themselves in court, they convert what otherwise would be an invasion of their person or their rights into a harmless or justified activity.[53]

The latter function (the creation of a new relationship), however, should not be ignored. Where A signals consent to the creation of a new relationship with B, this might be by virtue of some simple dynamic (such as the giving of a promise or agreement to an exchange of goods) or it might be by virtue of some more complex institutional set (as is the case, for example, if A invokes the law of contract or the law of marriage) or regulated scheme (such as one licensing assisted conception or physician-assisted suicide). Where A thus signals consent, then other things being equal A is precluded from asserting that B may not justifiably rely on, or hold A to, the agreed change of position or the terms of the new relationship. Whether A signals a change of position or agreement to the creation of a new relationship, consent functions as a *procedural* justification (for the benefit of the recipient, B) rather than as a full *substantive* justification of the kind that would be appropriate if one were arguing for a particular human right itself.[54]

[53] G.P. Fletcher, *Basic Concepts of Legal Thought* (1996), at 109.

[54] See, further, Brownsword, 'The Cult of Consent: Fixation and Fallacy' (2004) 15 King's College LJ 223; and D. Beyleveld and R. Brownsword, *Consent in the Law* (2007).

3. The Dignitarians

For the dignitarians, it is human dignity, not consent, that counts; and it is the interests of all members of the community that count, not merely those of the consenting community. The fact that all parties consent to participation in dwarf-throwing, or to playful killing at the Laserdrome,[55] is irrelevant insofar as these activities compromise human dignity. The duty to respect human dignity is not switched off by the consent of others any more than that it is switched on again by a withdrawal or a refusal of consent. Consent, for the dignitarian, is largely a sideshow.[56]

To this, and to avoid misunderstanding, we should add an important caveat. The dignitarian constituency within the bioethical triangle is a particular substantive articulation of duty-based ethics. It does not follow, however, that all duty-based thinking rejects the significance of consent. For example, in their important (obligation-driven) analysis of informed consent, Neil Manson and Onora O'Neill take consent just as seriously as rights-led thinkers,[57] treating consent as a waiver relative to a backcloth of rights, obligations, and expectations. Hence:

> In consenting, we *waive* certain requirements on others not to treat us in certain ways (sometimes this will include waiving rights), or we *set aside* certain expectations, or *license* action that would *otherwise* be ethically or legally unacceptable... Where an act would *otherwise* wrong an individual or disrupt their legitimate expectations, that individual's consent can waive their right, or modify their expectations in particular cases, and so justify an act that would *otherwise* be unacceptable.[58]

So, we should not make the mistake of thinking that a duty-driven ethic in itself entails that consent is downgraded. However, as Manson and O'Neill recognize, where autonomy-limiting dignitarianism is in play, actions that are judged to degrade others will be treated as unacceptable, the consent of those degraded others notwithstanding.[59]

C. Precaution

The precautionary principle has been developed in the context of environmental protection law, the need for sustainable development, and an emphatic support for the polluter pays principle. The principle is formulated in many different ways but the common denominator is that, in the face of scientific uncertainty,

[55] Case C-36/02 *Omega Spielhallen-und Automatenaufstellungs-GmbH v Oberbürgermeisterin der Bundesstadt Bonn* (14 October 2004); OJ C300/3, 04.12.2004.

[56] Compare Kadidal, 'Obscenity in the Age of Mechanical Reproduction' (1996) 44 American J of Comparative Law 353.

[57] N.C. Manson and O. O'Neill, *Rethinking Informed Consent in Bioethics* (2007).

[58] *Ibid*, at 72–3. [59] *Ibid*, especially at 20–1.

and possible damage to the environment, preventive steps should be taken.[60] This could be applied, say, to GM crops but there is no reason why it should not also apply to human genetics or similar science that impacts on humans. It might be said, for example, that it is not ethical to carry out clinical applications of stem cell research until we are confident that they are relatively risk-free, or that germ-line gene therapy should be put off limits.[61]

Cass Sunstein has written a damning critique of the precautionary principle for the simple reason that taking precautionary measures itself involves risk (sacrifice) and that this must be brought into what is otherwise a one-sided narrow-screen calculation.[62] Even if we adopt the wide-screen approach, however, the point is that precaution plays differently depending upon where we are coming from in the bioethical triangle.

1. *Utilitarianism*

Generally, precaution testifies to a utilitarian concern about safety. However, unless there is a rule-utilitarian justification for adopting a strict precautionary approach in an identifiable class of situations, the precautionary calculation should not be carried out in a one-sided way. For the utilitarian, each option should be reviewed. If it is proposed that we should give up on certain lines of research, it is not enough to plead precaution or risk-aversion or distress avoid-ance. The utilitarian will want to know what the costs of precaution are (quite literally, how much it is costing to buy a bit more safety, and what benefits are being foregone); and the anticipated net result of the precautionary option will have to be superior to that of any alternative option.

2. *Human Rights*

Because precaution relates to harm or damage, as conceived by a particular ethical viewpoint, when we adopt a human rights perspective, it is precaution for the sake of human rights that becomes focal. There are some tricky questions to be resolved here. However, suffice it to say that in a community of rights there

[60] For a general description, see Nuffield Council on Bioethics, *Genetically Modified Crops: The Ethical and Social Issues* (1999) at 162: 'This is the rule that permits governments to impose restrictions on otherwise legitimate commercial activities, if there is a risk, even if not yet a scien-tifically demonstrated risk, of environmental damage.' For a more specific elaboration, see Case T-13/99, *Pfizer* [2002] ECR II-3305, at para 143: 'a preventive measure cannot properly be based on a purely hypothetical approach to risk, founded on mere conjecture which has not been scientifically verified.' So, mere conjecture and hypothesis will not suffice. But, a precautionary measure may apply where the risk 'has not yet been fully demonstrated' (para 146). The underlying science must be consistent with principles of 'excellence, transparency and independence' (para 172).

[61] Compare the precautionary approach seemingly incorporated in Arts 16 (protection of future generations) and 20 (risk assessment and management) of the UNESCO Declaration.

[62] C. Sunstein, *Laws of Fear* (2005).

might be an argument that we should exercise caution when dealing with human embryos because we might be wrong in assuming that they are not bearers of rights;[63] or, again, it might be suggested that we need to be careful that treating human embryos as a research tool does not change our attitudes so that we suffer a loss of respect for fellow rights-holders. How far we might run with these particular arguments is moot; but there is certainly a precautionary form of argument available, and waiting to be fully articulated, within human rights thinking.

3. The Dignitarians

Dignitarians regard the compromising of human dignity as a self-standing reason for restraint—precaution does not enter into such an ethic in quite the same way. Crucially, dignitarianism is not interested in consequential calculation, whether of the likely costs and benefits (in the way that a utilitarian impact statement might be drafted) or of the effect on rights (including maximin-type precaution that might appeal within an ethic of human rights). Dignitarianism, it cannot be emphasized too strongly, is a red light not an amber light ethic; its credo is that we should not proceed at all (where activities are judged categorically to compromise human dignity) rather than that we should proceed only with care. For dignitarians, the proposition that we should exercise precaution against the risk that biotechnology might go wrong misses the point; the point is that, if biotechnology goes right, human dignity has already been compromised.

D. Taking Stock

The ethical plurality that sets the stage for the regulation of biotechnology brings together three key approaches: the consequential approach of the utilitarians; the rights-led approach of the human rights constituency (underpinned by a conception of human dignity as empowerment); and the duty-driven approach of the dignitarian alliance (pleading human dignity as constraint). In this setting, the rhetoric of harm, consent, and precaution is not neutral. These terms take on their own particular significance (or insignificance) when viewed from a particular point of the bioethical triangle.

With this groundwork in place, we can move on to reflect on the significance of the ethical plurality, together with its non-neutral rhetoric, in the context of the regulation of biotechnology. Whereas, in the next part of the chapter, the focus is on the ethical plurality as a force and a factor operating outside the law, in the following part it is on the plurality once incorporated within the law.

63 Compare the analysis in Beyleveld and Brownsword, *supra* note 29, especially Ch 6.

3. Ethics as an External Benchmark

Regulators might attempt to 'legitimate' their particular regime for the control or facilitation of biotechnology by various kinds of appeal. They might appeal to various public or national interest considerations, such as projected medical or economic benefits, or the alleviation of food shortages, or the protection of small farming interests, or the promotion of consumer choice, and so on. When they make such claims, regulators simply maintain that there is good reason for the regulatory position which, accordingly, merits respect. When regulators face a 'legitimation' crisis, it is because the reasons offered up are not persuasive.

However, even if there is no legitimation crisis as such, regulators might still be challenged more narrowly as to the ethical or moral 'legitimacy' of their regulatory position. So, for example, it might not be enough for regulators to present an intervention as one that promises to assist medical research (and, in due course, to deliver therapies for major human diseases); if the research in question is judged to violate moral or ethical standards—for example, by violating human rights or compromising human dignity—regulatees will seek to engage regulators as to the (moral or ethical) legitimacy of their actions. In other words, it might be conceded that regulators who seek to maximize the benefits of biotechnology may 'legitimate' their actions by reference to that general purpose, while insisting that regulation lacks 'legitimacy' if it violates the relevant moral or ethical standards.

Where the moral or ethical legitimacy of regulation is challenged, regulators might offer more than one kind of response. While stronger forms of legitimacy claim offer a substantive (or, on the merits) justification for the particular regulatory position, weaker forms of claim defend the regulatory position, not in itself, but as the outcome of what is claimed to be a fair regulatory process. The bottom line for the stronger kind of claim is that the regulation has legitimacy because, so it is argued, it is directly underwritten by relevant ethical standards—for example, that it is in line with respect for human rights and human dignity. Clearly, though, where legitimacy claims of this kind are advanced, and where they are advanced against a backcloth of ethical pluralism, it is difficult to speak convincingly to all constituencies. Faced with such difficulties, regulators might instead resort to the weaker (procedural) kind of claim. Again, however, there are serious difficulties in persuading those sections of the plurality whose views are not fully reflected in the regulatory outcome that their voices have been truly heard.[64] In what follows, I will speak principally to matters of substantive legitimacy.

[64] Compare Black, 'Regulation as Facilitation: Negotiating the Genetic Revolution', in R. Brownsword, W.R. Cornish, and M. Llewelyn (eds), *Human Genetics and the Law: Regulating a Revolution* (1998) 29.

In this part of the chapter, we can look at two hard cases for regulators who aspire to substantive legitimacy for their regulatory positions. The first case is that of therapeutic cloning and human embryonic stem cell research; and the second is that of the use of pre-implantation genetic diagnosis (PGD) and pre-implantation tissue-typing (PTT) for reproductive purposes.

A. Therapeutic Cloning and Human Embryonic Stem Cell Research

The ethics, and legitimacy, of so-called 'therapeutic cloning', or allied procedures involving the isolation and manipulation of human embryonic stem cells, are deeply controversial. Nations that are united in condemning human reproductive cloning and supporting its worldwide prohibition become divided when it is proposed that therapeutic cloning, too, should be covered by the prohibition;[65] and, around the world, within nation states, a patchwork of regulatory provisions is to be found. In some jurisdictions, there are outright prohibitions; in others, the position is permissive but heavily qualified (for example, restricting researchers to the use of supernumerary human embryos); in others, the regulation is relatively liberal (notably in the United Kingdom);[66] and, in yet others, we find a form of regulatory schizophrenia with prohibition coexisting with permission.[67]

Where a positive view is taken, the general perception is that therapeutic cloning and stem cell research represent a field of very considerable medical significance.[68] Whether we are thinking about bespoke stem cells and their derivatives for transplantation or generic lines or collections for drug-testing, it is conceivable that there could be major benefits resulting from this kind of research. There is a pressure on regulators to pave the way for such research initiatives. But, how might a permissive regulatory approach play relative to the bioethical triangle? After making some remarks, both general and more

[65] All 191 members of the UN agree in principle that there should be a declaration banning human reproductive cloning. However, only a minority agrees that the ban should extend to all uses of cloning technology in humans. See *supra* note 14.

[66] The regulatory Authority (the HFEA) has approved an application from Professor Ian Wilmut (who led the team that cloned Dolly the sheep) to undertake therapeutic cloning of human embryos with a view to understanding more about the early development of motor neurone disease: see Nuffield Council on Bioethics, *Bioethics News* (Issue 43, 8 October 2004) (on file with author).

[67] Such 'schizophrenia' might be reflected in funding restrictions that do not map on to the legislative scheme of permission and prohibition; or in prohibition subject to quite unprincipled permissive exemption. For a helpful review of the regulatory scene in Europe, see Halliday, 'A Comparative Approach to the Regulation of Human Embryonic Stem Cell Research in Europe' (2004) 12 Med L Rev 40; and, for the US, see Peterson, 'A Comprehensive National Policy to Stop Human Cloning: An Analysis of the Human Cloning Prohibition Act of 2001 with Recommendations for Federal and State Legislatures' (2003) 17 Notre Dame J of Law, Ethics and Public Policy 217.

[68] See Brownsword, 'Stem Cells, Superman, and the Report of the Select Committee' (2002) 65 MLR 568; and, even after many twists and turns, not to mention South Korean scandals, the optimism persists, see eg Sample, 'How Stem Cells Can Turn Back the Biological Clock', *The Guardian* (15 June 2006), 15.

particular, in response to this question, we can consider how the United Kingdom's permissive regulatory approach was defended against dignitarian objections.

1. *Why Not Take a Permissive Approach?*

For both utilitarians and human rights theorists, the promised benefits of human embryonic stem cell research are positive considerations. On the negative side, there seem to be no real counterweights: we assume that terminated embryos do not suffer and nor do they have rights; and, provided that those who donate eggs or embryos do so on the basis of free and informed consent, all is well.

Yet, are we looking hard enough for the negatives? First, are we being sufficiently careful to check for indirect damage to rights-holders? Maybe human embryos do not have rights; and, in this respect, they are comparable to other life forms that do not have rights. However, one concern about the mistreatment of such life forms is that it sets a bad example that leads, *indirectly*, to a weakening of respect for rights-holders. Granted, there is no straight-line connection. Notoriously, Hitler was a dog lover; and it might be that a person who is cruel to cats behaves perfectly properly when dealing with rights-holders. We really do not know. Secondly, there is the even less straightforward risk of the context that sustains a community of rights being compromised. Just as individual acts of pollution cumulatively corrode the physical environment, might it be that resort to therapeutic cloning cumulatively changes the social environment in a way that undermines the conditions for a flourishing community of rights? Some commentators believe that genetic engineering poses an insidious risk of this kind. However, although therapeutic cloning might involve some genetic manipulation, it is not carried through into social relations in the way that would occur in reproductive contexts. If there is a risk here it is that people perhaps live longer or that, thanks to stem cell treatments for Parkinson's and Alzheimer's disease, the quality of life for large classes of elderly people is considerably improved. These developments might have important economic consequences as well as consequences for the allocation of public health care resources; however, if there is some deeper threat to a rights-respecting community, it is unclear what it is. Accordingly, if the onus is on those who assert such indirect negatives, then the burden of proof is liable to be heavy (at any rate, for the time being). We might conclude, therefore, that therapeutic cloning seems to be ethically clean—unless, of course, one is a dignitarian.

Taking a dignitarian view (guided by a conception of human dignity as constraint), the ethical imperative is to avoid compromising human dignity; and the application is perfectly clear—the human embryo is to be directly respected. It matters not that the human embryo cannot experience distress or make its own choices. It is *not* like a rock or a stone. It is a living thing and a member of the human species. As such, it is protected by the overarching value of respect for human dignity. It has moral status and to treat it like a rock or a stone is to compromise human dignity. Once we have considered the details of a permissive regulatory approach, we can return to the dignitarian objection.

2. *The Devil is in the Detail (of Permissive Regulatory Positions)*

Translating these ethical judgments—the dignitarian apart—into preferred regulatory positions is more complex. Assuming that therapeutic cloning and human embryonic stem cell research is to be permitted, regulators might hedge their permission in various ways—for example, by permitting only supernumerary embryos that would be destroyed anyway to be made available for research. Even in the United Kingdom, where the regulatory framework allows for the use of supernumerary embryos as well as embryos donated for research *ab initio*, and (in defiance of Article 18(2) of the Convention on Human Rights and Biomedicine) allows for the use of embryos specially created for research, there are restrictions. In particular, research is only to be licensed where the regulatory Authority judges that it is necessary[69] (in the sense of it not being possible to do the research in any other way) as well as being necessary or desirable in relation to one of the approved statutory purposes.[70] Yet, why impose such restrictions? Would they (or more extensive restrictions) make sense from the perspective of utility or human rights?

Consider the logic of utilitarianism. Generally speaking, from a utilitarian perspective, the local context and culture will influence the particular contours of even a permissive regulatory regime. It follows that utilitarians might advocate some particular regulatory restrictions in some places at some times, but not in other places or in any place at other times. Nevertheless, how would we expect utilitarians to respond to the kind of restrictions mentioned?

First, utilitarians would surely accept efficiency arguments in favour of prioritizing the use of embryos that are surplus to requirement in IVF programmes and that will have to be destroyed anyway. Not only would it seem wasteful not to use such supernumerary embryos before turning to other sources, it might also chime in with public preference and sentiment. Against such utilities, there might be deficits if the embryos that are available to be sourced from IVF programmes are lower grade and not optimal for stem cell research. However, if the regulatory position merely prioritizes the use of supernumerary embryos provided that they are suitable for research, this does not hinder research and there is no real difficulty.

Secondly, if the regulatory position prohibits the creation of embryos specifically for research, this might prove a serious obstacle to stem cell research which

[69] Human Fertilisation and Embryology Act 1990, Sch 2, para 3(6).

[70] Originally, there were five such purposes: see *ibid*, Sch 2, para 3(2). To clear the way for human embryonic stem cell research and therapeutic cloning, three further purposes were added by the Human Fertilisation and Embryology (Research Purposes) Regulations 2001, SI 2001/188.

relies on specially designed embryos as its research tools. To be sure, stem cell research would not grind to a standstill: therapies might be developed using adult stem cells (or even foetal stem cells); stem cell lines might be imported; and there would still be supernumerary embryos available for research. Nevertheless, if the prohibition seriously interferes with research, utilitarians would need some convincing that it makes sense. In this light, utilitarians would want to take a hard look at Article 18(2) of the Convention on Human Rights and Biomedicine (as, indeed, they would want some convincing that research should be confined to adult stem cells).

Thirdly, what would utilitarians make of the twin constraints of *necessity* (meaning that the research cannot be carried out without using human embryos—in particular, that it cannot be conducted using non-human animals) and *proportionality* (meaning that human embryos should not be used for 'trivial' research purposes, such as, to take the standard example, research into baldness)? On the face of it, neither restriction would make immediate sense. If research on human embryos causes no distress, while research on animals causes distress, Peter Singer and other like-minded utilitarians surely would be quick to point out that the argument of necessity cuts in the wrong direction.[71] They would also resist the intuition that so-called 'trivial' research purposes should be excluded. If undertaking such research generates more utility than not doing so, the argument of proportionality tries to rewrite the basic arithmetic of utilitarian calculation.[72] Having said this, where local conditions reveal strong support for constraints of this kind (irrational though they might seem to be to any straight-thinking utilitarian), the disutility of opposing these preferences would need to be taken into account.

Turning to human rights theorists—first, there is no particular rights-related reason to prioritize the use of supernumerary embryos. To be sure, the persons for whom such embryos are now surplus to reproductive requirement should not be precluded from donating their embryos; but this is not an argument for their prioritization.

Secondly, to prohibit the *donation* of embryos specifically for research impinges on the autonomy of the donors; and to prohibit the *creation* of embryos specifically for research impinges on the freedom of researchers. Why should such persons not be permitted to make a useful contribution to medical research? Indeed, if prospective beneficiaries of the research have a right that the state at the

[71] See P. Singer, *Rethinking Life and Death* (1995) for Singer's thinking on the significance of sentience (sentient life to be valued more highly than non-sentient life) and purposive agency (purposive sentient life to be valued more highly then non-purposive sentient life).

[72] NB, the argument of proportionality applied to research on animals is in line with utilitarian thinking: researchers should not subject animals to great distress unless such disutility is offset by the utility produced for the human beneficiaries of the research. Compare the reasoning of the examiners at the European Patent Office in the Harvard Onco-mouse application, OJ EPO 10/1992, 590.

very least does not impede medical research unless there is a clear violation of right, patients' rights are infringed by such limitations.

What, then, of necessity or proportionality, or similar restrictions that are designed to reduce the use of human embryos or make them research tools of last resort? How do such constraints evince respect for rights? Under a standard articulation of human rights thinking, neither human embryos nor non-human animals have rights;[73] and, thus, there is no direct rights argument to favour one class of research resource over another. If there is some indirect argument here to support the necessity restriction, it needs to be teased out. As for the constraint imposed by proportionality, human rights theorists certainly hold that more pressing rights should take priority over less pressing rights—which perhaps has some purchase on the priorities that should direct the efforts of the research community. However, this is quite different from holding that non rights-bearing embryos should be used for research only where pressing human rights are at stake. Once again, unless an argument concerning the indirect impact on rights is to be teased out, proportionality makes no more sense as a restriction than necessity. Surprisingly, then, the logic of robust human rights thinking might be to take up an even more permissive regulatory position than that indicated by utilitarianism.

3. The Moral Status of the Human Embryo

Against a permissive regulatory approach, dignitarians stand on the protected moral status of the human embryo. On this question, dignitarians brook no argument; it is strictly non-negotiable; and there is little to be said other than that one accepts this view or starts somewhere else (as is the case with the utilitarian and human rights views). However, those who do start elsewhere often make a concession to the dignitarian position by recognizing the embryo as having a special status. Yet, if this concession does not resolve into the dignitarian position, what might it mean? More pointedly, what might it mean when it nevertheless allows for embryo research?

According to Karen Lebacqz:

> Researchers show respect towards autonomous persons by engaging in careful practices of informed consent. They show respect toward sentient beings by limiting pain and fear. They can show respect toward early embryonic tissue by engaging in careful practices of research ethics that involve weighing the necessity of using *this* tissue, limiting the way it is to be handled and even spoken about, and honoring its potential to become a human person by choosing life over death where possible.[74]

This is a brave attempt to translate respect for human embryos into a reluctance to treat embryos as research objects coupled with a recognition that this is a

[73] But NB, the position referred to in *supra* note 29.

[74] Lebacqz, 'On the Elusive Nature of Respect', in S. Holland, K. Lebacqz, and L. Zoloth (eds), *The Human Embryonic Stem Cell Debate* (2001) 149, at 160.

potential person. But, so long as we are eschewing the direct protective argument of human dignity as constraint, how can we render such a version of respect plausible? In other words, what kind of contingent considerations might be drawn into the utilitarian and human rights perspectives?

First, as indicated already, negative utilitarians seek to minimize the sum total of distress. Whether the source of my distress is what I regard as inappropriate action in relation to myself, or in relation to others, it matters not. Nor does it matter that the 'others' in relation to whom I am experiencing distress would not themselves figure directly in a utilitarian calculation. The crucial fact is that I do count; I am distressed; and this must be weighed. What this amounts to, therefore, is that utilitarians must respect the attitudes of its membership insofar as they hold negative views about embryo research. However, if we have no qualms about embryo research, and if there are no other negative utilities to be detected, we would have no problem with proceeding.

Secondly, in some contexts, the consequences of treating a human embryo as if it were no more than a rock or a stone might be damaging to the way in which we respond to the claims made by those who do have moral status. In other words, utilitarians might be concerned that lack of respect for human embryos tends towards lack of respect for those who are sentient; and rights theorists might worry that lack of respect for human embryos tends towards a weakening of respect for the entitlements of rights-holders. The fact that, in both utilitarian and human rights theories, there is a clear distinction between those, such as the human embryo, who do not have moral status and those who do is neither here nor there. If there is a reasonable belief that lack of concern for the one might lead to lack of respect for the other, there is a reason for acting with restraint in relation to human embryos.

Thirdly, there is the potentiality of the human embryo, mentioned by Lebacqz and argued about interminably in relation to both embryos and foetuses. On the one side, it is argued that potentiality is not actuality, that acorns are not yet oaks any more than students are already graduates or embryos already foetuses let alone born humans. From this it follows that whatever respect or recognition attaches to actual or achieved status cannot be read across to those still in the category of potentiality; and, thus, there is no reason why potential humans, merely by virtue of their potentiality, should be treated as though they are already actual humans. To this, the standard riposte is that we are dealing here, not with potentiality at large, but specifically with potential human life—and, this, so we are led to believe, is a special case. Clearly, this exchange is stuck in the mire. On the one side (the latter side as presented above), the potential for human life, the 'radical' potential, is seen as giving the embryo moral status.[75] It is not a matter of treating the embryo as if it had moral status; quite simply, it already has what it takes to be accorded moral status in its own right. Such is the position struck by

[75] See eg Keown, 'Restoring Moral and Intellectual Shape to the Law after *Bland*' (1997) 113 LQR 481.

those who subscribe to human dignity as constraint. On the other side, whether under utilitarianism or under human dignity as empowerment, a more restrictive view of moral status is taken. The embryo does not qualify and the question is why potential qualification should be treated as sufficient. From utilitarian and human rights perspectives, therefore, potentiality is a poor argument for moral status (which, itself, does not hinge on the potential for human life). Nevertheless, from such perspectives, is there anything more that can be said about potentiality as a contingent consideration?

The potentiality of the human embryo might figure as a factor in indirectly protective arguments of the kind already sketched. Over and above such considerations, the potentiality of the embryo might be linked into a precautionary approach that results in a degree of protection. In the case of utilitarianism, the relevant threshold is sentience. Quite where this threshold lies is moot. However, if there is any risk of inadvertently crossing the line, the utilitarian may think it right to err on the side of caution. By drawing the line on embryo research at 14 days, the utilitarian might judge that English law takes an excessively cautious approach. Nevertheless, the fact that the embryo (all being well) is on the way to becoming a being with moral status must generate some caution unless we are entirely confident about the point at which that status is achieved. In the case of human dignity as empowerment, the human rights standard for moral status is so far beyond the level of development achieved by the embryo that potentiality coupled with precaution seemingly adds little to the argument for protection. Indeed, on this approach to potentiality, the logic of human rights based on human dignity as empowerment is that it is not until we are dealing with children that we approach the threshold for moral status. There is, however, a more systematically precautionary line of argument. We make certain assumptions about the level of development and capacities achieved by the embryo, the foetus, the young child, the teenager, the adult, and so on. We make these assumptions in good faith and, provided that our particular assumptions about the embryo are correct, its potentiality for further development offers no argument against using it for research. However, what if our assumptions are wrong? If we take this thought seriously, we need to move carefully reminding ourselves, in the case of human embryos, that this might just be one who has moral status. In other words, precaution dictates that we operate in terms of possible and probable agents; and, within such a framework, the potentiality of the human embryo indicates that it (in contrast to rocks and stones) is at least a possible agent.

In the light of these comments, we can consider how, in the context of the amended terms of UK law, the House of Lords' Select Committee on Stem Cell Research[76] tackled the objection, strenuously put to it by a number of

[76] See House of Lords Select Committee on Stem Cell Research, *Stem Cell Research* (London: HMSO, February 2002) HL Paper 83(i) (Report) and 83(ii) (Evidence).

(dignitarian-minded) witnesses, that research on human embryos is, quite simply, unethical added to which any talk of 'respect for the embryo' by those who are licensing the creation and use of human embryos for research is incoherent nonsense.

4. *The Select Committee's Response to the Ethical Objection*

Dignitarians might advance two levels of ethical objection. In its stronger form, the objection is that the use of human embryos for research is unjustifiable and should be prohibited. On this view, hES cell research, whatever its supposed benefits, is categorically wrong. In its weaker form, the objection is that supporters of embryo research cannot have it both ways; that is to say, they cannot claim that they respect the embryo (or recognize its special status) and, at the same time, claim that the creation or use of human embryos for research is justifiable—in other words, the weaker objection challenges the coherence of a position such as that taken by the Warnock Committee (whose recommendations ushered in the Human Fertilisation and Embryology Act 1990).[77] What did the Select Committee have to say on these matters?

With regard to the objection in its stronger form, dignitarians can find significant support in Article 18 of the Convention on Human Rights and Biomedicine which, *inter alia*, categorically prohibits the *creation* of human embryos for research; and they can argue from the importance of human dignity (as constraint) which underpins Article 18. However, given that the United Kingdom has not yet committed itself to the Convention and (in view of the permissive state of its law) would have considerable difficulty in complying with the Convention, the Select Committee does not press for a fundamental change in policy.[78] As for arguments based explicitly on respect for human dignity, although the Select Committee notes that appeals to this idea increasingly feature in international frameworks designed to set ethical limits for science and medicine, they effectively reject it as an elusive notion which falls short of offering practical guidance when the aim of the exercise is to set limits to permissible research on human embryos.[79] This, however, is unconvincing because human dignity as constraint actually gives perfectly clear and precise practical guidance as to the limits of permissible research on human embryos—quite simply, no such research is permissible.

Clearly, the Committee is starting at another point of the bioethical triangle, essentially accepting the broadly utilitarian position struck by Warnock and

[77] Report of the Committee of Inquiry into Human Fertilisation and Embryology (the Warnock Committee Report) (Cm 9314, 1984).

[78] But cf the approach of the Human Genetics Commission, urging the Government to sign and ratify the Convention: see *Inside Information* (2002).

[79] Report, *supra* note 76, at paras 7.3–7.7. Interestingly, whilst the Committee fully supports the ban on reproductive human cloning in the Human Reproductive Cloning Act 2001, it does not rest on human dignity—rather it is the risk of physical abnormality coupled with the ambiguity of a cloned child's relationship with its parents that troubles the Committee: Report, para. 5.21 and Appendix 6.

enshrined in the 1990 legislation. This not only serves to shut out human dignity as constraint, it excludes any serious engagement with the ethical objection in its stronger form. What, though, of the objection in its weaker form? In what sense does the Committee treat the human embryo as having a special status?

As a measure of its respect for the human embryo, the Committee endorses the current 14-day limit for embryo research,[80] it underlines the sensitivity that is required where human tissue is handled,[81] and it argues that embryos, whether standard or CNR,[82] 'should not be created specifically for research purposes unless there is a demonstrable and exceptional need which cannot be met by the use of surplus embryos'.[83] The Report also makes a pair of mutually reinforcing recommendations, the aim of which is to minimize the use of embryos in research, namely: (i) that, where the Human Fertilisation and Embryology Authority (HFEA) grants licences for hES cell research, it should impose a condition requiring that any hES cell line generated in the course of the research should be deposited in a national cell bank; coupled with (ii) that, before granting any new licence for hES cell line research, the HFEA should be satisfied that no suitable cell lines are already available in the cell bank.[84] What are we to make of these concessions to the embryo?

From a utilitarian perspective, the 14-day limit, as we have remarked, might seem a touch overcautious—the embryo surely is still some way from being sentient. Nevertheless, if such a restriction serves to maintain public confidence in the integrity of embryo research and its regulation, this may well be a benefit worth taking. From a rights perspective, the attachment to a 14-day limit has no clear rationale. Even on a precautionary approach, the possibility that the embryo is already an agent does not markedly alter at this stage of what we understand to be its developmental path.

Why should the Committee emphasize the need for human tissue to be handled sensitively? At least three considerations spring to mind. First, the insensitive handling of such material can cause offence—one only has to recall the notorious *Kelly* case[85] where removed body parts were incorporated in so-called works of art. From a utilitarian standpoint, there is good reason to avoid occasioning such gratuitous distress. Secondly, if the way in which the tissue is handled deviates from the expectations of the donors, this will be a matter that concerns not only rights theorists (who will detect departures from the terms of the consent given)[86]

80 Report, para 4.22.

81 Report, para 4.25.

82 That is, an embryo produced by the process of cell-nuclear replacement and stimulation of the engineered egg.

83 Report, paras 4.28 (standard embryos) and 5.14 (CNR embryos).

84 Report, para 8.29.

85 *R v Kelly* [1998] 3 All ER 741.

86 Cf Skene, 'Proprietary Rights in Human Bodies, Body Parts and Tissue: Regulatory Contexts and Proposals for New Laws' (2002) 22 LS 102, at 117–18, for the proposal that there should be a personal, but not a proprietary, right to permit our bodies, body parts, and tissue to be used for *inter alia* medical or scientific research. Consent for such use may be subject to express or implied conditions. Where conditions of this kind are flouted by researchers (or other recipients), it is suggested that, in principle, damages should be available (at 123).

but also utilitarians who will fear that bad publicity might interfere with the donation of embryos needed for utility-yielding research. Thirdly, insofar as the tissue contains information about the genetic make-up of any individuals who do have moral status, then both utilitarians and rights theorists will be concerned, the former about the consequential disutility and the latter about breaches of privacy and confidentiality.

The Committee also insists that embryos should not be specifically created for research if surplus embryos are available and fit for the purpose; and it endorses the necessity principle written into the 1990 legislative framework. This amounts to the following two principles: (i) surplus human embryos are not to be used for research unless it is necessary (unless, in effect, the research cannot be carried out using animals); and (ii) human embryos are not to be specifically created and used for research unless the research cannot be carried out using surplus embryos. But why worry most about the creation of human embryos, then about the use of surplus human embryos, and least of all about non-human animals? As we have seen already, this order of protective priorities does not self-evidently cohere with either a rights perspective or a utilitarian approach.

Finally, it is worth observing that, although the Committee is strongly opposed to those who are sources of human embryos treating their embryos as commodities to be sold, it does not relate its view on this matter to the question of respect for the human embryo.[87] Yet, it would have been perfectly plausible to have identified a stand against commodification as one aspect of the respect due to human embryos. Or, at any rate, this would have been a natural move to be made by a Committee guided by the perspective of human dignity as constraint. That the Committee was not so guided is overwhelmingly clear. Indeed, as we have said, its remarks on respect for human embryos fit best with a utilitarian bioethics shaped by a political sense of how far and how fast the public would be prepared to go with the latest technology.

B. PGD and PTT[88]

During the last two decades, there have been rapid developments in the technology of embryo screening. Pre-implantation genetic diagnosis (PGD) can be used to check whether an embryo has a genetic marker for a particular disorder; and pre-implantation tissue-typing (PTT) is a technology that can be used to check whether an embryo is a tissue match with a particular person. How should regulators respond to such technologies?

For reasons that are now familiar, the dignitarian alliance views these developments with extreme concern. They see human embryos being commodified, instrumentalized, and unjustifiably set aside. Whether embryos are being used,

[87] See Report, para 8.32.

[88] In this part of the chapter, I am drawing on my paper, 'Happy Families, Consenting Couples, and Children with Dignity: Sex Selection and Saviour Siblings' (2005) 17 CFLQ 435.

abused, or (in the case of viable deselected embryos) not used, human dignity is compromised.

As one might anticipate, in the United Kingdom, where the Human Fertilisation and Embryology Act 1990 set out a cautiously permissive regulatory framework for the reproductive revolution, the dignitarian objections have little resonance. We have seen already how, in relation to stem cells, a largely utilitarian approach prevailed; and we will see this repeated in relation to PGD and PTT.[89] To this end, we can review two recent examples of the way that the ethical plurality plays in Parliament and the courts. First, there is the House of Commons Science and Technology Committee's report, *Human Reproductive Technologies and the Law*, in which the Committee recommends a light regulatory approach to both sex selection for the sake of family balancing and embryo selection for the benefit of a sick sibling.[90] Secondly, there is the decision in the *Quintavalle* case, where the House of Lords confirmed the view taken by the Court of Appeal, that the existing regulatory framework authorizes the Human Fertilisation and Embryology Authority to license PGD and PTT for saviour sibling purposes.[91]

1. *The Approach of the Science and Technology Committee*

The Committee's guiding principle is that 'the state should intervene only in carefully defined and justified circumstances, where there are specific harms.'[92] Alongside this governing harm principle, the Committee argues for an inverted version of the precautionary principle, recommending that regulators should permit reproductive technologies to proceed so long as 'there is no evidence of sufficiently serious harm or potential harm to outweigh benefit or potential benefit',[93] always

[89] Compare HFEA, *Sex Selection: Options for Regulation* (2003). There, the HFEA maintained that PGD 'involves some risk to both patients and embryos' (para 101) and that 'we cannot absolutely rule out the small chance of some long-term adverse effects for offspring' (para 121). Yet, in July 2004, the full 18-member HFEA announced its intention to relax the restriction that barred the way to authorizing free-standing PTT. The Chair of the HFEA, Suzi Leather, explained that, in the light of the latest evidence, the Authority was satisfied that there was no increased risk to the embryo arising from the tissue-typing procedure. See, Boseley, 'Green Light for "Designer Babies" to Save Siblings', *The Guardian* (22 July 2004), 1. And, for further signs of permissiveness, see Henderson, '"Designer Babies" Rule Changed', *The Times* (7 March 2005) 11 (this concerning the lifting of a ban on taking bone marrow from 'saviour sibling' children).

[90] Science and Technology Committee, *Human Reproductive Technologies and the Law*, HC 7-1 (Fifth Report of Session 2004–2005) (2005), at para 142 (sex selection) and 251 (saviour siblings). In fact, the Committee was badly divided, five of the eleven members of the Committee recording their disagreement with the Report.

[91] *Quintavalle v Human Fertilisation and Embryology Authority* [2005] UKHL 28.

[92] *Supra* note 90, at para 83 (under a discussion of GIFT and IUI).

[93] Paragraph 47. In a later discussion of the precautionary principle, at paras 273–6, the Committee (confusingly) runs together the risk of harm to persons who undergo reproductive treatment with the risk of harm to others or society. To maintain consistency with its governing qualified libertarian principle, the Committee should make it clear that consent covers the first kind of risk and the precautionary principle the latter. For further critique, see Somsen, 'Cloning Trojan Horses: Precautionary Regulation of Reproductive Technologies' in R. Brownsword and K. Yeung (eds), *Regulating Technologies* (2008).

being mindful of the risk of unintended and adverse consequences—in short, 'that alleged harms to society or [persons] need to be demonstrated before forward progress is unduly impeded.'[94] How does the Committee arrive at this, so to speak, qualified libertarian position?

In the key section of its Report,[95] the Committee devotes most of its discussion to a rights-based approach;[96] it dismisses dignitarianism as 'difficult to define and . . . extremely difficult to fashion into a foundation for legislation';[97] it touches briefly on a utilitarian approach;[98] and then it gathers up the essentials of its qualified libertarian view.

Starting with a rights-based approach, the Committee identifies 'the question of reproductive liberty'[99] as focal, the harm principle being relied on in defence of the negative rights or interests of citizens who wish to exercise their reproductive freedom in ways that are not harmful to others. Putting this in terms of the Human Rights Act 1998, the Committee suggests that reproductive liberty falls within Article 8(1) of the Convention and that it is only if an Article 8(2) reason is applicable (qua harm to others) that the state might legitimately restrict such liberty.[100] Quite correctly, though, the Committee notes that those who argue for limits on reproductive liberty appeal to reasons other than those listed in Article 8 (2)—for example, the protection of the embryo, the welfare of the child born, the exposure of those who undergo reproductive treatment to excessive risks, the wider impacts on society, and the argument that the elements of embryo-instrumentalization and commodification are intrinsically wrong as compromising human dignity.

Despite highlighting the importance of taking dignitarian claims seriously,[101] the Committee handles them poorly. First, there is the idea that the 'concept of dignity is difficult to define' and that, concomitantly, it would be 'extremely difficult to fashion into a foundation for legislation'—an idea that is correct insofar as it signals that the concept of human dignity is contested and that there would be problems with legislation drafted in terms that simply require 'respect for human dignity'. However, as we have said, one of the outstanding features of the dignitarian view is that it condemns unreservedly a particular catalogue of reproductive choices and practices. Nothing could be more straightforward than enacting legislation that translated these views into regulatory prohibitions—indeed, precisely as the UK government did when it legislated to prohibit human reproductive cloning. Secondly, the Committee points out that, with the exception of the Convention on Human Rights and Biomedicine, 'human rights

[94] *Ibid.* [95] Paragraphs 30–47. [96] Paragraphs 30–43.
[97] Paragraph 43. [98] Paragraph 44. [99] Paragraph 30.
[100] Paragraph 32.
[101] See eg paras 202–7. In para 206, for example, the Committee acknowledges that '[dignitarian] campaigners such as Josephine Quintavalle have made a valuable contribution to various debates and it could be argued that her application for judicial review in relation to preimplantation tissue typing is a valuable public service.'

conventions have in general avoided references to the human embryo'.[102] But this says no more than that dignitarian embryo concerns are not accepted elsewhere in the ethical plurality. Thirdly, the Committee says:

> In any event, in this section we are concerned with the fate of the embryo created for implantation; it is hard to argue that an embryo's dignity is in any sense negatively affected by being born.[103]

This would be a fair point but for the fact that the objections under review are supposedly non-rights-based objections to reproductive liberty. Amongst such objections, the dignitarians would argue that the deselection of perfectly viable embryos (because they are not of the desired sex or because they are not a tissue match) compromises human dignity. Whether or not one accepts the force of this objection, it will not do to say that it is not material because it concerns the treatment of embryos that are not implanted rather than those that are used.[104]

The objections to reproductive liberty that rest on the welfare of the child or the wider harm to society are elaborated a little but not in a way that advances our thinking. The problem is that such objections might be translated into more subtle rights-based considerations (in which case, they are internal to the rights-based paradigm) or they might be read as utilitarian arguments (in which case, they are external considerations). When, a few paragraphs later, the Committee turns to a utilitarian approach, it transpires that the Committee sees the libertarian and utilitarian views each supporting the regulatory principle that 'the state has no right to intervene in the choices of people to procreate unless evidence of harm can be shown.'[105]

Finally, with regard to the objection that voices concern about undue risk-taking, the Committee has no time for paternalism:

> The state will [should] only go so far to protect people from themselves, and will [should] intervene only when the risk is deemed unacceptably high or grave. The risks of assisted reproduction, if explained to and understood by, the individual concerned seem to us to sit firmly within those which can [may] be consented to in law.[106]

Gathering this together, we can say that the Committee's qualified libertarian approach advocates that regulators should not prohibit access to reproductive technologies unless there is an evidence-based risk of harm to others or to society.[107] Where the technology involves a risk of harm to the individual, any concerns for the welfare of the individual are met by a requirement of informed consent.

[102] Paragraph 43. [103] *Ibid.*
[104] And, in fact, the Committee goes on to rehearse (but not rebut) this objection in para 45.
[105] Paragraph 44. [106] Paragraph 40.
[107] In an earlier draft of para 46, the Committee, having accepted that 'the State's role in regulating assisted reproduction should primarily be guided by the human rights or libertarian approach', concludes that harm to other individuals or society only counts where it is 'clearly demonstrable': see HC 491, Science and Technology Committee, *Inquiry into Human Reproductive Technologies and the Law* (Eighth Special Report of Session 2004–05) (29 March 2005) Annex A.

The most instructive application of the Committee's general approach comes with its discussion of sex selection. Opening its discussion,[108] the Committee rightly points out that PGD apart, there are already ways and means of selecting the sex of a child (eg by sperm sorting)[109] and that selection for non-medical reasons might be for the sake of family balancing, or by way of replacement, or to fulfil a general preference for children of a particular sex.[110] Applying qualified libertarian thinking, the Committee should ask whether freedom to sex select for non-medical reasons would cause harm to others or society. In fact, the Committee does not pose the issue in quite this way. Instead, it lists a number of possible objections, including the dignitarian commodification objection.[111] However, it prefaces its general recommendation with a reminder of the essentials of its governing principle, and it concludes in the following terms:

> The onus should be on those who oppose sex selection for social reasons using PGD to show harm from its use. However, the use and destruction of embryos does raise ethical concerns and there are grounds for caution. The issue requires greater analysis . . . and we urge greater efforts to establish the demographic impacts across all sectors of society and the implications for the creation and destruction of embryos in vitro before new legislation is introduced. On balance we find no adequate justification for prohibiting the use of sex selection for family balancing.[112]

While much of this makes sense relative to the qualified libertarian perspective, the Committee seemingly yields to the dignitarians, not so much in conceding that the use and destruction of embryos raises (non-consequential) ethical concerns, but in accepting that there are therefore grounds for caution. Is this, one wonders, the same Committee that drafted the key section of the Report?[113]

That this concession is a hostage to fortune is evident if we pause over the confident recommendation that there should be no prohibition on sex selection for the sake of family balancing. From a qualified libertarian perspective, an important question is whether sex selection will have harmful demographic consequences. Sex selection for family balancing reasons seems to have positive consequences if anything because it compensates for the natural overproduction of one sex; and sex selection for replacement reasons (as with the Mastertons who, having lost their only daughter in an accident, campaigned to sex select for another girl)[114] is neutral because it preserves the demographic status quo. By contrast, sex selection as a matter of personal preference might lead to negative

[108] *Supra* note 90, at paras 131–42.

[109] Paragraph 131. [110] Paragraph 132.

[111] Paragraph 141. And see *supra* note 37.

[112] Paragraph 142.

[113] One suspects that this betrays the deep divisions within the Committee, see *supra* note 90; and see Montgomery, 'The Legitimacy of Medical Law', in S.A.M. McLean (ed), *First Do No Harm* (2006) 1, at 2.

[114] *Supra* note 90, at para 138 (Box 6).

demographic effects.[115] If it were faithful to the qualified libertarian perspective, the Committee should recommend the regulatory green light for sex selection, which is for family balance or replacement. However, once it is conceded that the creation and use of embryos for sex selection is a matter for concern and that caution is appropriate, it is wholly unclear why sex selection for reasons of family balance should be viewed in a more favourable light than sex selection for other non-medical reasons. The implicit dignitarian supplement, as some might put it, is a dangerous one.[116]

2. *The House of Lords in* Quintavalle

The background to *Quintavalle*[117] can be stated shortly. In February 2002, the Human Fertilisation and Embryology Authority granted a licence that would permit both PGD and PTT in relation to Mr and Mrs Hashmi's embryos. The purpose of such dual screening was to identify an embryo that would be suitable for implantation, being free of the beta thalassaemia gene that runs in the family as well as being a tissue match with young Zain Hashmi who suffers from the disease. All being well, Mrs Hashmi would carry one such embryo through to term and the newly born child would be fit to serve as a (cord blood or bone marrow) donor for the benefit of Zain. However, the Hashmis' project was stopped in its tracks when legal proceedings were brought by Comment on Reproductive Ethics challenging the power of the Authority to grant such a licence.

At first instance, Maurice Kay J ruled that the Authority had acted beyond its licensing powers;[118] but his decision was reversed by a unanimous Court of Appeal[119] and the latter decision was confirmed by a unanimous House of Lords. With the legal position thus clarified, the way was cleared for the Hashmis to continue with their attempts to create a saviour sibling for Zain; and, as we shall see, the way was also cleared for the Authority to license PGD and PTT for a much broader range of purposes (including simple sex selection) if, in its discretion, it judges that this is appropriate. However, this should not be read as a direct endorsement by the House of particular permissive licensing decisions that have already been made or that might be made by the Authority. For, the House is emphatic that the question to be determined in the appeal is not whether the Law Lords favour a conservative or a liberal approach to such questions as sex selection and saviour siblings but whether the allocation of responsibility set up

[115] For a strategy to cope with such a risk, see Jones, 'Sex Selection: Regulating Technology Enabling the Predetermination of a Child's Gender' (1992) 6 Harv J of Law and Technology 1.

[116] See eg D. Kairys (ed), *The Politics of Law* (1982).

[117] [2005] UKHL 28. For comment, see Freeman, 'Saviour Siblings', in S.A.M. McLean (ed), *First Do No Harm* (2006) 389.

[118] *R (Quintavalle on behalf of Comment on Reproductive Ethics) v Human Fertilisation and Embryology Authority* [2002] EWHC 2785 (Admin).

[119] [2003] EWCA Civ 667 (Lord Phillips MR, Schiemann and Mance LJJ). For discussion, see Brownsword, 'Reproductive Opportunities and Regulatory Challenges' (2004) 67 MLR 304.

by the 1990 legislation envisaged that the Authority should be empowered to deal with such matters or whether it put them off-limits. In other words, it is to settling matters of institutional design rather than the merits of designer children, or the like, that the interpretive efforts of the House are largely directed—and with that, the divisions in the ethical plurality are, to some extent, taken out of the spotlight.

With the general issue clarified, the House also has the gist of the challengers' claim clearly in focus. It is put very crisply by Lord Brown in the following terms:

> [T]he appellant's argument [is] that PGD screening is one thing, and properly licensable under the 1990 Act, tissue typing a completely different concept and impermissible. It is one thing to enable a woman to conceive and bear a child which will itself be free of genetic abnormality; quite another to bear a child specifically selected for the purpose of treating someone else. One can read into the statutory purpose . . . , that of 'assisting women to carry children', the notion of healthy children—only a genetically healthy embryo being 'suitable' . . . To read into [that statutory purpose], however, the notion that the child will be a suitable future donor for the health of another would be to stretch the statutory language too far.[120]

Notwithstanding such clarity of focus in relation to both the general question and the challengers' contention, it is arguable that the House gets things badly out of focus when it seeks to interpret the said statutory language.

To cut the interpretive story short, the principal question is what one makes of Sch 2, para 1 of the 1990 Act when it provides that, in the course of providing treatment services (treatment services being equated, broadly speaking, with medical services that are designed to assist women to carry children),[121] the Authority may license '(d) practices designed to secure that embryos are in a suitable condition to be placed in a woman or to determine whether embryos are suitable for that purpose.' This opaque drafting invites the following two questions:

1) In what sense might reproductive service providers be checking to ensure that an embryo is 'in a suitable condition' for implantation?
2) In what sense might reproductive service providers be checking to determine whether an embryo is, in the statutory language, 'suitable for that purpose'?

The answer to the first question is tolerably clear. Or, at any rate, it is not difficult to think of activities that fit the statutory description—for example, checking the condition of an embryo that, having been frozen and stored, is now unfrozen with a view to implantation. However, the answer to the second question is much less clear. Assuming that the activity is not concerned with checking the condition of the embryo, it seems that the check here is, as it were, for fitness for purpose rather than as to basic implantable quality. But, this begs the

120 [2005] UKHL 28, at para 51.
121 Section 2(1) of Human Fertilisation and Embryology Act 1990.

fundamental question: relative to which or whose purposes is an embryo to be judged fit—or, to which or whose purpose does the phrase 'that purpose' refer?

At first instance, Maurice Kay J read the relevant provisions as limiting the licensable activities to the overarching statutory purpose of assisting women to have children. Whatever particular activities are licensed they must occur in the course of providing treatment services—which is to say, they must be related to assisting women to have children. However, in the Court of Appeal, Lord Phillips MR devoted a great deal of his judgment to defending the proposition that the governing purpose was not simply assisting women to have children but assisting women to have healthy children (that is, children free of serious and detectable genetic disorders). By the time the appeal reached the House, it was conceded by the challengers that the broader reading defended by Lord Phillips must be correct. However, the view actually acted on by the Court of Appeal and confirmed by the House of Lords was far broader than that and quite differently focused, taking the particular reproductive purposes of the particular client woman as the governing reference point.[122] On this view, the range of licensable activities covers checking the suitability of the embryo relative to the particular purposes of the particular woman in question: in other words, 'suitable for that purpose' means 'suitable for whatever purposes the particular woman specifies'. How did the House account for this quite extraordinary reading of the legislation?

The central feature of Lord Hoffmann's leading speech (supported by Lord Brown) is to make and to reiterate the point that the word 'suitable' is an adjective that is context-sensitive, that it 'is an empty vessel which is filled with meaning by context and background.'[123] So, for instance, as Lord Scott apparently put it in argument, 'a suitable hat for Royal Ascot is different from a suitable hat for the Banbury cattle market.'[124] No doubt, this is the case. However, it is not the word 'suitable' that is the key to cracking the legislative code. Rather, the key is the phrase 'that purpose'. *If* 'that purpose' translates as 'whatever reproductive purpose the particular woman specifies' then suitability is indeed an empty vessel waiting to be filled by the particular context and background—namely, the particular context and background supplied by the particular reproductive purposes declared by the particular woman; and it might well be the case that the racing aristocracy at Ascot might have rather different criteria of suitability (both in relation to their hats and their offspring) from the country folk in Banbury, let alone the Hashmis from Leeds. On the other hand,

[122] See [2005] UKHL 28, at para 49 where Lord Brown neatly summarizes the three interpretations. These are that embryonic screening is permitted: (i) solely in order to ensure that 'the woman can carry the child successfully to full term'; (ii) not only in order to ensure that the woman can carry the child successfully to full term, but also 'to eliminate gene and chromosome defects such as may affect that child'; or (iii) in order to provide 'information about the characteristics of the embryo which is relevant to the woman's decision whether or not to carry the child.'

[123] *Ibid*, at para 33. [124] *Ibid*, at para 14.

if 'that purpose' refers generically to assisting women to have children (or healthy children), the context and background for suitability is already set; and for the women of Ascot, Banbury, and Leeds alike, the criteria are the same. It follows that the judgment handed down by the House passes legal muster only if it convincingly explains why 'that purpose' is to be read in the former (woman specific) rather than the latter (generic) way. Effectively, three arguments are directed against the latter interpretation.

First, once it is conceded that the Authority may license PGD not only to check the viability of an embryo (as Maurice Kay J held) but also to check its health prospects (as the challengers eventually conceded), it would be difficult to draw a line between those serious conditions for which PGD may be authorized and those less serious conditions that would fall beyond the Authority's licensing remit. Echoing Lord Hoffmann's reservations,[125] Lord Brown says:

> The fact is that once the concession is made ... that PGD itself is licensable to produce not just a viable foetus but a genetically healthy child, there can be no logical basis for construing the authority's power to end at this point. PGD with a view to producing a healthy child assists a woman to carry a child only in the sense that it helps her decide whether the embryo is 'suitable' and whether she will bear the child.[126]

But, to the contrary, once the concession is made, while there is a margin of interpretive doubt about the bounds of health (as in a healthy child), this does not entail that a generic reading of 'that purpose' must be abandoned. What this concession signifies is simply that the terms of the generic reading must be qualified so that the purpose in question is not simply to help women to have children but to do so in a way that avoids implanting an embryo that carries a serious disease. As Lord Brown said when rehearsing the gist of the challenge, '[i]t is one thing to enable a woman to conceive and bear a child which will itself be free of genetic abnormality; quite another to bear a child specifically selected for the purpose of treating someone else.'[127] And, similarly, it is one thing to modify the generic purpose as indicated and quite another to claim that this entails accepting that the governing purpose is supplied by whatever particular purpose the particular woman specifies.

Secondly, if the only purpose for which embryos may be screened is to avoid an implantation that cannot result in the birth of a genetically healthy child, then this would seem to mean that it would not be lawful for the Authority to license PGD to check an embryo where a mix-up of some sort is suspected—a *Leeds Teaching Hospitals* type of case, for example.[128] However, there is nothing in this point. If it is agreed that the legislation must allow for PGD to be used to confirm that the embryo to be implanted is that of the couple in question then it is easy enough to modify the generic purpose. To read 'that purpose' as covering

[125] *Ibid*, at para 27. [126] *Ibid*, at para 62. [127] *Ibid*, at para 51.

[128] The argument is considered by Lord Hoffmann at para 28. The test case is *Leeds Teaching Hospitals NHS Trust v A* [2003] 1 FLR 1091.

such cases in no way assists the argument that 'that purpose' hinges on the particular reproductive purposes of the particular woman.

Thirdly, Lord Hoffmann argues that the absence from the legislation of any reference to the use of PGD for sex selection is a telling indicator of the legislative intention to confer upon the Authority very broad licensing powers. Thus:

> [T]here was intense discussion, both in the [Warnock Report] and in Parliament, about selection for sex on social grounds. If ever there was a dog which did not bark in the night, this was it. It is hard to imagine that the reason why the Act said nothing on the subject was because Parliament thought it was clearly prohibited by the use of the word 'suitable' or because it wanted to leave the question over for later primary legislation. In my opinion the only reasonable inference is that Parliament intended to leave the matter to the authority to decide. And once one says that the concept of suitability can include gender selection on social grounds, it is impossible to say that selection on the grounds of any other characteristics which the mother might desire was positively excluded from the discretion of the authority, however unlikely it might be that the authority would actually allow selection on that ground.[129]

Once the persistent misdirection in this passage has been observed, however, it loses all persuasive force. The controlling phrase is 'that purpose' not 'suitable'. It is indeed implausible to suppose that Parliament thought that the word 'suitable' would signal that PGD for sex selection was prohibited. However, it is entirely plausible to suppose that Parliament thought that 'that purpose', by referring back to the general generic objectives of the legislation, signalled that PGD for sex selection was off-limits. On this analysis, the regulatory tilt is against conferring this discretion on the Authority; and, if PGD for sex selection is to be licensable, it needs to be explicitly so declared by Parliament. As for Lord Hoffmann's closing remarks, it is again the reading of 'that purpose' in a non-generic way, not the meaning of suitability as such, that extends the Authority's licensing powers.

Whatever we make of the merits of their Lordships' reasoning, the upshot of their decision, as Lord Hoffmann freely admits, is that the legislative scheme is to be understood as conferring on the Authority a much wider licensing remit than previously appreciated. As the challengers put it, and as Lord Hoffmann accepted, 'once one allowed the mother's choice to be a legitimate ground for selection, one could not stop short of allowing it to be based upon such frivolous reasons as eye or hair colour as well as more sinister eugenic practices.'[130] Yet, the House was not deterred. If particular women came along with preposterous reproductive purposes, the Authority would be expected not to license such activities; and, failing that, Parliament would intervene by exercising its reserve powers under section 3(3)(c) of the Act.

What *Quintavalle* boils down to therefore is a choice between two institutional designs. One design is that favoured by the House, under which the Authority

[129] [2005] UKHL 28, at para 29. [130] [2005] UKHL 28, at para 25.

has a broad discretion to deal with a wide range of controversial reproductive choices subject to Parliamentary reserve powers; the other design is one that confers upon the Authority a limited licensing discretion with Parliament having to expressly authorize each significant extension of the Authority's remit. The House does not take us into the kind of debate that might stake out the relevant considerations that bear on making a rational choice between these alternatives. If it were not for the fact that the Government cut across the process by reviewing the legislation and bringing forward proposals that are relatively restrictive in relation to the use of PGD, we might have thought that *Quintavalle* presaged a period during which the Authority would be invited to flex its full licensing powers.[131]

3. *Taking Stock*

Subsequent to *Quintavalle* the regulatory story has moved on apace. In an attempt, *inter alia*, to update the Human Fertilisation and Embryology Act, the Government first published the Human Tissue and Embryos (Draft) Bill. However, this wide-ranging bill did not survive pre-legislative scrutiny,[132] and the Government withdrew it in favour of a more focused bill designed simply to revise the terms of the 1990 legislative scheme. Subject to the outcome of this revitalized legislative process, however, we can say that the views expressed by the Science and Technology Committee, like the jurisdictional indications given in *Quintavalle*, are not helpful to the dignitarian cause.

In a context of ethical pluralism, the demand that the development and application of reproductive and therapeutic technologies should respect human rights and human dignity might delay or even bar access to human embryonic resources. However, in the United Kingdom, unlike in many parts of Europe, the dignitarian view has yet to register in a serious and systematic way with the regulators.

4. Ethical Pluralism and Internal Regulatory Standards

So long as ethical pluralism operates outside the regulatory tent, regulators can maintain a degree of consensus, even if the legitimacy of the regulatory regime is questionable. However, where the regulatory regime embeds an *explicitly* ethical test, this invites ethical pluralism inside the regulatory tent; and, even if ethics is

[131] Compare Department of Health, *Review of the Human Fertilisation and Embryology Act* (Cm 6989, December 2006), paras 2.41–2.43 (screening and selection of embryos). At the time of updating this chapter, a revised Human Fertilisation and Embryology Bill is making its way through Parliament.

[132] House of Lords House of Commons Joint Committee on the Human Tissue and Embryos (Draft) Bill, *Human Tissue and Embryos (Draft) Bill*, HL Paper 169-I, HC Paper 630-I (London: The Stationery Office, 1 August 2007).

not explicitly embedded, as we have seen in *Quintavalle*, it might implicitly shape legal reasoning where the regulatory position is open to interpretation.

A. Ethics as an Explicit Regulatory Standard

We can examine the phenomenon of explicit ethical incorporation in two contexts: first, that of patentability; and, secondly, that of moral exclusions against market access in free trade regimes.

1. Patentability and Moral Exclusion

The principal regimes of patent law in Europe, unlike in some other patent jurisdictions (for example, in North America), include an *explicit* provision against granting a patent where (roughly speaking) this would be contrary to morality.[133] In practice, of course, moral considerations can, and sometimes do, covertly influence the way that ostensibly technical notions, such as 'utility' or 'inventive step', are interpreted and applied. For the sake of argument, though, let us assume that there is a clean choice between a patent regime designed to make immorality a specific ground for exclusion from patentability and a regime that is designed to make patentability entirely independent of moral considerations. The general impression given by European patent practitioners is either: (a) that the latter is preferable to the former, the morality exclusion of the former constituting an unhelpful interference with the operation of the patent process; or (b) that the former is tolerable only so long as the exclusion is applied only in the most extreme circumstances, where the granting of a patent would be inconceivable.

Although the proposition that a patent regime should not include a morality exclusion takes us back to the ethical plurality as an external moderator, we can start by checking the strength of this claim relative to the approaches represented by the bioethical triangle. After that, we can review the proposition that the jurisprudence of moral exclusion against patentability should be developed in a way that tracks consensual ethical principles (rather than taking a partisan position within the plurality).

(a) The claim that a patent regime should not include a morality exclusion

For a utilitarian, it is far from obvious that a patent regime should include a morality exclusion. Utilitarians do not recognize categorically binding moral principles in the way that is characteristic of deontological approaches; and there might well be disutilities associated with morality exclusions if they discourage inventors from putting their inventive work in the public sphere and/or if they

[133] See Art 53(a) of the European Patent Convention and Art 6 of Directive 98/44/EC on the Legal Protection of Biotechnological Inventions.

consume significant resource at patent offices. On the other hand, if a morality check increases public confidence in patent systems, and if such a check is more efficient and effective when placed in the patent regime rather than outside it (as, say, in the political or wider legal regime), then a morality exclusion might have some utility. Indeed, Norway has taken the lead here (not necessarily, it should be said, for utilitarian reasons) by setting up a specialist ethics advisory committee as an adjunct to the local patent office.

Rejecting a utilitarian approach, a patent regime might explicitly exclude patents on inventions that are inconsistent with respect for human rights—indeed, this is the position in Belgium. Where states are already committed to human rights instruments, such as the European Convention on Human Rights, such an exclusionary provision is surely no more than the logic of their commitments.[134] Where states are not so committed, proponents of human rights will argue that patent regimes (like any other legal regime) must be subject to compatibility with human rights. Thus, the view from this corner is not simply that patent regimes should include a morality exclusion but, more specifically, that they should be human rights compliant.

From such a human rights perspective, the European Court of Justice (ECJ) erred badly in its response to the Netherlands challenge to the Directive on the Legal Protection of Biotechnological Inventions (Directive 98/44/EC).[135] One of the claims advanced by the Netherlands was that the Directive is morally deficient in that it fails to do enough to ensure that, where biological samples are taken, the donors give a full, free, and informed consent—crucially, whether provision is made to ensure that donors consent all the way through to possible patenting and commercial exploitation. The ECJ could have given a perfectly plausible response to this element of the Netherlands' challenge by pointing to the way in which Recital 26 of the Directive underlines the importance of informed consent. Far from relying on a robust interpretation of Recital 26, however, the ECJ effectively said that such a matter was not one for the patent regime. In other words, the ECJ sought to disconnect patent law from the ethics of patenting (even where the ethics at issue involved fundamental human rights values). On this approach, patent courts are technical courts, courts of law not morals, and if there is an issue about informed consent it needs to be remitted elsewhere.

For the dignitarians, the protracted campaign to entrench respect for human dignity in the EC patent directive eventually bore fruit.[136] The directive expressly

[134] Compare the argument in D. Beyleveld and R. Brownsword, *Mice, Morality and Patents* (1993).

[135] Case C-377/98. For commentary, see Beyleveld and Brownsword, 'Is Patent Law Part of the EC Legal Order? A Critical Commentary on the Interpretation of Article 6(1) of Directive 98/44/EC in Case C-377/98' (2002) Intellectual Property Quarterly 97.

[136] For analysis and discussion, see Beyleveld, Brownsword, and Llewelyn, 'The Morality Clauses of the Directive on the Legal Protection of Biotechnological Inventions: Conflict, Compromise, and the Patent Community', in R. Goldberg and J. Lonbay (eds), *Pharmaceutical Medicine, Biotechnology and European Law* (2000) 157.

ties its morality exclusion to the principle that human dignity should not be compromised. Moreover, the particular exclusions relating to cloning human beings and the use of human embryos[137] offer the dignitarians some hope of stemming the tide of human embryonic stem cell research. However, for the dignitarians, as for the human rights theorists, the argument is not simply that there should be a morality exclusion. Rather, it is that there should be an exclusion with a particular moral content, in this case the content prescribed by the dignitarian alliance.

(b) The claim that a morality exclusion should be applied only in the most exceptional of cases

When the European Patent Office (EPO) first gave serious consideration to the application of the morality exclusion in the European Patent Convention, it approached the matter (in the Harvard Onco-mouse application) in a distinctly utilitarian way, the examiners balancing the beneficial consequences (to humans) of the development of a genetically engineered test animal for cancer research against the negative effects on animal welfare and environmental hazard.[138] However, subsequently, the jurisprudence has become much more restrictive. In general, the position taken is that, in the absence of an overwhelming consensus that the exploitation of the invention would be utterly abhorrent (so much so that the granting of a patent would be inconceivable), the morality clause does not operate to bar patentability. Thus, in the more recent case of LELAND STANFORD/Modified Animal[139] (which involved a test animal for AIDS research), the Opposition Division at the European Patent Office said:

> [A]s long as a claimed invention has a legitimate use, it cannot be the role of the EPO to act as moral censor and invoke the provisions of Article 53(a) EPC to refuse on ethical grounds to grant a patent on legal research and directed to an invention indisputably associated with medical benefits. The technology underlying the present invention is undoubtedly controversial and the subject of intensive discussion in the media and among members of the public. However, there is at present no consensus in Europe[an] society about the desirability or otherwise of this technology, and public opinion is still being formed on this and related matters. It would be presumptuous for the EPO to interfere in this public debate. The provisions of Article 53(a) EPC are intended to exclude from patentability not subject-matter that is controversial, but rather that kind of extreme subject-matter (e.g. letter-bombs and anti-personnel mines) which would be regarded by the public as so abhorrent that the grant of a patent would be inconceivable.[140]

[137] Respectively, Art 6(2)(a) and (c). See Laurie, 'Patenting Stem Cells of Human Origin' [2004] EIPR 59, and Plomer, *Stem Cell Patents: European Patent Law and Ethics Report* (Nottingham, July 2006).

[138] OJ EPO 10/1992, 590.

[139] [2002] EPOR 2.

[140] *Ibid*, para 51.

Such a restrictive reading of the jurisdiction given by the morality clause reflects the patent community's view that the regulatory tilt of the patent regime is in favour of granting patents.[141]

What would our three bioethical approaches make of this view that morality exclusions are to be applied only exceptionally and on the basis of an overwhelming consensus?

From the utilitarian perspective, if a calculable patent regime encourages research and development that generates utility-producing inventive applications, it is important that a morality exclusion does not detract from these benefits. Accordingly, for a utilitarian, the relevant caveats in relation to the application of a morality exclusion are not so much that it should be invoked only exceptionally or on the basis of an overwhelming consensus; rather, the caveats are that the application of such a clause should be predictable and not so broad as to inhibit investment in utility-producing inventive work. Given such an approach, utilitarians would tend to favour clearly identified exclusions; and, if open-ended general clauses cannot be avoided, then so far as possible the patent community should be given specific advance guidance as to the application of such provisions. Thus, the attempt to give patent offices just such guidance (however inadequate) in the EC patent directive would be broadly welcomed by utilitarians.

Turning to a human rights perspective, matters will be viewed rather differently. Granted, not every application (in fact, relatively few) will directly raise human rights issues—not every application will raise issues of informed consent as directly as, say, the Relaxin Opposition at the European Patent Office.[142] However, where human rights issues are raised, whether directly or indirectly, they must be addressed. In this sense, applications of morality exclusions might be expected, in practice, to be relatively infrequent; but this is not to treat such applications as exceptional as a matter of principle. As for the idea that applications should be limited to cases where there is an overwhelming consensus in support of rejecting the application (or revoking the grant), this is way off track. Human rights applications are not always straightforward or uncontroversial; but those who operate the morality exclusions in a patent regime must apply human rights principles in good faith and conscientiously even if the result does not always command uniform support.

For the dignitarian alliance, the human rights view, albeit using the wrong currency, at least has the right sort of shape. In practice, it might be only occasionally that patent applications will raise issues relating to respect for human dignity. However, where such issues are presented, they must be addressed; and

[141] Cf Beyleveld, Brownsword, and Llewelyn, *supra* note 136.

[142] HOWARD FLOREY/Relaxin [1995] EPOR 541. For commentary, see Brownsword, 'The Relaxin Opposition Revisited' (2001) 9 Jahrbuch für Recht und Ethik 3.

they must be answered by reference to the value of human dignity rather than by reliance on any positive moral consensus. Indeed, from this perspective, the apparent willingness of the EPO to reject stem cell patents for dignitarian reasons,[143] even in the face of anything but a broad European consensus, must be seen as an important step in the right direction.

The lesson of this is that, in conditions of ethical pluralism, any attempt to base a moral exclusion on consensus or convergence will marginalize the role of the exclusion; and, while this seems to favour no particular constituency within the plurality, in practice it probably plays into the hands of the utilitarians. Indeed, when the EPO finally determines whether human embryonic stem cells are patentable, it will be surprising if it does not revert to the background utilitarian approach indicated by the weight of its jurisprudence.

2. *Moral Exclusion and Market Access*[144]

Article XX(a) GATT, and likewise Article XIV(a) GATS, recognize that a commitment to market access does not prevent a Contracting State from imposing trade-related restrictions that are necessary for the protection of 'public morals'. According to Thomas Cottier, these are examples of 'portals' in trade treaties that allow for human rights considerations to be brought into play.[145] Of course, even if it is conceded that human rights and human dignity fall within the compass of the 'public morals' exception, the chapeau to these Articles is designed to ensure that where the moral high ground is being taken, it must be taken transparently and in good faith—in other words, protective measures must not be 'applied in a manner which would constitute a means of arbitrary or unjustifiable discrimination between countries where the same conditions prevail, or [by way of] a disguised restriction on international trade.'

Now, if 'respect for human dignity' is recognized as a legitimate reason for the local imposition of trade restrictions, it makes a great deal of difference whether it is human dignity as empowerment or human dignity as constraint that is the operative conception. For, if it is the latter, while there is certainly some commitment to cosmopolitan prohibition, there also seems to be ample scope for local (and quite parochial) duty-driven standards to be set up against free trade. We should not expect to find the same exceptions everywhere. Whatever we make of such dignitarian exceptionalism as a matter of ethics, this

143 Compare the discussion in Laurie, *supra* note 137.

144 In this part of the chapter, I am drawing on Brownsword, 'Genetic Engineering, Free Trade, and Human Rights: Global Standards and Local Ethics', in Daniel Wüger and Thomas Cottier (eds), *Genetic Engineering and the World's Trading System* (2008) 287.

145 Cottier, 'Genetic Engineering, Trade and Human Rights', in F. Francioni (ed), *Biotechnologies and International Human Rights* (2007) 275.

is a prospect that might occasion some concern on the part of the sponsors of free trade.[146]

On the other hand, if it is human dignity as empowerment that is the operative conception, what then? If anything, human dignity as empowerment seems to support free trade (so long as it is not the slave trade or any other kind of commerce that obviously is incompatible with respect for human rights) and it remains to be seen how far there is scope for distinctively local restrictions within a framework of thinking that takes respect for human rights (human dignity as empowerment) as focal.

To focus some thoughts about dignitarian exceptionalism, consider the recent decision of the WTO Appellate Body in *United States – Measures Affecting the Cross-Border Supply of Gambling and Betting Services*.[147] The dispute in this case arose when Antigua and Barbuda complained that a set of US federal and state laws made it unlawful for suppliers who are located outside the United States to supply gambling and betting services to consumers within the US. After some filtering of the issues, the Appellate Body narrowed down the offending measures to three federal laws—the Wire Act, the Travel Act, and the Illegal Gambling Business Act[148]—these measures being targeted against the cross-border supply of gambling and betting services. Having found that these legislative prohibitions on the remote supply of gambling and betting services constituted a relevant restriction within the meaning of Article XVI of GATS, the Appellate Body then considered whether the US could avail itself of the general exception for 'measures . . . necessary to protect public morals or to maintain public order' under Article XIV(a). Following the view formed by the Panel, the Appellate Body held that the measures in question, being 'adopted to address concerns such

[146] Compare Charnovitz, 'The Moral Exception in Trade Policy' (1998) 38 Virginia J of Intl Law 689. In a very helpful discussion, Charnovitz draws a key distinction (at 695) between 'outwardly-directed' and 'inwardly-directed' applications of the moral exception, the former covering 'trade measures used to protect the morals of foreigners residing outside one's own country', the latter 'trade measures used to protect morals of persons in one's own country . . . ' Although it is arguable that inwardly-directed measures are less controversial, being an exercise in cultural self-defence, Charnovitz maintains that it is wrong to suppose that Art XX(a) GATT can never be applied in an outwardly-directed manner (738). Nevertheless, to confer an unqualified licence for such application would be problematic—as Charnovitz says (at 742): 'Although a unilateral determination of morality may be appropriate for inwardly-directed concerns, it is too open-ended for outwardly-directed concerns. Some method to determine the legitimacy of a moral claim is needed in order to ensure that the exception does not begin to swallow the rules. Allowing each government to restrict imports based on its own definition of morality could disrupt trade and allow imperialism by countries with market power.' By way of a solution, Charnovitz suggests that 'the WTO should use international human rights law to ascribe meaning to the vague terms of article XX(a). Thus, the moral exception could validate trade actions based on international norms while rejecting trade actions based on nationalistic aims. This solution would recognize the symmetry in our pursuit of both global commerce and global values' (742–3).

[147] WT/DS285/AB/R, April 7, 2005.

[148] Respectively, section 1084 of Title 18 of the US Code (the 'Wire Act'), section 1952 of Title 18 of the US Code (the 'Travel Act'), and section 1955 of Title 18 of the US Code (the 'Illegal Gambling Business Act').

as those pertaining to money laundering, organized crime, fraud, underage gambling and pathological gambling',[149] fell within the exception. In the event, the US argument failed to meet the chapeau requirement in Article XIV for it did not show that the domestic Interstate Horseracing Act[150] (which seems to permit certain types of remote betting on horse racing) does not amount to arbitrary or unjustifiable discrimination against extraterritorial suppliers. There is a great deal in this decision to be chewed over by international trade lawyers; but, for present purposes, it is the argument on the application of Article XIV(a) that is particularly relevant.

As we have seen, the US contended that the restrictive provisions were necessary measures in order to address a range of concerns. In response to this part of the US case, Antigua objected that the Panel had failed to determine whether the concerns were so serious as to touch one of the fundamental interests of society.[151] The Appellate Body saw little in this point, being satisfied that the Panel had taken account of the relevant matters and did not need to make a separate and explicit determination to the effect that the standard of a serious threat posed to the fundamental interests of the society had been met.

However, let us forget that Las Vegas is a global gambling metropolis. Counterfactually, let us suppose that the US treated all gambling and betting as immoral—indeed, as compromising human dignity—and that it had local laws prohibiting any such activities. Just as some societies might say, 'We do not want any reproductive cloning here; it offends our idea of civilized life', the hypothesis is that the US might say, 'We do not want any gambling or betting here; it simply is not civilized.' To permit such a dignitarian exception to be pleaded under Article XIV(a) is to invite each Contracting State to immunize its distinctive local cultural identity against global commerce. Where the particular cultural feature goes to the very identity of the society, it is plausible enough to treat it as touching and concerning a fundamental interest of *that* society (not of any society but of *that* particular society). However, if the restriction in question relates to a matter that falls short of that exacting standard (might we say this about the restriction in the *Omega* case?[152]), it would not qualify under provisions such as Article XIV(a)—or, at any rate, it would not qualify *qua* dignitarian exceptionalism (even if it might qualify under a rights-driven argument). To this extent, the objection raised by Antigua would become a focal point, and

[149] *Ibid*, at para 296. [150] *Ibid*, at paras 361–4. [151] *Ibid*, at paras 296–7.

[152] Case C-36/02 *Omega Spielhallen-und Automatenaufstellungs-GmbH v Oberbürgermeisterin der Bundesstadt Bonn* (14 October 2004); OJ C300/3, 04.12.2004. Compare Ackermann (2005) 42 CMLR 1107, at 1117: 'If respect for human dignity is a fundamental interest of society that can only be defined with reference to cultural conceptions of respect for personal identity, there is no alternative to taking national value judgments as a basis as long as the underlying conceptions vary between Member States.' However, national value judgments might be overridden where the interest is not of a fundamental nature. Or, to put this in the terms used by Advocate-General Stix-Hackl in the *Omega* case, the restriction must be directed at a genuine and sufficiently serious threat to a fundamental interest of society (para 100).

decisionmaking bodies would need to be satisfied (and to make it an explicit matter of record) that the dignitarian exception was eligible because it genuinely reflected the fundamental commitments of the community in question.

With these short remarks about dignitarian exceptionalism, we can turn to the question of how far an exception for human rights (and human dignity as empowerment) would be compatible with the cosmopolitan principle. Contrary to what we might suppose, a morality clause driven by human dignity as empowerment would not produce a standard set of exceptions. For, we are not dealing with a static moral community; a community of rights is not a finished product so much as an ongoing interpretive enterprise. Moreover, it is an enterprise that each community must take forward in its own way—each community, as it were, must have 'ownership' of its human rights project.[153] And, without attempting to be exhaustive, some of the more pressing and recurring types of questions to be addressed, debated, and (at least, provisionally) resolved within a community of rights are the following.

First, there is a large cluster of questions concerning which rights are to be recognized and what the scope of particular rights is. The international community of rights already recognizes a long catalogue of particular rights; but the world moves on and, as it does so, each community must decide which rights it will endorse.

Secondly, where positive rights and responsibilities are recognized, there needs to be a larger debate about how extensive these responsibilities are. While a community of rights will be sensitive to matters of historical injustice and second-best conditions (which, for instance, might require affirmative action), it will also be anxious to avoid compounding unfairness—for example, by imposing disproportionate burdens of positive responsibility or by creating opportunities for free-riding.

Thirdly, there are inevitably conflicts between rights: for example between the right to privacy and the right to freedom of expression. And, then, there are competing rights—that is, cases where two rights-holders present with the same right but there are not sufficient resources to satisfy both claims.

Fourthly, where a community of rights is committed to a will theory of rights, the dignity of choice includes making choices not only as to the way in which an option-giving right (such as the right to freedom of expression) is to be exercised but also as to the waiver of the benefits covered by the right. This latter kind of choice is expressed by the rights-holder's consent or, as the case might be, refusal of consent.[154] Consent is an extremely important dynamic in a community of rights and the community needs to debate the terms on which a supposed 'consent' will be recognized as valid and effective.[155]

[153] Compare C. Tomuschat, *Human Rights: Between Idealism and Realism* (2003).

[154] For the importance of giving a refusal its due weight, see Michalowski, 'Advance Refusals of Life-Sustaining Medical Treatment: The Relativity of an Absolute Right' (2005) 68 MLR 958.

[155] See, further, Brownsword, 'The Cult of Consent: Fixation and Fallacy' (2004) 15 KCLJ 223.

Fifthly, on which grounds may the state legitimately curtail the exercise of a recognized right? If a community of rights is to distinguish itself from a dignitarian community, it must reject conservative dignitarianism as any kind of justification for state encroachment on the right of self-determination. At minimum, we can take it that the state's intervention must be rationally connected to its legislative objective and, in some sense, rights-related. However, taking a stricter line, must the state's reasons be on a par with the grounds that we would demand in private dealings, namely that the intervention is essential in order to avoid a greater violation of rights? Which of these thresholds is the more plausible?

Finally, there is the vexed question of who has rights. Do young children, foetuses, or embryos have rights? What about the mentally incompetent or the senile? And, then, what about non-human higher animals, smart robots, and, in some future world, hybrids of various kinds? Each community of rights must debate such matters. Moreover, not only must it respond to the inclusionary question (who has rights?) but it must also determine its approach to those life forms that are to be excluded. What attitude does the community, as a community of rights, have to those who are judged not to qualify as rights-holders? Given the frailty of human reasoning, how precautionary should the community be? Is there a margin in which the benefit of doubt applies? If so, how wide is the margin to be?

In this catalogue of issues to be addressed by a community of rights, we see that recognizing human rights (or human dignity as empowerment) as a legitimate restriction on free trade is not recognizing an entirely predictable and unproblematic set of limitations. What is being globalized is not just the product of human rights, but equally importantly the process of deliberation and reflection that is characteristic of a community of rights. If each community is seeking, in good faith, to arrive at its best understanding of its commitment to human rights, there can be no guarantee that the answers arrived at will be the same everywhere—all of which allows for the possibility of distinctive human-rights based local restrictions on trade.[156]

Where does this take us? Let me suggest that exceptions for human dignity categorically must be recognized if the compromising of human dignity connotes the harming of the essential conditions for the existence of *any* moral community, including a community of rights. Indeed, if free trade agreements permit an exception to be made for threats that strike at the fundamental interests of society, my point is that for any society that aspires to be a moral community, there is no more fundamental threat than one that corrodes the conditions of

[156] Compare Scott, 'On Kith and Kine (Crustaceans): Trade and Environment in the EU and WTO', in J.H.H. Weiler (ed), *The EU, The WTO, and the NAFTA* (2001) 125 (especially with regard to *EC – Measures Concerning Meat and Meat Products (Hormones)*, Report of the Appellate Body, WT/DS26/AB/R, WT/DS48/AB/R, 16 January 1998, and the question of *scientific* uncertainty).

human dignity. While this suggestion does not serve to resolve the tension between those (moral) communities that articulate around human dignity as empowerment and those that articulate around human dignity as constraint, it points to the deep significance of respect for human dignity. On this analysis, to compromise human dignity is not to violate the principles of some particular moral code but to compromise the possibility of any form of moral life.

For a community of rights, this understanding of human dignity serves as a warning. *Inter alia*, it warns against excluding citizens from the standard-setting debates and, in general, from the ongoing interpretative deliberations that I have suggested are characteristic of a community of rights.[157]

In this context, why might a community of rights judge that, say, genetic engineering of human beings is an option to be pursued cautiously? One consideration that has been widely canvassed by human rights theorists is that children have a right to be born with an open future (preserving their lifestyle autonomy).[158] In other words, it is one thing to permit genetic engineering in order to avoid serious diseases but it is quite another matter to deny the child the opportunity of making its own lifestyle plans and developing its own projects. Cass Sunstein expresses this concern in a colourful way:[159]

> Imagine a situation in which cloning was freely permitted—in which individuals could clone themselves, and in which people interested in children with a certain genetic endowment could assure themselves of that very endowment. If a family wanted a terrific athlete, it could obtain a genetic equivalent of, say, Michael Jordan; if a family wanted a terrific musician, it could obtain a genetic equivalent of, say, Bob Dylan; if a family wanted a terrific scientist, it could obtain a genetic equivalent of, say, Ian Wilmut; if it wanted a terrific philosopher, it could clone John Rawls. Why would this be objectionable? The reason is that people would be treating their children-to-be as means, with prearranged agendas, rather than as ends, to seek their own path.

Clearly, the community would need to debate the distinction between negative and positive genetic selection and/or engineering and consider how the latter lie in relation to the right to an open future.

Again, following the reasoning sketched by Jürgen Habermas,[160] the community might fear that genetic enhancement (or positive genetic engineering), if not fundamentally altering the extent to which we have 'free will' and the extent to which the predicates for human responsibility are met, nevertheless might affect our perception of such matters. To the extent that the idea of free will rests on a

[157] See, too, the relevance of this warning in Part 5 of this chapter.

[158] See eg Davis, 'Genetic Dilemmas and the Child's Right to an Open Future' (1997) 28 Rutgers LJ 549, especially at 561–7 (arguing that a reflexive application of autonomy values will set limits to parents' reproductive autonomy).

[159] Sunstein, 'Is There a Constitutional Right to Clone?' (2002) 53 Hastings LJ 987, at 1000.

[160] J. Habermas, *The Future of Human Nature* (2003). See, too, B. McKibben, *Enough: Genetic Engineering and the End of Human Nature* (2003).

phenomenology of choice and responsibility, genetic modification could make a huge difference to the way in which we perceive ourselves, our 'achievements', and our 'failures'. If we come to see our genetic make-up as a programme that runs us, it will be tempting to think (against free will and responsibility) that we could not have acted otherwise; and, where our particular genetic make-up has been specified by others, if there is any responsibility left in the world, it seems to be with the specifiers rather than ourselves.[161] In other words, genetic enhancement threatens to alter the context by either undermining notions of free will and responsibility or by inviting the transfer of responsibility.

It follows that a particularly prescient community of rights, anticipating the corrosion of the conditions for moral community in general and a threat to a flourishing community of rights in particular, might impose some restrictions on the trade in biotechnological goods and services as an act of self-defence—as a measure designed to prevent the breakdown, not of a particular (and peculiarly local) moral order, but of the very conditions of moral order itself. Human dignity, so conceived, does indeed cut deep.

B. Ethics as an Implicit Regulatory Standard: The *Pro-Life Alliance* Case

There are a number of ways in which background ethical views can function, so to speak, as an invisible hand guiding the interpretation or application of legal standards. We have already seen one such instance in the *Quintavalle* case, where the judicial review, but not the House of Lords' determination, was inspired by (in that instance) a dignitarian ethic. Again, we might see ethics implicitly in play, albeit in a more complex way, in the recent *Biotech Products* (GM crops)[162] dispute at the WTO. Here, we might detect a number of subtle relationships between, on the one hand, scientific judgment and precautionary policies and, on the other, cultural and moral differences.[163] While the extent to which background ethics can operate in this way depends upon a number of factors—particularly, the governing culture of legal argument and adjudication, the level of the court, and the place of the tribunal in the national, regional, or international scheme of things—it is pretty clear that, in the common law world, and notwithstanding the frequent protestation that judicial proceedings take place in courts of law not courts of morals, ethics does have some influence; moreover, the

[161] Similarly, see R. Dworkin, *Sovereign Virtue* (2000), at 445.

[162] *European Communities – Measures Affecting the Approval and Marketing of Biotech Products* (WT/DS291/23 (United States), WT/DS292/17 (Canada), and WT/DS293/17 (Argentina), 8 August 2003), decided by the WTO in Autumn 2006. See, too, the chapter by Han Somsen in this volume.

[163] See, further, R. Brownsword, *Rights, Regulation and the Technological Revolution* (2008), especially Ch 7.

higher up the court system we go, the greater the chance that legal doctrine will be ethically tagged.[164]

To illustrate the possible implicit influence of the ethical plurality, we can examine in some detail the *Pro-Life Alliance* case,[165] another dignitarian-inspired challenge in relation to the scope of the Human Fertilisation and Embryology Act.

When the Act was enacted in 1990, two assumptions that might reasonably have been made were: (i) that a human embryo would necessarily be the product of a process of fertilization (of a human egg by human sperm); and (ii) that, if there was to be cell nuclear replacement for the purposes of cloning, then it would involve manipulation of an *embryo* (rather than replacement of the nucleus of an *egg*).

In line with the first of these assumptions, s 1(1) of the Act provides:

> In this Act, except where otherwise stated—
> (a) embryo means a live human embryo where fertilisation is complete, and
> (b) references to an embryo include an egg in the process of fertilisation, and, for this purpose, fertilisation is not complete until the appearance of a two cell zygote.

In line with the second assumption, s 3(3)(d) of the Act provides that no licence may authorize 'replacing a nucleus of a cell of an embryo with a nucleus taken from a cell of any person, embryo or subsequent development of an embryo'.

However, as embryologists developed processes for engineering eggs and then stimulating eggs to divide and develop into embryonic clusters, two nice questions of legal interpretation arose. One question was whether the 1990 Act applied to such embryonic clusters (to what were the functional equivalents of embryos produced by fertilization); and the other was how the prohibition in s 3(3)(d) related to such developments.

In the *Pro-Life Alliance* case, the principal thrust of the challenge was that these developments, having outstripped the terms of the legislative scheme, now

[164] Generally, see J.N. Adams and R. Brownsword, *Understanding Law* (4th edn, 2006), Chs 4–5. In particular, see Lord Steyn's judgment in *McFarlane v Tayside Health Board* [2000] 2 AC 59, at 82: '[T]o explain decisions denying a remedy for the cost of bringing up an unwanted child by saying that there is no loss, no foreseeable loss, no causative link or no ground for reasonable restitution is to resort to unrealistic and formalistic propositions which mask the real reasons for the decisions. And judges ought to strive to give the real reasons for their decision. It is my firm conviction that where courts of law have denied a remedy for the cost of bringing up an unwanted child the real reasons have been grounds of distributive justice. That is of course, a moral theory. It may be objected that the House must act like a court of law not like a court of morals. That would only be partly right. The court must apply positive law. But judges' sense of the moral answer to a question, or the justice of the case, has been one of the great shaping forces of the common law.' See, too, Lord Steyn's John Maurice Kelly Memorial Lecture, 'Perspectives of Corrective and Distributive Justice' (University College Dublin, 2002).

[165] *R v Secretary of State for Health ex parte Quintavalle (on behalf of Pro-Life Alliance)* [2001] EWHC 918 (Admin) (Crane J); [2002] EWCA Civ 29; [2003] UKHL 13.

needed to be brought back to Parliament for full discussion. However, the appeal courts ruled: (i) that CNR embryos (albeit not the product of a sperm-fertilized egg) fall within the terms of s 1(1); and (ii) that the prohibition against CNR in an embryo, as provided for by s 3(3)(d), does not cover CNR in an egg (because an embryo, literally, is not an egg). This ruling led to the triply convenient conclusion that: (i) the Act did not leave a worrying loophole in relation to human reproductive cloning;[166] (ii) there was no legislative obstacle in the way of engineering eggs for the purposes of therapeutic cloning; and (iii) there was no need for Parliament to revisit the vexed issue of research on human embryos.

In so ruling, the appeal courts rejected Crane J's literal reading of s 1(1). Instead, the appeal courts imaginatively read into the definition section the additional phrase '[if it is produced by fertilisation]'. With the benefit of this addition, s 1(1)(a) defined an embryo in terms of 'a live human embryo where [if it is produced by fertilisation] fertilisation is complete.' So transformed, s 1(1) simply specified what counted as an embryo for the purposes of the Act where the embryo in question happened to be the product of fertilization. However, this was no longer an exhaustive definition; for s 1(1) did not presume to exclude the possibility that an embryo might be produced by some process other than fertilization—indeed, by just such a process as that of CNR and stimulation, which became headline news almost a decade after s 1(1) was drafted.

Delivering the leading judgment, Lord Bingham asserts that the references to fertilization in s 1(1) cannot have been intended to signal a material distinction between 'live human embryos produced by fertilisation of a female egg and live human embryos produced without such fertilisation'; for, 'Parliament was unaware that the latter alternative was physically possible.'[167] Rather,

> the four words ['where fertilisation is complete'] were not intended to form an integral part of the definition of embryo but were directed to the time at which it should be treated as such . . . The essential thrust of section 1(1)(a) was directed to such embryos, not to the manner of their creation.[168]

Accordingly, the 1990 Act is read as applying to any entity that is *functionally* a human embryo. Because CNR embryos are functionally human embryos they are covered by the Act.

With regard to the interpretation of s 3(3)(d), the appeal courts were much shorter and much less imaginative. Following the literal approach taken by the Court of Appeal, their Lordships thought it sufficed to observe that, while an egg might develop into an embryo, an egg is not an embryo.

[166] Although, ironically, by the time the appeal courts had finally confirmed this, Parliament had rushed through the Human Reproductive Cloning Act 2001 precisely to create a new criminal offence to cover the maverick cloners.

[167] *R v Secretary of State for Health ex parte Qunitavalle (on behalf of Pro-Life Alliance)* [2003] UKHL 13, para 14.

[168] See [2003] UKHL 13, para 14.

There is a great deal that might be said about these judgments.[169] However, we can focus on just one question, namely: whether the conjunction of a purposive reading of s 1(1) with the literal reading of s 3(3)(d) betrays a degree of arbitrariness—which, in turn, implies an element of result orientation implicitly shaped by a rejection of the dignitarian agenda.

Let us suppose that the appellate court reading of s 1(1) is a plausible reconstruction of the regulatory intentions of Parliament—that is to say, let us suppose that it is plausible to treat Parliament as intending to regulate live human embryos, whether such embryos were produced by fertilization or by some other technique, not yet contemplated and, necessarily, yet to be developed. What, though, would a purposive approach of this kind suggest about the application of s 3(3)(d)? Clearly, it implies that no licence may authorize cell nuclear replacement in a human embryo, whether that embryo is produced by sperm fertilization of an egg or by a technique other than fertilization (such as CNR). If CNR is ruled out post-embryonic formation, does this not imply that CNR is also ruled out pre-embryonic formation? The thinking would be that, if Parliamentarians in 1990 saw fit to prohibit cloning at the embryonic stage, they would also intend to prohibit cloning at the earlier egg stage (even though the possibility of engineering an egg in this way was not anticipated). Copying across Lord Bingham's language as applied to s 1(1), we would say that the essential thrust of s 3(3)(d) was directed to cloning, not the particular technique of cloning, which Parliament (entirely understandably on the then current state of scientific knowledge) took for granted as involving *embryonic* manipulation. Yet, quite to the contrary, Lord Bingham reads s 3(3)(d) as having a very restricted ambit, literally to cell nuclear replacement in the human embryo and to nothing else.[170] Granted, the courts are presented with difficult cases when 'Parliament has to frame legislation apt to apply to developments at the advanced cutting edge of science.'[171] Even so, is there not some inconsistency and arbitrariness in the House's reasoning in the *Pro-Life Alliance* appeal?

To respond to the charge of arbitrariness in the Law Lords' reading of the Act, the standard move would be to argue that the interpretation adopted is consistent with the background legislative purpose. The notion of background legislative purpose, however, is a contested concept. Some would argue for a strictly historical reading of that purpose, insisting that the only purpose that counts is the one that the legislators subjectively had in mind (or that they committed to the record). Others would argue that the background legislative purpose must be read in a way that reflects such historical intentions but only insofar as such intentions are compatible with respect for fundamental values (such as respect for human rights and human dignity). But, of course, it will be appreciated that, in a

[169] See eg Beyleveld and Pattinson, 'Globalisation and Human Dignity: Some Effects and Implications for the Creation and Use of Embryos', in R. Brownsword (ed), *Human Rights* (2004) 185, at 195 *et seq.*

[170] [2003] UKHL 13, para 18.

[171] *Ibid.*

context of ethical pluralism, there is no clean answer to an inquiry of this kind. Hence, even if we agree about our particular conception of the legislative intent, we might run into further problems as we try to articulate a particular legislative scheme relative to a technology that has advanced beyond the imagination of the historic regulators.[172] In every sense, then, an attempt to elaborate the background legislative purpose takes interpreters into deep waters.

What can we say about the legislative purpose underlying the 1990 Act? As we have seen, the Warnock Report, which formed the basis of the legislation, was guided by the view that 'the embryo of the human species ought to have a special status' albeit less than that of a living child or adult.[173] Moreover, it is also part of the Warnock thinking that the use of pre-14-day embryos is less problematic than the use of post-14-day (post-primitive streak) embryos. However, what we make of these starting points rather depends upon where we are coming from in the ethical plurality. If we are coming from a utilitarian or human rights perspective, we might contend that the scheme of the 1990 Act and associated abortion legislation is consistent with the gradualist idea that pre-embryos merit no respect, that embryos at any stage merit a minimum degree of respect, and that the case for embryonic and foetal protection intensifies the further up the developmental track we go. However, if we are dignitarians, we are likely to reject Warnockian gradualism as incoherent and insist that the use of embryos at any stage compromises human dignity. In the *Pro-Life Alliance* case, the dignitarians seem to do little to advance their own cause by arguing that CNR embryos fall outside the regulatory framework (and might, therefore, be used for research without even a licence); but the longer-term strategy of forcing a fresh parliamentary consideration of the issue has, possibly fortuitously, come to pass.

[172] In both the *Pro-Life Alliance* and the CORE appeals, it is Lord Wilberforce's speech in *Royal College of Nursing of the United Kingdom v Department of Health and Social Security* [1981] AC 800 that is taken as seminal. According to his Lordship (at 822): 'In interpreting an Act of Parliament, it is proper, and indeed necessary, to have regard to the state of affairs existing, and known by Parliament to be existing, at the time. It is a fair presumption that Parliament's policy or intention is directed to that state of affairs... [W]hen a new state of affairs, or a fresh set of facts bearing on policy, comes into existence, the courts have to consider whether they fall within the Parliamentary intention. They may be held to do so, if they fall within the same genus of facts as those to which the expressed policy has been formulated... [or] if there can be detected a clear purpose in the legislation which can only be fulfilled if the extension is made. How liberally these principles may be applied must depend upon the nature of the enactment, and the strictness or otherwise of the words in which it has been expressed. The courts should be less willing to extend expressed meanings if it is clear that the Act in question was designed to be restrictive or circumscribed in its operation rather than liberal or permissive. They will be much less willing to do so where the subject matter is different in kind or dimension from that for which the legislation was passed.' Lord Wilberforce continued by emphasizing that what judges categorically must not do is gap-fill and second-guess Parliament's intention. But, of course, what constitutes gap-filling and second-guessing largely hinges on whether the present case can be subsumed under the same genus or the covering policy, which in turn depends on whether the court judges that a liberal or restrictive approach is called for.

[173] Report of the Committee of Inquiry into Human Fertilisation and Embryology (the Warnock Committee Report) (Cm 9314, 1984), para 11.7.

Similarly, the gradualist view nicely covers the literal reading of the protective effect of s 3(3)(d). This provision has no application to an egg, not so much because an egg is not an embryo as because the one has not yet crossed the threshold for even minimal protection—it is embryos that have a special status, not gametes. By contrast, if we take a dignitarian view, no such line can be maintained. Manipulation of gametes without more might be legitimate; but once the manipulation is designed to engineer a particular embryo product, human life is being commodified and human dignity is compromised.

Debates about the best reading of the *Pro-Life Alliance* judgments and, concomitantly, about the best interpretation of the legislative intent are liable to run on late into the night. Moreover, even when debate closes, it is unlikely to do so on a note of consensus or confidence. However, for our purposes, it suffices to remark that judges who wish to reach back into the ethical plurality can do so under the cloak of legislative purpose and that, to this extent, particular ethical viewpoints can come to bear implicitly on the judicial process. An adjudicative culture that operates in such a covert way wins no prizes for its transparency; but for judges to engage openly with a fractured ethical plurality is to invite a charge of partisanship and complaints about a lack of accountability.

5. The Revolutionary Sting[174]

In the foregoing parts of this chapter, the focus has been on the challenge of setting a legitimate regulatory environment for the development and application of biotechnology where the context is one of ethical plurality. We now switch the focus to the use of biotechnology as a regulatory instrument. The revolutionary sting to which I referred in my introductory remarks is precisely that the use of modern technologies, such as biotechnology, as a regulatory instrument might transform our culture and practice of criminal justice—ultimately with profound consequences for our valuation of both legality and human dignity.

Lawyers sometimes take the one-eyed view that law is the only way to channel human conduct. However, the reality is that regulators have a number of options: in addition to the criminal law (or other forms of law), regulators might rely on social or peer pressure, on moral exhortation, on market incentives or disincentives, and so on. Crucially, regulators might also rely on a technological fix, or architecture, or design of some kind. Intelligent regulation presupposes that the optimal mix of instruments (including various kinds of technical fix) is selected for the task at hand. Now, where the context is one of heightened concern about security and personal safety; where the ideology of the criminal justice system is

[174] In this part of the paper, I am drawing, in particular, on Brownsword, 'Neither East Nor West, Is Mid-West Best?' (2006) 3 Script 3 (available at <http://www.law.ed.ac.uk/ahrc/script-ed/vol3-1/brownsword.asp>) and 'Knowing Me, Knowing You—Profiling, Privacy and the Public Interest', in M. Hildebrandt and S. Gutwirth (eds), *Profiling the European Citizen* (2008) 362.

shifting away from due process to one of crime control; and where the culture views crime and security in terms of effective risk management, then the setting is ripe for the incorporation of new technologies that serve to improve the chances of criminal justice 'working'.[175]

As intelligent policing gains ground, the question arises: how far should we go with a (bio)technological approach to crime control? In *R v Chief Constable of South Yorkshire, ex parte LS and Marper,*[176] there was an unsuccessful challenge to the legal framework that now permits the police to retain full DNA samples and identifying profiles even though the person from whom the sample has been lawfully taken has not been convicted of any offence (indeed, not even brought to court). Appreciating the benefits of DNA profiling, the Lord Chief Justice, Lord Woolf, does not rule out the idea of a comprehensive national collection. Thus:

> So far as the prevention and detection of crime is concerned, it is obvious the larger the databank of fingerprints and DNA samples available to the police, the greater the value of the databank will be in preventing crime and detecting those responsible for crime. There can be no doubt that if every member of the public was required to provide fingerprints and a DNA sample this would make a dramatic contribution to the prevention and detection of crime. To take but one example, the great majority of rapists who are not known already to their victim would be able to be identified.[177]

With over four million profiles now held on the National DNA Database and with the average person being caught 300 times a day on CCTV, we might see the United Kingdom as a pilot for a technology-led approach to criminal justice. How far should we proceed with such a technological revolution?

The obvious riposte to this question (and the implicit suggestion that the development of a population-wide Database might be a smart move) is that it threatens a raft of values (privacy, autonomy, freedom, and so on) that are characteristic of liberal (human rights respecting) societies.[178] Moreover, it is important to review proposals for the deployment of biotechnological forensic resources, not in isolation, but in the larger context of an increasingly technological approach to social and crime control. Thus, as Mark Rothstein and Meghan Talbott have emphasized:

> The prospect of expanded use of DNA forensics needs to be placed in context. In a world in which personal privacy is difficult to maintain against an onslaught of computer file sharing, surveillance cameras, biometric imaging, thermal imaging, and other

[175] For an excellent scene-setting overview, see Bowling, Marks, and Murphy, 'Crime Control Technologies' in R. Brownsword and K. Yeung (eds), *Regulating Technologies* (2008).

[176] [2002] EWCA Civ 1275, [2004] UKHL 39.

[177] [2002] EWCA Civ 1275, at para 17; similarly Sedley LJ at para 87. And, for the mood of the moment, see Fresco and O'Neill, 'Strangler Conviction Triggers DNA Debate', *The Times* (22 February 2008), 1.

[178] As argued in Nuffield Council on Bioethics, *The Forensic Use of Bioinformation: Ethical Issues* (2007).

technological 'advances', for many people, the last 'off limit' area for access to personal information is law enforcement.

Assume that a hypothetical country routinely required all of its residents to submit the following items to the police: a DNA sample, a yearly photograph, handwriting exemplar, voiceprint, fingerprints, hair samples, retinal scans, bank statements, credit card information, health records, and other details of their personal life. Obviously, ready access to this information by police would help solve crimes. Nevertheless, such comprehensive information submission to law enforcement would be widely viewed as hallmarks of a repressive, totalitarian state.[179]

The point, while well made, is not the one that I want to make. My point is not about striking the right balance between competing moral considerations—for example, between the interest in security of person and property and the interests in freedom, the integrity of families, and equal respect and non-discrimination.[180] Rather, my point reflects a very particular concern about human dignity, a concern that, as I have already suggested, cuts much deeper than the stock dignitarian objections that we have discussed earlier in this chapter.

A. The Chaplain's Concern

We can start by reminding ourselves of a key moment in Anthony Burgess's remarkable novel, *A Clockwork Orange*.[181] Observing the successful results of young Alex's treatment at the State Institute for the Reclamation of Criminal Types, the chaplain laments:

He [ie Alex] has no real choice, has he? Self-interest, fear of physical pain, drove him to that grotesque act of self-abasement. Its insincerity was clearly to be seen. He ceases to be a wrongdoer. He ceases also to be a creature capable of moral choice.[182]

To which Dr Brodsky, who has pioneered the process of aversion therapy (the so-called Ludovico's Technique) to which Alex has been subjected, responds:

These are subtleties . . . We are not concerned with motive, with the higher ethics. We are concerned only with cutting down crime.[183]

[179] Rothstein and Talbott, 'The Expanding Use of DNA in Law Enforcement: What Role for Privacy?' (2006) 34 JLME 153, at 160–1.

[180] See eg Etzioni, 'A Communitarian Approach: A Viewpoint on the Study of the Legal, Ethical and Policy Considerations Raised by DNA Tests and Databases' (2006) 34 J of Law, Medicine & Ethics 214; Williams and Johnson, 'Inclusiveness, Effectiveness and Intrusiveness: Issues in the Developing Uses of DNA Profiling in Support of Criminal Investigations' *ibid* 234; and Greely *et al*, 'Family Ties: The Use of DNA Offender Databases to Catch Offenders' Kin' *ibid* 248. In their concluding remarks, Greely *et al* nicely capture the dilemma about the discriminatory impact of DNA databases where the criminal justice system already operates with a bias against ethnic minorities. Thus, at 260, they suggest that 'it may be wise to consider an expressly population-wide database rather than ending up with something that, through the extension to family members, becomes a large but racially biased database.'

[181] (1962). My page references are to the Penguin Classics edition, 2000.

[182] *Ibid*, at 94.

[183] *Ibid*.

However, for the chaplain, the higher ethics really do matter. In his assessment, the principal objection to Ludovico's Technique is not that it might misfire, but that it works; and that, when it works, it succeeds in depriving the reclaimed person of the capacity for moral choice.

Of course, in *A Clockwork Orange* the technology is applied only *ex post* (Alex is a serial offender). However, the chaplain's point can be generalized: too much technology can be bad for moral choice. If we assume that some forms of technological control *ex post* are acceptable (although this is certainly a matter for further consideration), we can concentrate on: (i) the development of panopticon technologies that guarantee that any criminal types are certain to be detected; and (ii) the development of technologies that can be employed to ensure either that Alex-types simply do not get born or that they are never released into the larger community. The question, broached already in this chapter, is whether reliance on such technology might have a corrosive effect, not simply on a particular interest such as privacy, but on the underlying interest in moral community itself.

B. Panopticon Technologies

Imagine that the state has access to technologies of surveillance, location, and identification that are so sophisticated and reliable that there is no chance that where a criminal offence is committed, the offending agent will not be detected. Even if such a community has a fairly limited criminal code, the fact that all crimes will be observed, all offenders detected, might be a cause for concern.[184] But why? If the code penalizes violations of just those interests that are fundamental to agency, what is the problem with such an effective detection strategy?

The problem, stated shortly, is that moral agents expect to make a choice between compliance and non-compliance with their legal-moral criminal code. To be sure, panopticonism presents agents with the paper option of non-compliance, but the reality is that agents who do not obey most certainly will pay. Echoing the chaplain's concern, this real state of affairs might be thought to interfere with the development of agent virtue, particularly the virtue of choosing to do the right thing for the right reason. For, if agents comply only because they fear certain detection and punishment, there is little room for the promotion of the desired virtue.

Against this, it might be argued that such a view is 'idealistic': the fact of the matter is that agents rarely do the right thing for just the right reason. In practice, for many agents it is the background (albeit uncertain) threat of penal sanctions that deters the commission of crime. On this view, where agents rarely do the right thing for the right reason, panopticonism (by converting a low-risk threat of

[184] Compare Tadros, 'Power and the Value of Privacy', in Claes, Duff, and Gutwirth (eds), *Privacy and the Criminal Law* (2006) 105 (especially Tadros's discussion of what he calls Camerania).

punishment into a high-risk threat of punishment) simply extends the logic of the existing arrangements—this is no real change in kind, simply a change in degree. Even in a morally disposed community, there has to be some sanction to compensate for weakness of the will.

Against this injection of realism, those who hope for a morally progressive evolution will protest that panopticonism makes it much more difficult—in fact, nigh impossible—for that evolution to take place. For, when the threat of detection and punishment is relatively low, there is still space for moral reason to play a part in influencing the decision to comply or not. Indeed, in the Platonic fable of the Ring of Gyges,[185] it was precisely where there was no risk of detection and punishment that moral reason came into its own. If, by extreme contrast, the threat of detection and punishment is overwhelming, this prudential consideration will dominate practical reason—inevitably, in most cases, the right thing will be done, but not for the right reason.

The spirit of the chaplain's view that informs the idealist approach is that the doing of the right thing for the right reason speaks to what it is to be human. As Terry Eagleton puts this way of thinking:

> Being human is something you have to get good at, like playing snooker or avoiding the rent collector. The virtuous are those who are successful at being human, as a butcher or jazz pianist are successful at their jobs. Some human beings are even virtuosi of virtue. Virtue in this sense is a worldly affair; but it is unworldly in the sense that success is its own reward.[186]

But, if we argue against panopticonism on the grounds that it compromises what it is to be essentially human, or by asserting that the moral life is its own reward, this is liable to be rejected as mystical, or metaphysical, or just a fresh outbreak of dignitarian dogma. The danger with arguing that the virtue of doing the right thing has an expressive value is that we reduce what a moral community prizes to little more than a *modus vivendi*, leaving advocates of this way of life simply to preach to the converted. Is it possible to put the case more convincingly?

Moral agents are committed to the ideal that their regulatory framework should present them with the option of doing the right thing for the right reason. In turn, agents should learn to act in other-regarding (fellow agent-respecting) ways because they understand that this is morally required. In a sense, agents who do the right thing for the right reason self-regulate. Now, it is arguably irrational for an agent to deny being bound by other-regarding principles;[187] and, whilst such rationalist foundationalism is deeply unfashionable, this is what it takes to reveal the commitment to doing the right thing as more than a *modus vivendi*: doing the right thing is rationally required. What is more, although this argument develops the idea of an infrastructure for agency, its advocates may legitimately reject the

[185] Plato, *The Republic* (1974), at 105–6 (Book II).
[186] T. Eagleton, *After Theory* (2003), at 125.
[187] Seminally, see A. Gewirth, *Reason and Morality* (1978).

claim that the reason for doing the right thing is merely instrumental. To the contrary, the reason for doing the right thing is not that this is in the longer-term interest of all agents (even though it might well be)[188] but that it is, so to speak, in the agent's own interests as a rational being—an agent who does the right thing for the right reason maintains its own integrity as a rational being.

C. Exclusionary Technologies

Elsewhere,[189] I have used Lawrence Lessig's terminology to sketch two regulatory models. One, the so-called East coast model, is an approximation of the way in which regulators operate in modern societies; the other, the so-called West coast model, is a thought experiment, purely ideal-typical. In the former, regulators rely on a mix of regulatory strategies—traditional legal command and control techniques, informal social and peer pressure, market adjustment, and so on;[190] in the latter, regulation is of an entirely technological kind—risk management is handled exclusively by technological design of one kind or another. Whereas, in the East coast community, Alex and his friends would not be out of place, in the West coast community, there would be no place for people like Alex. Not only would the West coast regulators either design-out people like Alex or design-in a control mechanism that eliminates the risk that Alex otherwise would present, they also would design-out the characteristic East coast choice between doing right and doing wrong. Even though the pattern and product of the use of exclusionary technologies is a community that is regulated for compliance and agent security, the West coast model invites a range of objections.

First, as Lessig himself has perceptively pointed out, a technology that codes for compliance has its price. Thus:

> Code is an efficient means of regulation. But its perfection makes it something different. *One obeys these laws as code not because one should; one obeys these laws as code because one can do nothing else. There is no choice about whether to yield to the demand for a password; one complies if one wants to enter the system.* In the well implemented system, there is no civil disobedience. Law as code is a start to the perfect technology of justice.[191]

Imagine a convergence between cybertechnology, biotechnology, nanotechnology, and neurotechnology and we can conceive of West coast regulators simply

[188] Compare the strategy in D. Gauthier, *Morals by Agreement* (1986).

[189] Brownsword, 'Code, Control and Choice: Why East is East and West is West' (2005) 25 LS 1. See, too, Brownsword, 'Red Lights and Rogues: Regulating Human Genetics', in H. Somsen (ed), *The Regulatory Challenge of Biotechnology* (2007) 39.

[190] L. Lessig, *Code and Other Laws of Cyberspace* (1999), at 53–4; and 'The Law of the Horse: What Cyberlaw Might Teach' (1999) 113 Harv L Rev 501, at 533–4.

[191] Lessig, 'The Zones of Cyberspace' (1996) 48 Stan L Rev 1403, at 1408 (emphasis added).

managing their regulatees, bypassing practical reason to design-in a solution to a problem of which regulatees might not even be aware.[192] But, the import of such exclusionary technologies is that West coast regulatees are not only excluded from the option of criminality, they are also excluded from the discourse and debate of regulatory standard-setting. Modern public law virtues, such as those of transparency, accountability, and participation, are conspicuous by their absence.

Secondly, the objection to panopticonism that we have rehearsed already applies *a fortiori* to the exclusionary approach. With panopticonism, there is no real chance of doing the right thing for the right reason; with the exclusionary approach, there is literally no chance of doing anything other than the right thing (as pre-coded by the regulators). It is true that an exclusionary technology might leave some space for minor moral decision-making; and the possible significance of leaving such space merits further consideration. However, by excluding the most important moral matters, the West coast yields only an ersatz moral community.

In this light, we might view some of the concerns about the administration of methylphenidate (Ritalin) and amphetamine (Adderall) to children whose conduct is outside the range of acceptability, as prefiguring concerns that we might have about an exclusionary strategy. In its report, *Beyond Therapy*, the President's Council on Bioethics expresses just such a concern very clearly:

> Behavior-modifying agents circumvent that process [ie the process of self-control and progressive moral education], and act directly on the brain to affect the child's behavior without the intervening learning process. If what matters is only the child's outward behavior, then this is simply a more effective and efficient means of achieving the desired result. But because moral education is typically more about the shaping of the agent's character than about the outward act, the process of learning to behave appropriately matters most of all. If the development of character depends on effort to choose and act appropriately, often in the face of resisting desires and impulses, then the more direct pharmacological approach bypasses a crucial element... By treating the restlessness of youth as a medical, rather than a moral, challenge, those resorting to behavior-modifying drugs might not only deprive [the] child of an essential part of this education. They might also encourage him to change his self-understanding as governed largely by chemical impulses and not by moral decisions grounded in some sense of what is right and appropriate.[193]

In other words, with the onset of an exclusionary strategy, we see a dramatic corrosion of moral community. As the President's Council graphically warns, once we take an interventionist biotechnological approach to respond to (or manage) our social problems, there is a danger that 'we may weaken our sense of

[192] A point made forcefully by Tien, 'Architectural Regulation and the Evolution of Social Norms' (2004) 9 Intl J CLP 1.

[193] *Supra* note 1, at 105–6.

responsibility and agency.'[194] Technological advances might be morally progressive, but not necessarily so.[195]

Thirdly, when imagining an exclusionary strategy, I have assumed: (i) that regulators would be well-motivated; and (ii) that the technology would be safe and reliable. For a thought experiment, this is all well and good; but, in practice, these are reckless assumptions. Even if the precautionary principle has taken a beating recently,[196] there is good sense in being cautious about handing over irreversible powers of this kind to future generations of regulators.[197]

6. Conclusion

The rapid developments in modern biotechnology have elicited a distinctive strand of dignitarian concern, creating a three-way ethical plurality. Without doubt, the emergence of the dignitarian alliance has made the task of regulators more difficult.[198] There is less common ground and ethical conflicts are more complex than they were. Some regret the appearance of human dignity on the scene. However, even if we do not share the approach of the dignitarian alliance, the potential of biotechnology to corrode the basis of moral community, including the dignity of moral choice, is a serious cause for concern.

Indeed, to return to Baldi's remarks on the significance of the acceleration in both information and biotechnology, some might express this concern in more dramatic terms. For, the convergence of several streams of modern technologies might be thought to presage not just the end of moral community but the end of human community. As Bill Joy explained the seductive power of the new technologies:

> Each of these technologies also offers untold promise: The vision of near immortality... drives us forward; genetic engineering may soon provide treatments, if not outright cures, for most diseases; and nanotechnology and nanomedicine can address yet more ills. Together they could significantly extend our average life span and improve the quality of our lives. Yet, with each of these technologies, a sequence of small, individually sensible advances leads to an accumulation of great power and, concomitantly, great danger.[199]

[194] *Ibid*, at 106.

[195] Compare Touraine, 'The Crisis of "Progress"', in M. Bauer (ed), *Resistance to New Technology* (1995) 45.

[196] See, in particular, the powerful critique in C. Sunstein, *Laws of Fear* (2005).

[197] See, further, Brownsword, 'Genetic Databases: One for All and All for One?' (2007) 18 King's LJ 247.

[198] Compare L.R. Kass, *Life, Liberty and the Defense of Dignity* (2002), at 22: 'the new biotechnologies threaten not so much liberty and equality as something we might summarily call "human dignity"'.

[199] Joy, 'Why the Future Doesn't Need Us' (2000) 8(4) *Wired* (<http://www.wired.com/wired/archive/8.04/joy.html>) at p 3 (last accessed 10 July 2008).

The moral is clear. As agents who value our autonomy, we need to be aware that the cumulative effect of seemingly benign individual choices could represent a threat to the conditions that make those choices meaningful. And, as agents who aspire to moral community, we need to be particularly mindful of the corrosive impact of a (bio)technological approach to social control.

3

Regulating Human Genetics in a Neo-Eugenic Era

*Han Somsen**

No one really has the guts to say it... If we could make better human beings by knowing how to add genes, why shouldn't we do it? (James Watson)

1. Human Genetics: The Mountain Comes to Mohammed

Human civilization appears to be at imminent risk. The decline of nature is occurring with such frightening speed that a single generation is witnessing the accelerated disappearance of thousands of animal and plant species, and freak changes in global weather patterns with disastrous consequences. Fossil fuel reserves have also been all but exhausted in a matter of decades, and fresh water supplies have become similarly scarce. For the cause of this crisis we need not look far: *Homo sapiens* is not as smart a species as we might have hoped.

This view finds increasing support, and not just from 'usual suspects', such as green politicians, anti-globalists, or climatologists looking for research money. Bill Clinton, Al Gore, Rupert Murdoch, Sir Richard Branson, Bill Gates, Sir David Attenborough: it is an unlikely alliance of liberals, ultra-conservative multi-millionaires, progressive entrepreneurs, and nostalgic nature conservationists who will no longer sit idle until their political leaders take effective action to protect our civilization from certain ruin. This eco-pessimism is reminiscent of the wave of cultural pessimism and anti-democratic sentiment that engulfed Europe at the start of the last century. At that time, concerns focused on the social degeneration of the growing working classes, a process that was believed to be funnelled by inbreeding of genetically inferior individuals. It proved an ideal atmosphere for the rise of a eugenic movement.

In short, eugenics advocates improvement of the (human) race by the application of genetic laws.[1] It was around 1880 that the term 'eugenics' was first coined by a

* This is an amended, updated, and extended translated version of an inaugural lecture delivered in Tilburg, 10 November 2006. The author wishes to thank the editor, Vivian Carter, and Annemarie Sprokkereef (TILT) for their useful comments and suggestions.

[1] *Encyclopaedia Britannica* (11th edn, 1910).

distant cousin of Charles Darwin. Based on Darwin's evolutionary theory and Mendelian laws of inheritance, eugenics was to avert the progressive 'degeneration' and 'regression' of the human race that was believed to threaten civilization.[2] Eugenics had prominent supporters from different backgrounds, including the Rockefellers and John Kellog. Eugenic ideas appealed to nationalists, racists, and socialists, but also to academics.

Whether there exist parallels between European cultural/democratic pessimism at the start of the previous century giving rise to the eugenic movement, and current eco-pessimism/anti-globalism to the extent that it may foreshadow a revival of (neo-)eugenic thinking is speculative, but I believe it is plausible.[3] Politicians, scientists, and organized groups of citizens more and more frequently trumpet a gospel in which technological progress in the fields of ICT, biotechnology, nanotechnology, and neurosciences will free us from the yoke of our natural surroundings and the random whims of body and mind.[4] Thus, the logic goes, since the human race has proved unable to adapt to the demands of nature or the society in which it functions, nature must be adapted to the demands of humans, and humans to the demands of society. Modern biotechnology plays a pivotal role in this process: it allows for the reconstruction of plants, animals, and humans precisely to this end.[5]

After the dark ages, rather than respect for Creation, lack of scientific understanding and technological capability dictated that 'the climate', 'nature', or 'human nature' remained beyond our regulatory reach.[6] The rise of modern biotechnology, in combination with other technologies, however, has given rise

[2] In 1883, Francis Galton defined 'eugenics' as 'the science of improving stock—not only by judicious mating, but whatever tends to give the more suitable races or strains of blood a better chance of prevailing over the less suitable than they otherwise would have had'. F. Galton, *Hereditary Genius* (1869) and *Human Faculty* (1883).

[3] This is how I read Jasanoff's comment: 'If collective defense and welfare goals remain intransigent problems, as the "war on terror" clearly demonstrates, then the mood of the moment seems all the more hospitable to state-supported advances in the life sciences, which promise citizens fulfillment on an intimate, personal scale, through longer, healthier, more liberated lives for themselves and, in time, their genetically tailored children.' S. Jasanoff, *Designs on Nature: Science and Democracy in Europe and the United States* (2005), at 36.

[4] The transhuman movement offers a good example of this view, see <http://extropy.org/>.

[5] Here, the word 'process' was chosen deliberately. Biotechnology can be seen as a 'technological process' or as a totality of the 'products' obtained through that process. Opting for one or the other perspective will have regulatory consequences. In the United States, the second perspective is followed, which means that a number of fundamental human values that biotechnology could change or erode are less likely to be protected in the regulation framework and therefore are less likely to give rise to limitations in the technology. In Europe, biotechnology is seen as much more of a 'process', and explicitly ethical considerations even feature in the regime that regulates GMOs. See also Jasanoff, *supra* note 3, at 54.

[6] Arguably, the eugenic movement in the 1950s suddenly disappeared because 'the science of genetics progressed, leaving genetic scientists increasingly dubious of the central factual claims of the movement.' At the same time, '[s]ome would even maintain that the eugenicists did not in fact abandon ship, although they said they did. In this view, the Nazi connection motivated eugenicists to refuse to endorse eugenic ideas under that name, and to support eugenic beliefs and projects in other guises.' See A. Buchanan *et al*, *From Chance to Choice* (2000).

to ambitions to reshape our natural world.[7] And since, despite many warnings about environmental degradation, Mohammed has not come to the mountain, the hope now is that humans are smart enough to move the mountain to Mohammed. This involves changing nature, human nature included.

Naturally, freedom to reconstruct our world generates widespread feelings of fear, which is reflected in an emotional debate on the regulation of genetically modified organisms (GMOs). However, fear alone does not justify the elevation of the status quo as policy objective, particularly if the status quo does not appear viable for much longer. To be sure, the consequences of failures to take radical technological measures to reconstruct our world may raise a more fearful image than a technological voyage in sometimes unchartered territory.[8]

It therefore has become necessary critically to assess the regulation of our technological capabilities in a way that suits deliberative democracies, and for the principles that arise from that process to be anchored in a regulatory framework. In this respect, regulators have gained most experience with GMOs, where scientific advances have allowed their large-scale commercial introduction.[9] Within the European Union (EU), fears about unknown consequences of GMOs ultimately led to a moratorium.[10] The implications of this moratorium were global, as third countries were directly affected by this EU policy. This necessitated judicial review of the EU measures within the context of the global trading regime presided over by the World Trade Organization (WTO).

Although this chapter concerns the regulation of human genetics, I will draw upon this global experience regarding the regulation of GMOs to gain insights about the regulation of genetic reproductive technologies. In particular, in the case of GMOs, risks put forward to justify restrictions on the introduction of GMOs that impact on third countries must pass a formal legitimacy test, and there are no reasons why a regulatory framework restricting human reproductive technologies on the basis of ethical considerations should not also satisfy some form of procedural test. I will use pre-implantation genetic diagnosis (PGD) as an example of a modern genetic reproductive technology. PGD is controversial, in part precisely because it is regarded as a tool for neo-eugenic projects. As is the case with GMOs, decisions regarding PGD are taken in the absence of (global)

[7] For the first European bureaucratical thoughts on this matter, see *Eurofutures: The Challenges of Innovation, The FAST Report* (1984), which contains a chapter entitled 'bio-society'.

[8] See C. Sunstein, *Laws of Fear: Beyond the Precautionary Principle* (2005). We do not have to immediately fear an ecological Armageddon; simply feeding an ever-increasing world population without reverting to GM crops could well prove to be an untenable situation in the long term.

[9] Article 2.2 of the Council Directive 2001/18/EC of the European Union defines a GMO as 'an organism, with the exception of human beings, in which the genetic material has been altered in a way that does not occur naturally by mating and/or natural recombination'; in other words, it expressly excludes mankind from this regulatory framework.

[10] The moratorium lasted from October 1998 until May 2004, when the European Committee approved Syngenta's genetically modified BT-11 corn.

ethical consensus, and it will be impossible for legislators to develop a regulatory framework that is to everyone's satisfaction.

The challenge is to articulate a regulatory framework that is based on a clear and consistently applied ethical perspective (which I will term a 'meta-regime'), and in which clinical risks are addressed transparently. Only then will regulation enjoy both internal legitimacy (with regard to nationals) and cosmopolitan legitimacy (with regard to third country citizens). The WTO's recent judgment in *EC Biotech*, although concerning the legitimacy of EU policy regarding *bio-risks*, yields useful insights that are relevant for domestic regulatory frameworks concerning PGD.[11] The case is of direct relevance for the way in which legislators must weigh (clinical) *risks* in their decisions to restrict access to PGD. In addition, the judgment implies that the legitimacy of a domestic (or EU) regulatory framework for PGD is best guaranteed by a cosmopolitan perspective, which in turn will direct us towards a human rights paradigm.

The remainder of this chapter is in four parts. Part 2 provides a brief introduction to PGD and its uses, and maps some of the most pressing ethical controversies to which it gives rise. It pays attention to the question of whether PGD amounts to a neo-eugenic project, and the implications of an affirmative answer. Part 3 concerns what I term the cosmopolitan legitimacy of regulatory decisions, and draws on literature and case law on risk regulation in the context of GMO regulation. Part 4 builds upon the 'bioethical triangle' elaborated by Roger Brownsword, and argues that, from a perspective of cosmopolitan legitimacy, a regulatory regime inspired by human rights is preferable to a regime based on local morals. Fears that such a 'universal' meta-regime will usurp flexibility to accommodate local variation will also be briefly addressed. Finally, Part 5 applies some of these insights to recent Dutch policy regarding PGD, and concludes that a consistent policy inspired by a human rights perspective would give rise to different outcomes than those advocated by the Dutch Health Minister.

2. PGD and 'Neo-Eugenics'

A. PGD: Applications, Clinical Risks, and Ethical Dilemmas

Annually, close to eight million children are born worldwide with serious illnesses with a genetic cause.[12] At present, medical research laboratories are testing for at least 1228 specific genetic defects on an estimated total number of between 3500

[11] See WTO Biotech case *European Communities—Measures Affecting the Approval and Marketing of Biotech Products*, in which the WTO decided against the EU moratorium on 7 February 2006 (preliminary report) and on 10 May 2006 (final report). See also CIEL (Centre for International Environmental Law)'s analysis: *EC-Biotech: Overview and Analysis of the Panel's Interim Report* (2006) and more recent information on the Internet at <http://ciel.org/Tae/Trade_WTO.html>.

[12] Christianson, Howson, and Modell, *The March of Dimes Global Report on Birth Defects: The Hidden Toll of Dying and Disabled Children*, available at <http://www.marchofdimes.com/professionals/871_18587.asp>.

and 5500.[13] Without the possibility of conducting such prenatal tests, options for prospective 'risk parents' would be limited.[14] However, parents and doctors can decide to test an embryo or foetus prior to birth. Prenatal genetic tests have been conducted for several decades on full pregnancies, for instance through chorionic villus sampling. Such prenatal tests make it possible to assess the severity of the risk that a child will be born with a genetic defect or illness, to detect chromosomal defects so as to improve the birth rate and decrease the number of miscarriages, or to choose the sex of an embryo. Many parents make use of the ability to conduct prenatal genetic tests. Some defects can be treated *in utero*, although in many cases this will not be possible. If parents in such cases decide to carry a child with a genetic defect to term, it is valuable from both a psychological and a practical perspective to be aware of the condition of the unborn child at an early stage.

Pre-implantation genetic diagnosis (PGD) involves prenatal tests on embryos created by means of IVF *prior* to placing them in the uterus.[15] PGD is therefore an alternative to 'traditional' prenatal genetic testing: embryos are genetically tested prior to implantation, so that a healthy embryo can be implanted in the uterus through a process of intentional (de)selection. Two types of pre-implantation genetic tests are distinguished: a molecular analysis aimed at discovering mono-genetic defects (defects caused by damage to a single gene); and a chromosomal test aimed at discovering potential defects caused by chromosome mutation (a 'wrong' number of chromosomes) and translocations (where a part of a chromosome has broken off and has attached itself to another chromosome).[16]

Without becoming more technical than is needed, it is immediately clear that PGD carries fewer clinical risks for mothers than an invasive post-implantation test, especially if such a test is followed by an abortion in cases where a genetic defect or illness of the foetus is discovered. The fact that PGD renders abortion unnecessary may also offer an important psychological and moral advantage for the prospective parents.[17]

[13] Online Mendelian Inheritance in Man (OMIM), 'OMIM Statistics', available at <http://www.ncbi.nlm.nih.gov/Omim/mimstats.html>.

[14] Prior to the development of prenatal testing, such parents could choose between: (1) accepting the risk of conceiving a child with a genetic defect; (2) using donor sperm (donor egg cells were not technologically possible yet); (3) adopting a child with no genetic connection to either parent; or (4) having no children.

[15] The Planning Decision (*Planningsbesluit*) defines PGD as: 'research of the ovum or in vitro embryo for diagnosis of constitutional and hereditary conditions.' Usually, this is done by removing one or two individual cells (by biopsy) from an IVF embryo consisting of eight cells in total. These cells are then tested cytogenetically (ie chromosomally) or molecularly (ie for DNA sequences). Biopsy is prohibited in Germany, where the analysis is carried out in a different way (by analysing polar bodies).

[16] Aneuploidy screening is mainly applied with 'old' mothers, and mothers with an anamnesis of miscarriages and/or failed IVF procedures.

[17] The advantages for patients seem to be confirmed by the statistics on the experience of patients, see Lavery *et al*, 'Preimplantation Genetic Diagnosis: Patients' Experiences and Attitudes' (2002) 17 Human Reproduction 2664–7. According to this study, about 76% of 20 patients who had used both PGD and prenatal diagnosis would opt for PGD with a next child, 16% for prenatal diagnosis, and 8% would not use any test. Such preferences are important to a regulatory regime inspired by utilitarian ethics, see *infra*.

The (relatively minor) risk associated with PGD predominantly concerns the IVF phase of the treatment, which can lead to hyperstimulation[18] of the ovaries and a more frequent occurrence of multiple birth pregnancies. It is worth stressing that these risks must be faced by anyone choosing to undergo IVF treatment anyway, so that PGD does not carry any *additional* risks with it for this group. However, for prospective parents without fertility problems who still choose PGD (usually for reasons based on family history) these risks are additional. An important implication is that, if attention is focused on the interests of the child, there do not seem to be any compelling reasons, if risks associated with IVF are considered acceptable, for prohibiting PGD.[19]

Genetic tests are not yet one hundred per cent accurate. Studies have reported five misdiagnoses in 754 live births,[20] although more recent numbers are more encouraging, with no wrong diagnoses in 382 babies.[21] Even though PGD is a young technique, and at this moment only a few thousand children have been born with the aid of the technique, the best available evidence suggests that the technique carries with it no greater risks to mother or child than other basic techniques such as IVF or intracytoplasmic sperm injection.[22]

As stated, simple chromosomal tests make it possible to determine the sex of an embryo, so that PGD makes sex determination possible for social reasons. In western countries, preferring one sex to the other primarily plays a part in relation to a growing desire for a balanced family make-up (boy(s) and girl(s) in one single family). Sex selection of an embryo for this sort of reason is prohibited in most European countries but not, for instance, in the United States. In cultures where there is an *a priori* preference for boys, rather than using PGD, cheaper and quicker ultrasonic techniques are used. In addition to sex selection, controversies surrounding PGD also arise from the possibility of

[18] According to a recent study, this complication occurred in 2% of a total of 2500 cycles, to a slight to serious degree. In the United Kingdom, the syndrome only once led to death in a total of 500 000 cycles. Papanikolaou *et al*, 'Incidence and Prediction of Ovarian Hyperstimulation Syndrome in Women Undergoing Gonadotropin-releasing Hormone Antagonist In Vitro Fertilization Cycles' (2006) 85 Fertil Steril 122, at 20.

[19] PGD differs from IVF: (a) by the biopsy; and (b) by the genetic test. In only 1% of all cases the biopsy is not successful. See *infra* note 20. Concerning the genetic test: according to the latest data from the European Society for Human Reproduction and Embryology (ESHRE) 91% of the embryos are successfully genetically tested, *ibid.*

[20] Verlinsky *et al*, 'Over a Decade of Experience with Preimplantation Genetic Diagnosis: A Multicenter Report' (2004) 82 Fertil Steril 292, at 4.

[21] Harper *et al*, 'ESHRE PGD Consortium Data Collection V: Cycles from January to December 2002 with Pregnancy Follow-up to October 2003' (2006) 21(1) Human Reproduction 3, at 21.

[22] See *The Science and Clinical Utilisation of Pre-Birth Genetic Testing with Particular Focus on PGD* at 17: 'Current data suggest that PGD is a safe procedure for mother and child, with risks such as congenital malformation no greater than those for foundation technologies such as IVF and intracytoplasmic sperm injection (ICSI). There is a need for more reporting clinics providing PGD services to enable monitoring of technology, including for usage and for long-term follow-up of children born through PGD.' Available at <http://www.otago.ac.nz/law/genome/resources/GenomeTeaser_06.pdf>.

selecting an embryo of which the tissue matches that of a seriously ill older sibling (so-called HLA-typing).[23] As a result of this technique, bone marrow transplantation for the benefit of the sick child has a far greater chance of success, as a rejection of 'foreign' tissue can usually be avoided. Parents might therefore seek access to PGD, not because of fertility problems or because they want a child, and even if under normal circumstances they would prefer *not* to have a child, but for the sole reason of securing effective treatment for an existing child who is seriously ill. Such children conceived after HLA-typing are referred to as 'saviour siblings'.

Another ethically problematic implication of PGD is that, although most carriers[24] of damaged alleles will never exhibit the related defect, it is also used to exclude such carrier embryos. This strongly reduces the chance that the defect will manifest itself in future generations, but at the same time implies that embryos which could have developed into healthy children are not implanted. Believers in 'slippery slope theories' are therefore concerned about the current use of PGD for defects that only manifest themselves in later life,[25] that are not fully penetrating, or that imply only a certain predisposition for the defect.[26]

B. PGD: A Eugenic Project?

Eugenics will always be associated with the racial policy during the Nazi era, constituting one of the most barbarian chapters in human history.[27] Legal manifestations of the term 'eugenics' reflect its reputation: Article 3.2 of the EU Charter of Fundamental Rights prohibits 'eugenic practices', which could be interpreted in such a way that it includes PGD. This complicates a well-informed debate about PGD.

To be sure, PGD is a diagnostic technique that is not used for modifying ('enhancing') embryos: embryos are selected, but this is accomplished through negative selection—certain embryos are not implanted in the uterus. In addition, 'desirable' traits, if it were to become known with which genes these should be

[23] Although media attention mainly focuses on the phenomenon of the saviour sibling, matching tissue can also protect the health of a healthy foetus itself in case of incompatibility with its mother's blood, which will make the mother's immune system threaten the foetus' life.

[24] A carrier of a hereditary disease does not necessarily display the specific serious illness symptoms related to the condition. Some symptoms can manifest themselves in carriers, but never as seriously as with patients who have the diseases in full. There is a large risk that carriers will pass on the disease to their children.

[25] PGD is, for instance, used for Huntington's disease, which is a dominant, fully penetrant, very serious disorder manifesting itself at a later age.

[26] Recently, it was advocated in the United Kingdom to use PGD for BRCA1 and BRCA2 mutations for breast cancer and cancer of the ovaries. Breast cancer caused by BRCA mutations manifests itself later in life and it is 80% penetrant in women at age 70 and 6% in men. PGD is also used for a type of Alzheimer's that manifests itself at an early age.

[27] See *supra* note 6.

associated, would require the testing of far more than a single gene. In practice this means that an embryo will always test either positive or negative for one of the genes looked for. PGD will then lose its usefulness. For the foreseeable future, influencing behavioural qualities such as criminality or intelligence, for this and other reasons, will not be possible. Many human traits are of course also (especially) influenced by environmental factors, and the effectiveness of genetic manipulation can be questioned.[28] In any event, the question of whether or not a certain trait is 'desirable' depends on context, and it is difficult if not impossible *in abstracto* to pronounce upon this matter.[29] Having said this, it is submitted that a reproductive technology that excludes the implantation of certain embryos after inheritable genetic defects are discovered, or prefers them on the basis of desirable traits such as sex, *does* meets accepted definitions of eugenics. Also, it cannot be ruled out that genetic technologies will further develop in a direction that makes several of the above-mentioned scenarios realistic in the future. More important still is the fact that, besides the technique, society's perception of the concept 'serious genetic defect' evolves with time, quite possibly in ways that leads to exclusion of embryos based on questionable (individual) preferences. Although PGD is therefore a form of eugenics, in contrast with the eugenic policies and laws of the twentieth century, the initiatives for such eugenic procedures seem now to reside with private individuals. I will therefore refer to PGD as a manifestation of 'neo-eugenics'. The sole fact that the state does not play a guiding role in deciding whether or not couples should use PGD does not free it from its eugenic character, nor does it offer any guarantees that the reprehensible type of eugenics in recent history will not be repeated. Conceivably, the state will have to act in order to tame the eugenic instincts of private individuals which, as a cumulative of individually expressed preferences, could systematically damage the interests of carriers of undesirable alleles.[30]

[28] Depression is, for instance, associated with the 5-HTT allele, but only if the carrier has had a traumatic childhood. See Caspi *et al*, 'Influence of Life Stress on Depression: Moderation by Polymorphism in the 5-HTT Gene' (2003) 301 Science 386, at 9.

[29] It is often said that there is a correlation between body length and social success. However, being a tall person, my chances of being successful as a jockey or Formula 1 driver are nil because I am too heavy for a thoroughbred or the cockpit of a Ferrari.

[30] Buchanan *et al*, *supra* note 6, at 53, illustrate how cumulative individual preferences can lead to social injustice:

'1a. I favour a genetic intervention because I want my child to have the "best" (healthiest, etc.) genes.
1b. We favour genetic interventions (on behalf of each of us) because we want our children to have the "best" (healthiest etc.) genes.
1c. I favour genetic interventions (for each person in our group) because I want our children to have the "best" (healthiest etc.) genes.

If 1a is morally acceptable, it doesn't become wrong when voiced by several people (in the form of 1b). And how can one person be faulted for endorsing that group's hope (1c)? 1b and 1c are merely the aggregate of many instances of 1a. One might expect to hear 1c uttered by, say, a health official or a legislator who sponsors a measure that would provide genetic services to a large number of people.'

It is therefore understandable that PGD and the eugenic period of the last century are often mentioned in one breath. Still, this label in itself should not colour our assessment of the morality of PGD as such. Just as Hitler's construction of the *autobahn* does not shed light on the morality of our present-day network of motorways, Nazi eugenics should not affect our *a priori* assessment of the morality of PGD. The pursuit of improved individual or public health through the use of a genetic reproductive technique that meets with our definition of eugenics should therefore not be rejected solely for that reason. The morality of PGD will have to be assessed from an ethical perspective that, regardless of the question of whether PGD seeks to 'improve the race', offers consistent guidance to support or oppose the application of PGD in individual cases.[31]

C. Interim Summary

I have argued that in a globalized world in which national and EU regulatory actions often produce important extraterritorial effects, domestic policies regarding PGD must possess cosmopolitan legitimacy. Within a European context, this follows from a meaningful interpretation of 'European citizenship', and from the fact that PGD is a medical service falling within the ambit of the EC Treaty. As a result, limitations do not only affect, for example, Dutch citizens but all other EU citizens as well.[32] Such policies, therefore, should be subject to a 'European' legitimacy review. This argument equally applies to a global context. Some form of review is appropriate, if only because 'world citizens' will be prepared to travel any distance to conceive a saviour sibling through PGD to save another child's life, which they will be forced to do if the technology happens to be unavailable or unlawful in their home jurisdiction. The WTO carries out such review for trade-related national and EU regulatory measures.

A brief analysis of selected WTO and EU cases will help shed light on the nature of these cosmopolitan minimal standards. It will be seen that WTO and EU judicial review mechanisms, leading to the articulation of formal and procedural conditions for domestic policies, have both a cosmopolitan and an internal legitimizing function. Put simply, WTO and EU discipline ensures that states rationalize policies in a manner that surpasses a parochial logic, by virtue of which they acquire not just domestic, but also European and global, legitimacy. *EC – Biotech* is an example of the WTO disciplining the EU, which pursued

[31] *From Chance to Choice, ibid*, offers an excellent discussion on the morality of eugenics.

[32] See Joerges and Godt, 'Free Trade: The Erosion of National and the Birth of Transnational Governance', in S. Leibfried and M. Zürn (eds), *Transformation of the State* (2005), who phrase this problem very powerfully: 'Compelling normative reasons which militate in favour of such co-operative commitments can be derived directly from the post-national constellation in which Member States of the EU find themselves. *Their interdependence has become so intense that no state in Europe can take decisions of any political weight without causing "extra territorial" effects for its neighbours. Put provocatively, but nonetheless brought to its logical conclusion, the Member States of the EU have become unable to act democratically.*' (emphasis added)

policies that harmed third countries without being able to offer sufficient justifications in terms of the protection of health or the environment. In that sense, EU policies regarding GMOs lacked cosmopolitan legitimacy.

As the EU and WTO practices that we will discuss have been responses to national policies focusing on *risk*, it may be asked how they can be expected to offer support for principles pertaining to the regulation of human genetics which, after all, raises mostly *ethical* questions. However, if we agree that EU and global judicial review of national risk policies are necessary (and possible), it becomes hard to deny that the (ethical) principles on which domestic PGD regulation is modelled likewise should be made subject to a degree of international oversight. That this is no easy task is manifestly true, but not sufficient reason to leave the domestic regulation of vital genetic reproductive technologies entirely unchecked.

I will argue that the most promising avenue to follow is an ethic grounded in human rights, as articulated in numerous general international treaties and several specific bioethical instruments.

3. Regulation of Technology, Deliberative Democracy, and Cosmopolitan Legitimacy

A. International Judicial Review and Domestic Deliberative Democracies: A Legitimacy Paradox

Even for mature liberal democracies, biotechnology forms a challenge of immense proportions. This is largely because biotechnology gives rise to new formal and actual centres of power that often elude democratic control and that are, by their nature, not easily accessible to citizens. Technical experts analyse and regulate bio-risks, on the basis of data provided by a globalized life-science industry that is subject to trade secrets laws; in the back rooms of patent agencies ethically charged patent applications are assessed and decided; and bioethicists elaborate ethical principles without having to consult citizens.

A solution to part of this problem is found in the rule that legislators are not bound to follow scientific or ethical advice.[33] In the sphere of risk regulation this has resulted in a separation between a scientific risk assessment and a politically controlled risk management phase.[34] The European Court of Justice (ECJ) has

[33] It is difficult to be sure if regulators need to have better reasons to depart from 'scientific' advice than from ethical advice. If the authorization of releases of GMOs is representative, past practice suggests that politicians do not feel compelled to offer countervailing evidence to ban GMOs that have been judged to be perfectly safe. Similarly, environmental impact assessment procedures culminate in advice to planning authorities that is used fairly lightheartedly.

[34] This is not a full answer to the legitimacy issue if only because the demarcation between risk analysis and risk management is partly cosmetic. For instance, scientists will have to decide on the type of risk that plays a role in their assessment, and this is a policy issue, not a scientific one.

underlined these respective roles for science and politics for the EU institutions.[35] It has happened frequently (arguably too frequently) that even though on the basis of a thorough scientific risk assessment it was concluded that introduction of a GMO into the environment was safe, the political branch nonetheless decided to deny an authorization. As we shall see, the Dutch Health Minister (Ross) similarly disregarded the National Health Council's (ethical) advice regarding PGD on most points. The precautionary principle, by authorizing regulatory action under circumstances of scientific uncertainty, appears to provide politics with additional discretion to disregard scientific advice.

Such outcomes, which evidence the primacy of deliberative democracy over technocracy, by no means always survive the disciplines of the EU and the WTO, however. And even when policy does reflect the best available scientific advice, it is possible that it does not stand up to a review by the EU or WTO. This may be the case, for instance, when the same policy objective can be achieved with less restrictive measures, or when measures arbitrarily discriminate. It is understandable that such interventions by supranational or international organizations, of which the democratic credentials are dubious to boot, are generally regarded as a threat to the democratic legitimacy of domestic or EU policy. However, in order to judge whether such assertions are valid, we should give a moment's thought to the manner in which these organizations exercise their jurisdiction.

B. Cosmopolitan Legitimacy of Risk Regulation: An Alternative Perspective to the Dilemma

Both the EU and the WTO enjoy jurisdiction to review national policies embedded in legislation and administrative practice for compatibility with EU or WTO law respectively. For the EU, judicial review is carried out by the Court of First Instance (CFI) and the European Court of Justice (ECJ), and for the WTO by Dispute Settlement Panels and the Appellate Body (AB). In cases of incompatibility between domestic law and such law of higher order, the latter has priority. Similarly, in case of a clash between EU and WTO law, the latter prevails. This is no different if disputed rules or practices are outcomes of domestic or European democratic processes.

From the perspective of legitimacy, such supranational or international interventions would be problematic in a world in which the effects of national

[35] See Case T-13/99, *Pfizer* [2002] ECR II-3305 para 199: 'That finding can also be justified on grounds of principles relating to the political responsibilities and democratic legitimacy of the Commission. While the Commission's exercise of public authority is rendered legitimate, pursuant to Article 155 of the EC Treaty . . . by the European Parliament's political control, the members of SCAN [Scientific Committee on Animal Nutrition], although they have scientific legitimacy, have neither democratic legitimacy nor political responsibilities. Scientific legitimacy is not a sufficient basis for the exercise of public authority.'

policies remained limited to national territory. After all, in such a world only the *internal* legitimacy of a policy would need to be safeguarded. Within the EU and in a globalized world more generally, however, nearly every national policy of any significance will have European or global implications. Besides internal legitimacy, national or EU policy must therefore also enjoy *external* or *cosmopolitan* legitimacy. This in fact implies that a 'deliberative supranationalism' must be constructed, which has the interdependence of states as its point of departure and simultaneously offers a solution to the problem of extraterritorial effects of national and, *a fortiori*, EU policy.[36] The EU and the WTO can be regarded as authors of such a deliberative supranationalism, primarily by exercising their respective jurisdictions and by demanding transparency of decision-making procedures. Obviously, transparency not only serves cosmopolitan legitimacy, but also greatly enhances internal legitimacy.

Although the institutional differences between the EU and the WTO are enormous (for instance, the WTO does not have any legislative authority), there are parallels in how they work. Both bodies review policies applying a 'meta-norm', ie a norm of higher order that trumps other rules. The meta-regime, importantly, tends to be predominantly procedural rather than substantive. Insofar as the yardstick to review substantive domestic and European law is procedural, objections based on the democratic legitimacy of those substantive rules thus reviewed are much less persuasive. The EU employs the principle of mutual recognition, for example, which the ECJ formulated in *Cassis de Dijon* as such a meta-norm. The SPS agreement performs a similar role:

> the SPS Agreement does not invoke some supranational legislative authority. It provides a framework within which WTO Members are to seek a resolution of conflicts arising from the extra-territorial impact of their regulatory policies.[37]

Joerges and Godt refer to 'proceduralized policy coordination through conflict-of-laws methodologies.'

With respect to GMOs, EU and WTO meta-regimes stipulate, *in abstracto*: (a) the objectives to be pursued by members of the WTO and EU Member States, and institutions on the basis of which the WTO and EU promulgate norms; (b) how the objectives are to be put into practice and the role of the actors involved; (c) the enforcement machinery that is triggered when Member States do not comply with (a) or (b).[38]

[36] See Joerges and Godt, *supra* note 32. See also Keohane who argues that the chasm between jurisdiction and impact gives rise to an 'external accountability gap' in: 'Global Governance and Democratic Accountability', in D. Held and M. Koenig-Archibugi (eds), *Taming Globalization: Frontiers of Governance* (2003).

[37] Joerges and Godt, *supra* note 32.

[38] Scott, 'European Regulation of GGOs: Thinking about "Judicial Review" in the WTO', Jean Monnet Working Paper 04/04, New York University School of Law.

For the EU regime on GM food and feed, this concretely means that: (a) only the health of humans, animals, plants, environmental protection, and consumers may play a role in the approval procedure;[39] (b) the different actors each play a clearly defined role in a complex system of 'bio-governance';[40] (c) violation of this regime will trigger the enforcement procedures as laid down in the EC Treaty and secondary legislation.[41]

If the EU regime of approval is used to protect, for instance, ethical or cultural values, then the ECJ will therefore declare such measures unlawful on the basis of (a). If the objective of a measure is compatible with the EU regime but, for instance, no sound risk analysis has been carried out, or if the Commission dissents from the opinion of the European Food Safety Agency (EFSA) without providing reasons, then the ECJ will reverse a decision on a GMO on the basis of (b). In both cases, the ECJ derives its authority from (c).

This exercise can also be used for the WTO as regards GMOs. Focusing on the SPS Agreement, this means that: (a) WTO members may take phytosanitary measures to protect the health of plants, animals, and humans;[42] (b) such measures have to be proportional, may not arbitrarily discriminate, and must be based on sound scientific evidence;[43] (c) in case of conflict with (a) or (b), Member States have to align their legislation with a WTO decision, and in case of failure to do so, Article 23 of the Dispute Settlement Understanding (DSU) comes into effect.[44]

[39] See Art 2(1) of Council Regulation 1829/29 OJ 2003 L268. However, Art 7(1) allows the Commission to include 'other legitimate factors to the matter under consideration' in its draft decision in addition to EFSA's opinion. This is a strange provision as Art 2(1) regulates the objectives exhaustively.

[40] The European Commission draws up a draft decision for an approval decision. The final decision is taken according to a comitology procedure in which Member States exercise their power through their representatives in the 'Standing Committee on the Food Chain and Animal Health' and their representatives in the Council of Ministers.

[41] In an EU context, the consequence of such a violation could be that the national rule loses its legal effect entirely. The consequences of a WTO verdict do not normally go this far, but it does pave the way for far-reaching punitive measures.

[42] See Annex A(1) to the SPS Agreement which provides that a phytosanitary measure is a measure:

'(a) to protect animal or plant life or health within the territory of the Member States from risks arising from the entry, establishment or spread of pests, diseases, disease-carrying organisms or disease-causing organisms
(b) to protect human or animal life or health within the territory of the Member from risks arising from additives, contaminants, toxins or disease-causing organisms in foods, beverages or feedstuffs
(c) to protect human life or health within the territory of the Member from risks arising from diseases carried by animals, plants or products thereof, or from the entry, establishment or spread of pests; or
(d) to prevent or limit other damage within the territory of the Member from the entry, establishment or spread of pests.'

[43] See Art 2.2 SPS, *infra* note 45.

[44] Article 23(1) DSU reads: 'When Members seek the redress of a violation of obligations or other nullification or impairment of benefits under the covered agreements or an impediment to the attainment of any objective of the covered agreements, they shall have recourse to, and abide by, the rules and procedures of this Understanding.'

It is important to note that this framework is procedurally/formally worded in the SPS Agreement: phytosanitary measures have to be based on sufficient scientific evidence,[45] and therefore on a sound risk analysis.[46] If insufficient relevant scientific evidence is available, members may take provisional measures.[47]

The WTO will not pass judgment on the legitimacy of the actual level of protection pursued by the measures; that substantive power falls within the primacy of (democratic) governments. In brief, states may set the level of protection as high or low as they wish, as long as such measures are scientifically grounded, no less restrictive measures that would result in the same level of protection (proportionality) are possible, and they do not discriminate arbitrarily. The requirement of a sound scientific basis does not apply to limitations that fall under Article XX(a) GATT or Article XIV(a) GATS.[48]

It is of course true that 'science' offers no satisfactory answer to all our questions of legitimacy, especially when ethical or normative controversies are at issue. However, the alternative to science, ie a substantive meta-regime that would determine the level of protection that states are allowed to pursue, or which ethical choices are legitimate, surely is an unacceptable alternative. In other words:

> It is submitted that, all of these difficulties notwithstanding, a conflict-of-laws approach to regulatory differences offers an often viable alternative to a search for substantive transnational rules. This alternative is less intrusive and therefore easier to accept. Even where meta-regimes remain indeterminate, they may nevertheless help to structure the controversies among the parties to a conflict by re-opening political, potentially deliberative, processes.[49]

[45] Article 2.2 SPS reads: 'Members shall ensure that any sanitary or phytosanitary measure is applied only to the extent necessary to protect human health, animal or plant life or health, is based on scientific principles and is not maintained without sufficient scientific evidence, except as provided for in paragraph 7 of Article 5.'

[46] Article 5.1 SPS: 'Members shall ensure that their sanitary or phytosanitary measures are based on an assessment, as appropriate to the circumstances, of the risks to human health, taking into account risk assessment techniques developed by relevant international organizations.'

[47] Article 5.7 SPS: 'In cases where relevant scientific evidence is insufficient, a Member may provisionally adopt sanitary or phytosanitary measures on the basis of available pertinent information, including that from the relevant international organizations as well as from sanitary or phytosanitary measures applied by other Members. In such circumstances, Members shall seek to obtain the additional information necessary for a more objective assessment of risk and review the sanitary or phytosanitary measure accordingly within a reasonable period of time.'

[48] Article XX(a) GATT reads: 'Subject to the requirement that such measures are not applied in a manner which would constitute a means of arbitrary or unjustifiable discrimination between countries where the same conditions prevail, or a disguised restriction on international trade, nothing in this Agreement shall be construed to prevent the adoption or enforcement by any contracting party of measures: (*a*) necessary to protect public morals;' and Article XIV(a) GATS reads: 'Subject to the requirement that such measures are not applied in a manner which would constitute a means of arbitrary or unjustifiable discrimination between countries where like conditions prevail, or a disguised restriction on trade in services, nothing in this Agreement shall be construed to prevent the adoption or enforcement by any Member of measures: (a) necessary to protect public morals or to maintain public order.'

[49] Joerges and Godt, *supra* note 32.

Recently, a WTO Dispute Settlement Panel ruled on the cosmopolitan legitimacy of EU GMO legislation.[50] The case concerned the EU moratorium on GMOs, product-specific EU delays in the assessment of authorizations, and a number of trade-limiting 'safety measures' that Member States applied to GMOs in individual cases.[51] The Panel reached the conclusion that (a) the objectives of the EU regime, relative to the SPS Agreement, were legitimate, but that (b) the *operationalization* of those objectives was not sound. Without having to examine the ruling further, the only matter that is important to us for the moment is that none of the three disputed EU measures passed the procedural test. The Panel commented on the moratorium:

If procedural delay could be used, directly or indirectly, as an instrument to manage or control risks, then Members could evade the obligations to be observed in respect of substantive SPS measures, such as Article 5.1, which requires that SPS measures be based on a risk assessment.[52]

And:

application of a prudent and precautionary approach is, and must be, subject to reasonable limits, lest the precautionary approach swallow the discipline imposed by Annex C(1)(a), first clause. Indeed, if a Member could endlessly defer substantive decisions on the grounds of perceived need for caution and prudence in the assessment of applications, Annex C(1)(a), first clause, would be devoid of any meaning or effect. . . . it is therefore important always to bear in mind that Annex C(1)(a) first clause, implies a core obligation to come to a decision on an application.[53]

The safety measures applied to a number of GMOs by France, Germany, Italy, and Greece were all rejected. The scientific risk analyses that preceded the measures (which took the form of prohibitions) did not reveal any safety risks that could justify a ban on the GMOs. The EU did indeed present documents to justify the safety measures, but those were not risk analyses.[54]

From a perspective of democratic legitimacy, these conclusions by the Panel can be considered problematic because they undermine the primacy of politics

[50] Council Directive 90/220/EEC, OJ 1990 L117/15, Council Directive 2001/18/EC, OJ 2001 L106/1 and Council Regulation 258/97, OJ 1997 L43/1. This regime was changed drastically after the case had been brought before the WTO: see Council Regulation 1829/2003 OJ 2003 L268/1 and Council Regulation 1830/2003 OJL 268/24.

[51] *European Communities—Measures Affecting the Approval and Marketing of Biotech Products.* Interim Rapport of 7 February 2006. See *supra* note 11.

[52] Paragraph 7.1510. [53] Paragraph 7.1516.

[54] The documents assessed the possibility, but not the probability, of adverse effects on human health, etc, and often were limited to arguing the need for more research. Annex A(4) of the SPS Agreement defines risk analysis as follows: '*Risk assessment*—The evaluation of the likelihood of entry, establishment or spread of a pest or disease within the territory of an importing Member according to the sanitary or phytosanitary measures which might be applied, and of the associated potential biological and economic consequences; or the evaluation of the potential for adverse effects on human or animal health arising from the presence of additives, contaminants, toxins or disease-causing organisms in food, beverages or feedstuffs.'

over science, and possibly also because they disregard the precautionary principle which, within the EU, enjoys *quasi*-constitutional status.[55] Let us first see if the WTO review of EU and national GMO legislation disregards the primacy of politics over science. This would indeed be the case if the WTO obliged political bodies always to adhere to scientific opinion in its decision-making on GMOs. However, all WTO legislation does is oblige the political branch to offer reasons for deviating from scientific advice, and attaches conditions to the quality of such reasons. Moreover, as to the latter, a departure from scientific advice also requires scientific substantiation. The WTO therefore explicitly allows members to follow a scientific minority opinion, as long as such a view actually exists.[56] What is not accepted, and this should not surprise us, is that members take measures without giving sound reasons. This is fair, given that the effects of such measures are usually felt also, and at times even especially, in third countries. This rule, hence, is justifiable by virtue of a need to secure cosmopolitan legitimacy, but no less to satisfy requirements of internal legitimacy. If democracy amounts to more than translating populist anxieties and prejudices into policy, then a meta-regime that raises the publication and use of sound information to a central procedural requirement and facilitates deliberation is democracy reinforcing.[57]

The disciplinary function carried out by the WTO, which is predominantly procedural, hence guards the cosmopolitan *and* internal legitimacy of national and EU policy. It does so by forcing politics to rationalize and articulate policy in a manner that allows for *ex ante* and *ex post* democratic deliberation and control. If legislators wish to prohibit a GMO, a justification based on fear, the position of the stars, or public opinion[58] is insufficient. If this were allowed, the state's power would become total in a way completely incompatible with a deliberative democracy. If all this sounds exaggerated: let us not forget that the eugenic movement of the last century had *scientific* pretensions. If the legitimacy of eugenic legislation had been procedurally reviewed along the lines employed by the WTO, ie if measures that cannot be based on scientific evidence had been struck down, it might have spared the world one of the worst chapters in its history.

For the regulation of risk, the *lingua franca* is science, and it is for the sake of safeguarding cosmopolitan legitimacy that the WTO requires members to phrase risk policy in that global language. In this latter respect, it is telling that in

[55] The principle derives its quasi-constitutional status from its place in the EC treaty, which takes up the highest position in the hierarchy of the EC legislation sources. Secondary legislation (eg directives or regulations) that are in conflict with the precautionary measure can therefore be struck down by the ECJ.

[56] See *EC – Hormones* (WT/DS26/AB/R).

[57] In this context, see also Howse, 'Democracy, Science and Free Trade: Risk Regulation on Trial at the World Trade Organization' (2000) 98 Michigan L Rev 1, at 26.

[58] On the dangers of using public opinion as the basis for policy, see Sunstein, *supra* note 8, and Scott, *supra* note 38.

EC – Biotech the Panel made the following observation about the phrase 'appropriate to the circumstances' in the SPS Agreement:

> We need not determine whether relevant scientific evidence was or is insufficient *for Austria*, and if so, whether this would be a relevant circumstance. Even if this were the case, the flexibility which the phrase 'as appropriate to the circumstances' may in some situations provide does not relieve Austria from the requirement in Article 5.1 to base its safeguard measure on a risk assessment which meets the definition of Annex A (4).[59]

In *Hormones*, the AB decided that science can *never* provide *absolute certainty* that there will *never* be any health risks, and that this kind of 'theoretical uncertainty' can therefore not offer a basis for measures that can be justified by Article 5.7 SPS.[60] This brings us to the tricky subject of the precautionary principle which, after all, sanctions regulatory action in situations where scientific uncertainty prevails.[61] From the cited example it becomes clear that precaution, depending on how it is manifested, is a risky principle in light of the requirements of democratic legitimacy. This is illustrated by the judgment of the CFI in *Pfizer*.[62] The ECJ decided that a ban on an antibiotic in cattle feed was justified, despite the fact that no thorough risk analysis had been conducted, that there was little proof of a risk of resistance, that the antibiotic was not being used for human beings, and that the relevant EU science commission (SCAN) advised against the ban. The justification: precaution.

Even though WTO law has no direct effect within the EU, EC legislation must be interpreted in accordance with WTO law.[63] This means that 'extreme' versions of precaution cannot inform EC policy and case law. There are indeed several indications that EU institutions are influenced by WTO guidelines. Although the Court's verdict in *Pfizer* leads one to suspect otherwise, it did state:

> a preventive measure cannot properly be based on a purely hypothetical approach to risk, founded on mere conjecture which has not been scientifically verified.[64]

Another example is an Italian ban on genetically manipulated corn that contained between 0.04 and 0.30 particles of transgenetic protein per million, even though, according to the *Instituto Superiore di Sanità*, there was no scientific evidence of risks to human health or animals. The ECJ decided that a mere possibility of a threat was insufficient justification for the measure: 'The risk must be adequately sustained by scientific evidence.'[65]

[59] *EC – Biotech*, *supra* note 11, at 930, para 7.3043 (emphasis added).

[60] *Supra* note 47.

[61] On precaution see, also, the chapter by Laurence Boisson de Chazournes in this volume. See, in addition, Somsen, 'Cloning Trojan Horses: Precaution in Reproductive Technologies' in R. Brownsword and K. Yeung (eds), *Regulating Technologies* (2008).

[62] *Supra* note 35.

[63] Case C-149/96, *Portugal v Commission* [1999] ECR I-8395.

[64] *Pfizer*, *supra* note 35, para 142.

[65] Case C-236/01, *Monsanto Agricultura Italia v Presidenza del Consiglio dei Ministri* [2003] ECR I-08105.

Scientific uncertainty can justify measures based on precaution as long as that scientific uncertainty continues. Scientific uncertainty will exist only insofar as a risk assessment cannot be carried out, in which case precaution is a sound justification for regulatory action. Again, the Panel is spot on in this regard:

> We note that the Appellate Body in *Japan – Apples* referred to the insufficiency of available scientific evidence to perform an 'adequate' assessment of risks. The European Communities appears to rely on the Appellate Body's use of the term 'adequate', for it argues that an 'adequate' assessment of risks is one which is '*adequate for the purposes of the legislator.*' The Appellate Body failed to explain the term 'adequate'. Moreover, the term 'adequate' nowhere appears in Article 5.1, Article 5.7 or Annex A (4). In these circumstances, we are not convinced that we should attach much significance to the term. Indeed, the term 'adequate' may have been intended as nothing more than a reference to the definition in Annex A (4). In this view, *a risk assessment would be 'adequate' if it meets the standard and definition provided in Annex A (4).*[66]

In summary, it can be concluded that scientific evidence is 'sufficient' or 'adequate', not if a certain quantitative threshold value has been reached, but if it makes a qualitatively meaningful dialogue possible within the process of democratic deliberation.[67]

C. Lessons for Regulating PGD

How can we make use of these lessons in internal and cosmopolitan legitimacy for the development of a regulatory framework for PGD? Are the differences between GMOs and embryos far too profound to enable us to apply insights obtained from the regulation of GM corn to the regulation of PGD? Not necessarily.

First of all, it is simply not logical to place higher demands on the legitimacy of a policy on GM corn than on a policy on embryos. On the contrary, given the ethical implications of PGD, we should strive for a debate on PGD that is as clear and transparent as possible, in a manner befitting modern deliberative democracies. WTO and EU case law regarding GMOs offers a number of useful guidelines for such a transparent and clear debate. Insofar as this debate concerns the *clinical risks* associated with PGD, the lessons from the WTO and EU jurisprudence are in any case of direct relevance. This conclusion is important, because risk appears to have played an important role in the formulation of, for example, the Dutch position. In addition, the precautionary principle has started to surface in the debate about human genetics, and the insights derived from WTO case law are instructive here as well. Nonetheless, judicially reviewing ethical decisions is much more difficult and controversial than reviewing risk regulation. Unlike the risks surrounding GMOs, not a single international

[66] *EC – Biotech*, *supra* note 11, para 7.3226 (emphasis added).
[67] See also Howse, *supra* note 57, at 16.

organization has *prima facie* jurisdiction to formally review ethical choices. Also, at first glance we lack the support of a cosmopolitan meta-norm, such as the SPS rule that dictates that decisions have to be based on a scientific assessment that makes such review possible in the first place. All the same, as argued above, there is a need to articulate such a rule. I will suggest that a human rights perspective offers the most suitable candidate for such a meta-regime. By reviewing a national PGD policy in light of that regime, which again is predominantly procedural in nature, the internal and cosmopolitan legitimacy of that policy can be better guaranteed than without such a review. Using the human rights perspective is not a matter of course: there are alternative meta-regimes out there, and each has its advocates. In what follows, I will use Roger Brownsword's 'bioethical triangle' to illuminate the most important of these different perspectives.[68]

4. The Bioethical Triangle

A. Utilitarianism, Human Rights, and Human Dignity

1. Utilitarianism

The Australian philosopher and bioethicist Peter Singer recently commented disapprovingly on the increasing use of PGD.[69] His reflections were reported by the world press, since Singer is a contemporary figurehead of utilitarianism. Utilitarians believe that the law must be informed by the principle that the final result of a rule should yield the best possible net result for the group affected by this rule.[70] This leads to a comparison of advantages and disadvantages. Utilitarianism is therefore widely used in the legal sub-discipline known as law and economics.

Ask a utilitarian 'should we clone people' or 'should we use PGD to make better people', and his/her answer will depend solely on the advantages and disadvantages involved in that kind of activity. He/She will calculate, weigh up the pros and cons, and arrive at a provisional conclusion. The answer is provisional, because whether PGD will get a green, amber, or red light for a utilitarian depends on time and place. When the technology improves so that the advantages come to outweigh the disadvantages for a certain community, the utilitarian

[68] See Brownsword, 'Ethical Pluralism and the Regulation of Modern Biotechnology' in F. Francioni (ed), *The Impact of Technologies on Human Rights* (2006).

[69] Singer, 'The Costly Appliance of Science', *The Guardian* (19 September 2006).

[70] Utilitarianism is associated with philosopher Jeremy Bentham, and his work *An Introduction to the Principles and Morals of Legislation* from 1781 in particular (republished 2000). His first famous words read as follows: 'Nature has placed mankind under the governance of two sovereign masters, *pain* and *pleasure*. It is for them alone to point out what we ought to do, as well as to determine what we shall do. On the one hand the standard of right and wrong, on the other the chain of causes and effects, are fastened to their throne.'

will change a red light to green. The group that the utilitarian will consider in his/her assessment consists of humans or animals that can feel pain or pleasure, which excludes foetuses or embryos, simply because they are not supposed to experience pain, sadness, or pleasure, but does include animals with a developed nervous system.

Utilitarians are well disposed towards technology. Unless PGD damages the utility of a specific group, there are no principles or rules of a higher order (religious or otherwise) that stand in the way of technology's development. So what can explain Singer's disapproval of PGD? PGD can prevent a great deal of human suffering by offering parents the prospect of a 'suitable child', and it can probably also bring about a great deal of joy to childless couples, which should please utilitarians. Of course, PGD also has its drawbacks. Relying on specific views on 'human dignity' (discussed later), some consider the destruction of unused embryos an insurmountable disadvantage. In itself, however, this is not a point of concern for utilitarians, although they do take these objections by others into consideration in their analysis of pros and cons. The same goes for the possibility of sex selection: for some an unacceptable violation of human dignity, but not necessarily a problem to utilitarians. The creation of a human being either through negative selection (the exclusion of embryos with undesirable genetic traits) or through positive selection (conscious choice of an embryo with desired genetic traits, as with saviour siblings) is also controversial. Both positive and negative genetic selection indeed reek of past eugenic times in which certain groups, directed by the state, were encouraged to procreate, whereas others were discouraged to do so, and in the most extreme cases involuntarily sterilized or, as happened during the Nazi regime, simply murdered.

Singer's objections to PGD, however, have nothing to do with the destruction of embryos or foetuses, or with the eugenic scent of PGD. Rather, they concern the social inequality caused by the disparity between wealthy prospective parents who can afford this useful technology, and the poor who cannot. However, should PGD be covered by health insurance, thus avoiding social inequality, then Singer will switch the light to green. The key point is that utilitarians do not recognize any meta-regime, other than utility, that provides an *a priori* answer to the question of whether or not PGD should be permitted: advantages and disadvantages will simply have to be weighed. In this latter respect, the utilitarianism that Singer represents differs crucially from two other important and partly competing ethical paradigms that greatly impact on the policy and law informing human genetics. These perspectives are associated with human rights and human dignity respectively.

2. *Human Rights*

This second school is not so much interested in the outcome of welfare-maximizing regulation, but focuses on respect for individual autonomy, as outlined in human rights incorporated in several regional and international legal instruments. A human rights centred ethic is aimed at the holders of human rights. Embryos and foetuses are not considered to be beneficiaries of human rights, a point of view that has been confirmed by the European Court of Human Rights.[71] UNESCO, the international organization that has taken the lead in developing bioethical principles, seems to embrace this body of ideas and emphasizes 'the importance of taking international human rights legislation as the essential framework and starting point for the development of bioethical principles'.[72]

If a medical technology leads to an infringement of such a (universal) right of a higher order, such as for instance the right to privacy, ownership, or bodily integrity, the technology should in principle be banned.

The reverse is of course also true: if there is no infringement of a right of a rights-holder there is, from a human rights perspective, no reason to prohibit an activity. A crucial exception to the rule that a technology may not infringe the rights of rights-holders can be found in the rule that individuals can decide to alter their own legal position. This rule results directly from the primacy of personal autonomy, and is put into practice in a medical context by means of consenting to, for instance, a medical operation after having been informed about the consequences.[73] Provided that informed consent has been given by the holders of the relevant rights, abortion or the destruction of embryos does not infringe the right of any rights-holder. That aspect of PGD, therefore, does not generate any insurmountable objections from a human rights perspective.

All in all, human rights may therefore form an obstacle to the unhindered exploitation of technological potential, but not necessarily an insurmountable obstacle, if only because the procedural instrument of prior informed consent plays a crucial role.[74]

[71] *Vo v France* [2005] 40 EHRR 12. The Court decided here that Art 2 (protection of life) of the European Convention on Human Rights did not apply to a six-month old foetus.

[72] UNESCO, Explanatory Memorandum the Elaboration of the Preliminary Draft Declaration on Universal Norms on Bioethics (Paris, 21 February 2005 SHS/EST/CIB-CIGB/05/CONF.204/4).

[73] For instance, recital 26 of Directive 98/44/EG reads: 'Whereas if an invention is based on biological material of human origin or if it uses such material, where a patent application is filed, the person from whose body the material is taken must have had an opportunity of expressing free and informed consent thereto, in accordance with national law.' The Dutch government argued that the Directive should be annulled because it was inadequate with regard to prior informed consent, but they lost the case. The ECJ denied that prior informed consent should have a place in patent law, and attached no value to the recital at all: see Case C-377/98, *Netherlands v Council* [2001] ECR I-7079.

[74] It is relevant that the prior informed consent is not a material human right, but a procedural instrument that can change existing relationships between the giver and recipient of the consent. See Brownsword, 'The Cult of Consent: Fixation and Fallacy' (2004) 15 King's College LJ 223. For utilitarians the instrument does not have any discrete value; it will simply be included in the sum of advantages and disadvantages. With biobank research, for which keeping a system of prior consent is complicated and expensive, utilitarians will much sooner be inclined to renounce a strict regime of prior consent than with, for instance, individual genetic research. A less strict regime, one of presumed consent, was introduced for the biobank in Iceland.

3. Human Dignity

The third school, which has been strongly on the rise recently, uses the far less well-defined principle of 'human dignity' to answer the sort of questions that we have just put to utilitarians and advocates of an ethics based on human rights. The contours of the concept of human dignity can only be determined with the greatest difficulty. The interpretive problem is first of all caused by the fact that there are two prevailing versions of human dignity which, in addition, are used with little consistency in international and national legal instruments. The first version sees human dignity as the final justification for the existence of articulated human rights: human rights arise from the fact that each individual possesses intrinsic human dignity. Human dignity in the first version plays an emancipating role (Brownsword refers to 'human dignity as empowerment'). Human dignity is a precondition for the existence of human rights, and respect for human dignity is therefore completely compatible with respect for human rights.[75] In a world based on respect for human rights, the foundation of human rights (human dignity) will have to be protected from erosion by new technologies all the time. This means, for example, that it has to be considered whether PGD endangers children's autonomy to shape their future selves, or that any erosion of that autonomy is compatible with the foundations of the community of rights.[76] Modern thinkers such as Ronald Dworkin and Jürgen Habermas share this concern,[77] which Brownsword phrases as follows:

[75] Article 1 of the Universal Declaration of Human Rights 1948 states: 'All human beings are born free and equal in dignity and rights.' The Preamble reads: 'recognition of the inherent dignity and of the equal and inalienable rights of all members of the human family is the foundation of freedom, justice and peace in the world.' A biomedical example is couched in Art 1 of the International Declaration on Human Genetic Data, 2003, the main purpose of which is: 'to ensure [respect for] human dignity and protection of human rights and fundamental freedoms in the collection, processing, use and storage of human genetic data, human proteomic data and of the biological samples from which they are derived.'

[76] See Harris's explanation (speaking of euthanasia) of the importance of such an autonomy: 'The point of autonomy, the point of choosing and having the freedom to choose between competing conceptions of how, and indeed why, to live, is simply that it is only thus that our lives become in any sense our own. The value of our lives is the value we give to our lives. And we do this, so far as possible at all, by shaping our lives for ourselves. Our own choices, decisions and preferences help to make us what we are, for each helps us to confirm and modify our own character and enables us to develop and to understand ourselves. So autonomy, as the ability and the freedom to make choices that shape our lives, is quite crucial in giving to each life its own special and peculiar value.' Harris, 'Euthanasia and the Value of Life', in J. Keown (ed), *Euthanasia Examined* (1995), quoted by Brownsword, 'Bioethics Today, Bioethics Tomorrow: Stem Cell Research and the "Dignitarian Alliance"' (2003) 17 Notre Dame J Law, Ethics & Public Policy 15–51, at 22.

[77] See R. Dworkin, *Sovereign Virtue* (2002) at 444: 'If we were to take seriously the possibility that we are now exploring—that scientists really have gained the capacity to create a human being having any phenotype that they or their prospective parents choose—then we could chart the destruction of settled moral and ethical attitudes starting at almost any point. We use the chance/choice distinction not simply in our assignments of responsibility for situations or events, for example, but in our assessments of pride, including pride in what nature has given us.' See also J. Habermas, *Die Zukunft der menschlichen Natur. Auf dem Wege zur liberalen Eugenetik?* (The Future of Human Nature. Are We on our Way to Liberal Eugenics?) (2001).

In other words, genetic enhancement (by way of reproductive cloning for desired and desirable characteristics) threatens to alter the context by either undermining notions of free will and responsibility or by inviting the transfer of responsibility. If individuals can no longer make much sense of their personal responsibility for their own actions and omissions, then surely this will weaken, too, their sense of collective responsibility.[78]

The second version of human dignity has a repressive and constraining effect, and it is this version that is strongly on the rise, including in my native country the Netherlands.[79] Human dignity as a restriction is rooted in Kantianism,[80] but it also has considerable support from the influential religious conservative constituency. Advocates of this second version of human dignity are referred to as 'dignitarians' by Brownsword. Dignitarians will categorically oppose any technology or intervention that harms their constraining notion of human dignity. To a dignitarian, the right to life and physical integrity mean that any intervention in the embryonic stage should be prohibited. The principle that human life should not be instrumentalized to dignitarians means that sex selection, reproductive cloning, and the phenomenon of saviour siblings must be halted. A ban on the commercialization of the human body implies a ban on the patentability of human genetic material, surrogate parenthood, and so on.

Clearly, dignitarians have major objections to PGD, and no amount of consultation, education, or deliberation is likely to change that position, as they are uncompromising in their belief that the destruction of embryos is categorically

[78] Brownsword, 'Stem Cells and Cloning: Where the Regulatory Consensus Fails' (2005) 39 New England L Rev 535–71, at 561.

[79] Brownsword calls this version 'human dignity as constraint'. In the Convention for the Protection of Human Rights and Dignity of the Human Being with Regard to the Application of Biology and Medicine: Convention on Human Rights and Biomedicine, 1996, countries oblige themselves 'to take such measures as are necessary to safeguard human dignity and the fundamental rights and freedoms of the individual with regard to the application of biology and medicine.' UNESCO's Universal Declaration on the Human Genome and Human Rights, 2003, declares that, although 'research on the human genome and the resulting applications open up vast prospects for progress improving the health of individuals and of humankind as a whole... [it is imperative]... that such research should fully respect human dignity, freedom and human rights.' An example on the EU level is Art 6(2) of Council Directive 98/44/EC which settles the patentability of biotechnological inventions. The directive stipulates that on account of clause 1, the following is expressly considered unpatentable: methods for cloning people; methods for changing the germinal genetic identity of man; the use of human embryos for industrial or commercial purposes; methods for changing the genetic identity of animals that are likely to make them suffer without considerable medical use for man or animal, including the animals obtained from such methods.

[80] See I. Kant, *The Metaphysics of Morals* in M. Gregor (ed & trans), (1797) at 209: 'Every human being has the legitimate claim to respect from his fellow human beings and is *in turn* bound to respect every other. Humanity itself is a dignity; for a human being cannot be used merely as a means by any human being... but must always be used at the same time as an end. It is just in this that his dignity (personality) consists, by which he raises himself above all other beings in the world that are not human beings and yet can be used, and so over all *things*. But just as he cannot give himself away, for any price (this would conflict with his duty of self-esteem), so neither can he act contrary to the equally necessary self-esteem of others, as human beings, that is, he is under obligation to acknowledge, in a practical way, the dignity of every human being. Hence there rests on him a duty regarding the respect that must be shown to every other human being.'

wrong.[81] The rise of the restrictive notion of human dignity therefore does not bode well for the further development of clinical genetics, including PGD. To the extent that deliberation and debate require preparedness to compromise, constructive dialogue on the desirability of such a development is also bound to become more difficult.

It is tempting to dismiss the human dignity paradigm for the regulation of PGD for this reason, but that would leave us with one crucial question: on what grounds can governments intervene if PGD starts to erode human dignity in its emancipating role, thus eating away at the very foundations of the human rights community? Assuming that parents give prior informed consent for such use of PGD, a human rights response (in which, as we have seen, 'human dignity as empowerment' is subsumed) offers little solace. Nor can a solution to this dilemma be expected to come from utilitarians, at least not for as long as the advantages of PGD outweigh its disadvantages. Unless we wish to turn to the kind of uncompromising dignitarianism that we have just concluded falls short of a cosmopolitan meta-regime enabling a transparent process of deliberation, we therefore would appear to be left with a serious problem.

It has been suggested that the precautionary principle could serve to combat the gradual erosion of the foundations of a community of rights. If precaution is part of an ethics inspired by human rights, governmental restrictions of such rights may be justified by appealing to the uncertain future consequences of, for instance, PGD for the viability of the community of rights itself:

> On this analysis, the State is authorised to act in a precautionary way for the sake of the integrity of the Community—for example, where an act that is permitted under a procedural justification has a negative impingement on *possible* rights-holders, on *arguable* rights or on the viability of the community of rights itself. In such cases, the State's case for intervention does not rest on the need to intervene for the sake of a clear and settled overriding right, but for reasons of risk-avoidance relative to the regime of rights and the community of rights-holders.[82]

The next section looks at which of the three ethical paradigms described would serve best as a meta-regime giving direction to the regulation of clinical genetics and what role precaution has to play.

B. Human Rights as Meta-Norm for the Regulation of PGD

Before arguing in favour of using human rights as the meta-regime for the regulation of PGD, it is useful to remind ourselves of the function of such a regime. The basic assumption is that policy expressed in legislation has to enjoy both internal and cosmopolitan legitimacy when those rules produce extraterritorial consequences,

[81] Brownsword, *supra* note 78, at 543. [82] *Ibid*, at 451 (emphasis added).

which is often the case in the European and globalized context in which we live. Policy on GMOs clearly has extraterritorial effects, mainly because it has implications for international trade. The legitimacy review developed by the WTO is procedural in nature, and expresses itself in principles such as non-discrimination, proportionality, and, with respect to the SPS Agreement, in the requirement that restrictive measures must be based on a sound risk analysis. This last requirement enables a transparent and constructive assessment of laws, which does not only benefit the cosmopolitan legitimacy of the measures, but also serves the quality of the internal democratic deliberative process. The regulation of medical genetic services, such as PGD, also has extraterritorial impacts. This is of course the case within the EU on the basis of the free movement of services, and the concept of European citizenship. However, the regulation of PGD also has global implications for potential suppliers and recipients of this service outside Europe. Therefore, there is no reason to subject the regulation of PGD to this form of review, whereas in the case of GMOs such review, for good reasons, is relatively rigorous.

However, legitimizing a policy with far-reaching ethical implications poses an enormous challenge. First of all, one cannot attribute the same legitimizing effect to 'science', as is possible in the case of the regulation of risk. Furthermore, at first sight there is no worldwide institutional machinery offering a platform for the type of procedural policy coordination as advocated by Joerges and Godt. To put it more simply: we are looking for an ethical perspective with cosmopolitan appeal, and a platform where differences can be settled on the basis of this cosmopolitan criterion, like the WTO and the ECJ. Such settlements will by necessity be temporary, since science and moral insights will constantly evolve. With this goal in mind, I will assess the three ethical paradigms that were briefly presented above.

1. Utilitarianism

For fundamental and practical reasons, a utilitarian ethical paradigm cannot provide answers to the question whether a state measure allowing or prohibiting PGD or any other clinical genetic technique is legitimate. A fundamental and insurmountable objection to a utilitarian review is that it sacrifices the rights of the individual for the good of the many. This objection especially holds true in a discussion on PGD, a technology that posesses neo-eugenic potential. Buchanan *et al* quote the following passage from *The Eugenics Catechism* by the American Eugenics Society:

> Q. How much does segregation cost?
> A. It has been estimated that to have segregated the original 'Jukes' for life would have cost the State of New York about $25,000
> Q. Is that a Real Saving?

A. Yes. It has been estimated that the State of New York, up to 1916, spent over $2,000,000 on the descendants of these people.
Q. How much would it have cost to sterilize the original Jukes pair?
A. Less than $150.

On this kind of utilitarian logic, Brownsword comments: 'the problem with utilitarian calculations is that they are capable of outraging our sense of justice particularly by authorizing the pursuit of the larger benefit at the cost of individual human rights.'[83]

A second fundamental and related objection to a utilitarian legitimacy review is that it offers the right answers to the wrong questions. The danger of PGD or any other technology involving human genetics is not so much in its safety, costs, etc (for which a utilitarian review is adequate) but at least partly in its heralding a 'posthuman' era.[84] Utilitarianism does not offer sound protection against such risks. The idea that an ethics that uses utility as a benchmark can serve as a meta-regime for guarding the legitimacy of policy also runs aground because of practical reasons. Because the utility of a rule can only be determined *ex post*, the question arises whether a utilitarian model can ever offer firm guidance on whether a proposed or existing rule is legitimate. As commented previously, utility depends on time and space. If a utilitarian model were to be used as legitimacy review, this would give rise to situations in which one and the same rule would first be acceptable and a little later it would not. In addition, the rule would be acceptable for jurisdiction A but not for jurisdiction B. From the viewpoint of the desired transparency in a global context this is an undesirable outcome, which would also wreak havoc for international trade. Evidently, this does not affect the usefulness of utilitarianism as a whole. It has important roles to fulfil in making ethical decisions, for instance about the use of scare public funds to improve public health in national and global contexts. It merely does not provide support in deciding whether a rule or activity is acceptable or not from a cosmopolitan perspective.

2. *Dignitarianism*

The fact that one may question what a deliberative democracy exactly amounts to is at the same time the best answer to that question. In a democracy, difference in opinion is approached through a reflexive debate that is as transparently and rationally structured as possible. When it comes to PGD, dignitarians are not much interested in reflection and debate; human dignity must *never* be compromised. The sanctity of life is paramount (as long as it is embryonic, at least).[85] For that reason alone, the dignitarian paradigm does not meet our demands.

[83] *Ibid*, at 543.
[84] The term is derived from F. Fukuyama's *Our Posthuman Future* (2002).
[85] Dawkins draws our attention to the paradox that dignitarians often are stern supporters of the death penalty: see R. Dawkins, *The God Delusion* (2006) at 291–2.

When the paradigm has cosmopolitan pretensions, it assumes a totalitarian character. In its local form, it leads to local and sub-local 'PGD-free zones'. Whatever guise the perspective assumes, it will lead to the infringement of individual liberties, free trade, or both.[86] The restrictive version of human dignity can therefore never yield a useful meta-regime that enables us to review a national policy's cosmopolitan legitimacy.

We are stuck, then, with the problem formulated at the end of the previous section: emancipating human dignity must be protected as it is precondition for the existence of articulated human rights. Any measures that a state might wish to take to protect this constructive conception of human dignity will, however, not often pass a utility review. The same goes for a review framework that only encompasses articulated human rights and attributes key importance to the instrument of prior informed consent. Take the example of reproductive cloning after informed consent by parents. A state could decide, in order to protect a constructive concept of human dignity, that the treatment should be prohibited. On the basis of a utility review, utilitarians may conclude that a green light should be given, for example because cloning offers the only chance of a child with a genetic relationship with the parents. A human rights perspective will protect the autonomy of the parents as soon as a link to an articulated human right can be found. It is not beyond dispute that the principle of reproductive autonomy represents an articulated human right, but assuming that this is in fact the case, a state will have to offer very good reasons if it is to limit that right. There is a need for such 'good reasons', however, because the autonomy that forms the basis of every articulated human right is at stake. Neither a utility review, nor a human rights review can offer those good reasons. Dignitarians will not understand why reproductive cloning is not allowed where therapeutic cloning, IVF, and PGD are, and will rightly feel that the policy is inconsistent or even discriminatory.

Does the precautionary principle provide a solution? Brownsword thinks it does:

> there is a plausible role for the State exercising a precautionary jurisdiction where the long-range consequences of individual choice are unclear or where we cannot be entirely confident that an act has no negative impingement on rights-holders or on arguable rights. Some States may be more risk-averse (more precautionary in their approach) than others and we must be careful that this stewardship jurisdiction is not a hostage to

[86] See Brownsword, 'Genetic Engineering, Free Trade and Human Rights: Global Standards and Local Ethics', in T. Cottier (ed), *Genetic Engineering: Challenges Posed by a New Technology to the World's Trading System* (forthcoming): 'if we think that, where morality exceptions allow for dignitarian principles, this will not result in the same rules everywhere, we are quite right—but it surely is not a good thing that we are right. To permit the cosmopolitan principle to be destroyed by dignitarian exceptions, or to allow dignitarian imperialism to push back human rights, is a poor result—and it is a poor result whether one looks at it by reference to free trade or the morality of individual freedom.'

dignitarian fortune. Nevertheless, my suggestion is that we should not rule out the possibility of such a legitimate jurisdiction which would then add to the reasons for denying the sufficiency of the couple's consent.[87]

The attraction of this solution is that it upholds an ethics based on human rights, while at the same time protecting humanity against dehumanization, which would eventually be fatal to a community based on human rights. But is this really true? Brownsword also seems to have his doubts. At the very least, he is aware of the fact that the precautionary principle could prove to be dangerous in the hands of dignitarians. And he is right. In the regulation of GMOs, precaution for the regulation of risk in situations of scientific uncertainty is a tricky matter. We have seen from the case law of the WTO and the EU that it is difficult to maintain the integrity of a procedural legitimacy requirement based on science, if it is accepted through recognition of the precautionary principle that science is not sacrosanct. However, it may prove impossible to maintain the integrity of a legitimacy requirement based on autonomy if through recognition of the precautionary principle autonomy becomes a relative value. Without wanting to lose myself in listing the dangers hidden in embracing the precautionary principle in this context, a few of them are self-evident.

First, there is a big difference with regard to the *subject of the uncertainty*, in comparison to GMO regulation, where it always concerns scientific uncertainty. However, for Brownsword, uncertainty about two questions can trigger precautionary regulation, and therefore compromising a human right:

1) *who* holds a right;
2) *which* rights are relevant.[88]

Both conditions (1) and (2) have been met, according to the current state of the law, and the scope for precautionary action becomes almost limitless. There is absolutely no consensus, let alone a universal consensus, on the moment when a human being can be a rights-holder: from the moment of conception, after three, six, or nine months, or not until after birth. Even the question of which rights are relevant from a human rights perspective (never mind their implications!) has its degree of uncertainty; the right of reproductive autonomy is an example here. A government can therefore, at all times, and on the basis of either of the two uncertainties, call on the precautionary principle to justify restrictions of human rights.

But the dangers of using precaution in the context of human rights review go further. In addition to obscurity about the *subject of uncertainty*, the *subject of risk* against which precaution should protect is also not clearly defined. In GMO regulation, this is about risks to the health of plants and animals and to the environment. It is clear that Brownsword is thinking of risks to the emancipatory/constructive form of human dignity. But what keeps us from giving the

[87] Brownsword, *supra* note 78, at 472.
[88] *Ibid* (emphasis added).

precautionary principle a different twist, and to focus on the risks for individual autonomy, the general good, or a restrictive version of human dignity? The fact that this is an acute danger is confirmed in practice. In the United Kingdom, the House of Commons Science and Technology Committee reached the following conclusion in its important report *Human Reproductive Technologies and the Law*:

> We do not see why the area human reproductive technologies should do anything other than proceed under a precautionary principle currently prevalent in scientific, research and clinical practice. This means . . . that alleged harms to society or to patients need to be demonstrated before forward progress is unduly impeded.[89]

It is not entirely clear which social or clinical risks the Select Committee is thinking of here, but precaution at least justifies a *laissez-faire* approach that Brownsword precisely wanted to limit with an appeal to the principle.

The fact that precaution in this context can turn into a travesty is shown when we note the similarities between the Select Committee's conclusion based on the use of the precautionary principle and the definition of the 'proactionary principle' developed by transhumanists:

> People's freedom to innovate technologically is highly valuable, even critical, to humanity. This implies several imperatives when restrictive measures are proposed: Assess risks and opportunities according to available science, not popular perception. Account for both the costs of the restrictions themselves, and those of opportunities foregone. Favor measures that are proportionate to the probability and magnitude of impacts, and that have a high expectation value.[90]

The charming young lady who introduced herself as 'Lady Precaution' in poor light earlier in the evening, turns out to be 'Mr Proaction' upon return to the hotel room; a particularly unwelcome surprise.[91] A principle that was supposed to protect us against the product of transhumanists leads to results that are identical to the principle that is actually Article 1 of the constitution of those same transhumanists.[92]

[89] House of Commons Science and Technology Committee, *Human Reproductive Technologies and the Law*, Fifth Report of Session 2004–5, Vol 1, para 47.

[90] Available at <http://www.extropy.org/proactionaryprinciple.htm>.

[91] This is a classical example of a 'nightmare scenario'. After all, the reverse would have less serious consequences. If, on the basis of 'Mr Proaction's' reputation, an eventful night in the hotel room had been anticipated, but the travesty reveals itself as 'Lady Precaution', the disappointment would undoubtedly remain, but the consequences would be far less dramatic.

[92] Should doubts remain about this conclusion, section 66 of the same report may remove these: 'The ethical status of hybrids and chimeras is complex. While there is revulsion in some quarters that such creations appear to blur the distinction between animals and humans, it could be argued that they are less human than, and therefore pose fewer ethical problems for research than fully human embryos. We recognise concerns that hybrids could be used for reproductive purposes and recommend that new legislation (a) defines the nature of these creations, (b) makes their creation legal for research purposes if they are destroyed in line with the current 14-day rule for human embryo cultures, and (c) prohibits their implantation in a woman.'

It is interesting to compare this British use of precaution with the Dutch Health Minister's conclusions:

Although children born with PGD do not seem to have a higher risk of defects than children born with 'normal' IVF, the Indication underlines the importance of a long-term follow-up to gain more certainty on the matter. To me, this is a reason to be extra cautious in case of a broadening of the assessment for the need of PGD.[93]

Although precaution (taking restrictive measures justified by scientific uncertainty) is not actually mentioned, it would lead to a result that is the opposite of that of the Select Committee.

Irrespective of the ethical perspective that it is supposed to serve—utilitarian, human rights, or human dignity—the precautionary principle is a potential Trojan Horse. Precaution does not offer any protection, not even in the hands of an independent judiciary, and can turn against those who seek justice without so much as a preamble. In summary, precaution should not have a place within a meta-regime that is supposed to yield predictable outcomes where the legitimacy of measures is reviewed. The charge is not that precaution leads us in a *certain* wrong direction. The accusation is much more serious: precaution does not give any direction.

3. Human Rights

There are several reasons for submitting the regulation of human genetics, including PGD, to a review based on human rights. First of all, this connects it to UNESCO's point of departure and to the evolving international practice. At the same time, it must be recognized that UNESCO and other international organizations are far from univocal in the ethic that inspires the international regulatory framework: in one and the same text there are references to human rights side-by-side with dignitarian and utilitarian thoughts. That ambiguity has enabled formation of an apparent international consensus between diametrically opposed positions, for instance on cloning, but carries as a price that the regime will become either irrelevant or, at a later time, a source of conflict. Because, as we have seen, the three ethics can lead to completely different outcomes, such ethical pluralism is not an option: choices will have to be made. Secondly, an ethic focused on autonomy is not only the most suitable for a liberal deliberative democracy, but according to many also provides the only answer to the globalized reality.[94] This argument is partly based on the universal claim made by

[93] Minister Ross's viewpoint, dated 10 May 2006, is available at <http://www.minvws.nl/kamerstukken/pg/2006/premplantatie-genetische-diagnostiek-en-premplantatie-genetische-screening.asp>.

[94] See F. Fukuyama, *The End of History and the Last Man* (1983); Brownsword, 'What the World Needs Now: Techno-Regulation, Human Rights and Human Dignity', in R. Brownsword (ed), *Global Governance and the Quest for Justice* (2004), at 203–34.

human rights. Extreme local variation, such as would be caused by a regulatory framework based on local applications of human dignity or utility, would be hard to justify from a perspective of cosmopolitan legitimacy, and would be incompatible with the global trade system.

At the same time, the universality of human rights could also be considered a problem. One of the *leitmotivs* of Brownsword's work is that a community of rights constantly has to guard the viability of that community by using a reflexive process of question and answer to protect the emancipatory/constructive version of human dignity. The outcome of this deliberative process will and must allow local variation. In other words: a review framework based on human rights always has to offer room to local measures aimed at protecting the foundations of articulated human rights. How to meet this demand without opening the door to arbitrariness or dignitarianism is a formidable challenge, and in this paper I will limit myself to the following observations and considerations. First of all, a regime that caters for the need for local variation by providing room for the precautionary principle should be rejected. Precaution leads to arbitrariness and provides governments with a blank cheque for limiting individual liberties. This is an important conclusion, for precaution is more and more often considered an easy-fix for difficult problems. Just as the WTO cannot tolerate such use of precaution, we must now guard against the principle in the context of human genetics.

Furthermore, the system of human rights offers much flexibility and room for local variation. A rule that, for instance, prescribes 'protection of life' has hardly the same pre-emptive effect as an EU rule stipulating that confectionery merits the predicate 'chocolate' only if it contains 43 per cent cocoa. On the question of when the life to be protected commences, for instance, there can and may be differences of opinion, as was shown in the ruling in *Vo v France*. Similarly, whether this rule only protects citizens against states (vertical effect), or whether it also has horizontal effect, ie also regulates the behaviour of citizens amongst themselves, is also open to debate. In addition, enforcement of the rule remains to a high degree outside the scope of the rule and there, too, different communities can pursue different policies. Finally, it should not be forgotten that articulating human rights is an open and never-ending project. New human rights are continuously codified, and if the need to protect the emancipatory/constructive version of human dignity from erosion increases, the task is to translate that need to an articulated human right. The fact that reaching consensus on new rights often seems nearly impossible (especially in comparison to a simple embracing of precaution) is the very reason from which an articulated human right derives its externally legitimizing effect. What remains to be done at this point is to apply these insights to a concretely formulated policy in order to illustrate the consequences of a review based on human rights. The position on PGD that the Dutch Health Minister outlined in 2006 will serve this purpose.

5. The Minister's Opinion Reviewed

A. Brief History

On 7 November 2003, the Minister requested that the National Health Council advise on PGD and pre-implantation genetic screening (PGS). The immediate cause for the request was that the criterion regarding the availability of PGD used in the Regulation of Clinical Genetics of January 2003 raised questions within the medical profession. What this criterion basically amounts to is that there has to be an individual risk of conceiving a child with a serious genetic disorder or disease. In her letter, the Minister inquires as to what this criterion precisely means for the permissibility of the various applications of PGD. The National Health Council's report was presented on 18 January 2006. With regard to the availability of PGD, the National Health Council focused on the question of whether PGD should be available on the basis of:

1) being a carrier;
2) diseases that occur late in life;
3) defects with variable expression or with incomplete penetrance;
4) HLA-typing for the benefit of donor creation (saviour sibling);
5) sex selection.

The National Health Council advised more or less positively for all indications. The Council did place conditions on the use of PGD: depending on the indication they placed major or minor restrictions on parents who wish to make use of the technique. The Minister formulated her Opinion on 10 May 2006.

B. The Importance of Transparency

The Opinion differs from the advice of the National Health Council on all counts. We saw earlier that this separation of specialist and political responsibilities fits within our deliberative democratic establishment. The WTO and the EU explicitly accept this separation, but they do press for procedural demands for the motivation of measures that deviate from scientific advice. Within the framework of risk regulation the demand is, simply put, that a departure from scientific advice also requires a *scientific* motivation. This motivation may be based on a scientific minority opinion. The aim of the demand 'fight science with science' is to introduce transparency and in doing so to facilitate (global) deliberation, possibly within the framework of a WTO or EU dispute settlement. For the regulation of a technology of which the risks are primarily of a bioethical nature, this necessary transparency can be safeguarded by indicating the ethical approach that has been adopted. This makes a process of deliberation possible, which is the minimum condition for obtaining both internal and cosmopolitan legitimacy of decisions. The articulation of the meta-regime explicates:

a. the objectives that can be aimed for within that framework (protection of human rights, maximizing utility, protection of human dignity);
b. the operationalization of those objectives (invoking articulated human rights and/or the procedural instrument of prior informed consent, a cost-benefit analysis, or prohibiting activities or rules that compromise human dignity).

With regard to GMO legislation, it has been seen that the meta-regime of the WTO and the EU also regulates the machinery that is put into motion when Member States do not comply with either (a) or (b). This will not normally apply to national measures that restrict access to PGD on ethical grounds. However, if a measure has, for instance, European or global trade implications, then the EU or the WTO will claim jurisdiction. It is also possible that a measure has to be judged within the framework of, for instance, the European Convention on Human Rights, which, incidentally, also has a strong disciplinary function for the EU and the WTO.

I will first and foremost pay attention to the transparency of the ethical framework that provided the basis for the Minister's Opinion. This provides clarity on the objective that can be pursued, and the manner in which those objectives can be put into operation. Furthermore, in each case the Opinion will be assessed against a review based on human rights.

C. The Ethical Plurality Employed by the Minister

The Minister's letter articulates the ethical angle chosen for assessing the five specific questions on the indication assessment of PGD. Her language does not always fit in with Brownsword's triangle, but the translation to the model is not very difficult. The Minister employs a multitude of ethical perspectives:

> With PGD, there are usually no fertility problems with the future parents, but still an IVF procedure must be undergone for which several embryos are created, whilst the parents, in most cases, can have children in a natural way. This moral aspect needs to be part of the consideration whether the situations in which PGD is permitted are to be extended. In this connection, it is of importance that the IVF procedure is in itself a burden for a woman and that the chance of success of IVF is relatively small. Furthermore, complications for woman and child occur more often. With IVF there is an increased risk of morbidity for the child.[95] . . . In any case, the future parents have to be well informed about the disadvantages that these methods may involve, before this IVF-PGD course may be followed.[96]

[95] *Supra* note 93, at 2. [96] *Ibid.*

The first moral aspect to which the Minister refers is the fact that in the case of IVF with a PGD, procedure embryos are used (and the surplus embryos are not used), whilst a pregnancy could also be achieved in a natural manner. Utilitarians do not have any *a priori* problems with creating and destroying embryos. If the word 'this' refers to creation and destruction of embryos, then that is where the ways of the utilitarians and the Minister part. In their assessment, utilitarians will include the fact that an expensive IVF procedure is employed, whilst a pregnancy can also be achieved naturally. At the same time, they will want to ascertain the other advantages and disadvantages of PGD in this sort of situation. If there are no compensating advantages, a utilitarian will indeed consider immoral the use of finite public means to fund IVF without there being a necessity to do so. In practice, such advantages of course do always exist, and the question for the utilitarian then is whether they outweigh the costs. The second part of the quote provides some information about this, but it is limited to the disadvantages: a great burden on the woman, a small chance of success of IVF, a chance of complications for woman and child, and an increased risk of morbidity. The instrument of informed consent is not essential to utilitarians. It can be useful, especially in a society where the requirement is highly valued, but it is not a 'solid' component of utilitarian ethics.[97] All in all, it cannot be ruled out that the Minister's Opinion is based on a utilitarian ethic, but this conjecture is weakened by the emphasis on the *unnecessary* creation and destruction of embryos, and the reference to the importance of informed consent.

In an ethical paradigm informed by human rights, scant attention is given to the protection of embryos: embryos, after all, are not holders of rights. The precautionary principle, which Brownsword supports so it may provide this protection, is employed in the Opinion, but not with regard to the protection of possible holders of rights, or to the protection of possible human rights. The Minister's viewpoint employs the principle to prevent risks of damage to the embryos that have not yet been scientifically proven to exist.[98] Precaution, therefore, is rather traditionally used here to regulate risks in the absence of scientific certainty. An invasive PGD procedure would violate several articulated human rights if it were not preceded by informed consent. From a human rights perspective, prioritizing the procedural instrument of informed consent is crucial

[97] See Brownsword, *supra* note 76, at 8: 'There is no golden rule requiring that consent of those upon whom an action of decision impacts should be obtained. For example, requiring researchers or doctors to deal on an informed consent basis with research participants or others is not necessarily an improvement on compulsion, ignorance or paternalism. The calculation always depends on context, convenience, contingency and circumstance. Having said this, in a culture where preferences strongly favour the currency of consent, even if there is no golden rule requiring consent, utilitarians might well accept the sense of a general rule to this effect.'

[98] See *supra* note 93, at 3: 'What little research has been carried out into biopsy to date seems to indicate that it does not lead to damage by virtue of itself. However, there is no certainty about this yet, as the Health Council also indicates . . . To me, this is a reason to be extra cautious in case of a broadening of the assessment for the need of PGD.'

indeed. Once this consent has been granted, and no rights of third parties are at risk, there is no reason at all to limit PGD.

Hence, the Minister's Opinion offers advocates of an ethic based on human rights some support. From the quoted passage, however, it appears that individual autonomy is not the only basic assumption. Even if parents wanted to give their informed consent for PGD, it is possible, taking the other considerations mentioned into account, that permission would still not be granted. Neither is the ethic in the Opinion fully compatible with the viewpoint of dignitarians. The destruction of an embryo, for dignitarians, must be prohibited in *all* cases. In this connection, it does not matter whether these embryos are destroyed in a procedure that enables infertile couples to have children, or whether it occurs in a PGD procedure. Even if PGD carried no risk at all for the mother or the future child, the fact that surplus embryos are destroyed or are used for research is sufficient reason for dignitarians to veto PGD.

D. Answers to the Questions Regarding Availability of PGD

1. Carriers

The implications of being a carrier were briefly explained earlier. Being a carrier is not usually manifested by disease, or in any case not by the often extremely serious symptoms of the disorder. Being a carrier does imply a risk that the disease is passed on to children or grandchildren, which can be a heavy psychological burden. Prospective parents can therefore wish to undergo PGD treatment to rule out the possibility that they will conceive a child who is carrier. The Opinion distinguishes two situations. In the first, PGD is prescribed for a serious hereditary disease. In such cases, it is often not only revealed which embryos will develop the disease to its full extent, but also which embryos are carriers of the disease. No additional procedure, therefore, is needed to establish whether embryos are carriers. In the second situation, an additional procedure is required, and usually this is IVF. The objective of the PGD treatment is solely to identify carrier embryos.

The Minister considers PGD permissible in the first case, but not in the second:

> With regard to PGD, I want to adhere to the criterion that there has to be an individually increased risk that a child will suffer a serious genetic disorder. Given the moral considerations, restraint in establishing carriers is appropriate. This especially applies if extra procedures are required in order to establish which embryos are carriers.[99]

This conclusion would not pass a review based on human rights. After prior informed consent, the PGD treatment should be permitted. There is no reason at all to limit the parents' autonomy, which they express by means of informed consent. Even an 'emancipating' dignitarianism will not lead to another

[99] *Ibid*, at 5.

conclusion as a matter of course. The open future of the child is not jeopardized by a PGD procedure to identify carrier embryos. The risk that availability of PGD for carriers can undermine the collective feeling of responsibility that feeds a community of rights is therefore far-fetched.

One can never predict with absolute certainty the outcome of a *utilitarian* assessment, but because both parents and their future children will benefit from exclusion of carriers, it is likely that in such cases PGD will be permitted rather than denied. The cost of the treatment could be a consideration, but because excluding carriers brings with it future cost savings, this too is an argument *in favour of* permitting PGD. The rights and interests of the future child—the Opinion's basic principle—are not harmed either. Thanks to PGD treatment, the future child will not be the carrier of a hereditary disease, and it is difficult to substantiate that this harms the child's interests in any way. The Minister's decision can therefore only be understood as an expression of restrictive dignitarianism. What is protected are surplus embryos.

2. *Diseases that Occur Later in Life*

The question of whether PGD should be permitted for hereditary conditions that will occur later in life, like Huntington's disease, raises several ethical issues. The focus here is on the dilemma when prospective parents know that the disease runs in the family, but, because it is a late-onset disease, they cannot know for certain whether they themselves have the disease. Many parents do not *want* to know, because this knowledge would be an unbearable psychological burden to them. They then call upon the so-called 'right not to know'.[100] The dilemma arises when such parents want a PGD treatment so that a future child will in any case not develop the disorder, but at the same time they wish to remain ignorant regarding their own condition.

Two different types of genetic test can be used, with far-reaching consequences for the right not to know: a mutation test (non-disclosure test), or an exclusion test. A mutation test determines whether the foetus carries the damaged gene: if the test turns out positive, then the mother has the disease. A negative test result does not say anything about the status of the mother. In the case of a positive test result, the right not to know can only be respected if the attending physician does

[100] The status of this right is not yet clear. However, it is strongly represented in recent international bioethical legal instruments. Article 10 of the Convention on Human Rights and Biomedicine, 1997, says: 'the wishes of individuals not to be so informed [on their health] shall be observed.' Article 5 of the UNESCO Universal Declaration on the Human Genome and Human Rights, 1997, stipulates: 'The right of every individual to decide whether or not to be informed of the results of genetic examination and the resulting consequences should be respected.' Lastly, Art 10 of the International Declaration on Human Genetic Data, 2003, recognizes that: 'When genetic data, human proteomic data or biological samples are collected for medical and scientific research purposes, the information provided at the time of consent should indicate that the person concerned has the right to decide whether or not to be informed of the results.'

not disclose the information provided by the test. This possibility was rejected by the National Health Council. An exclusion test works differently, and *neither* in the case of a positive test result *nor* in the case of a negative test result does it provide any new information about the mother. If priority is given to the right not to know, the exclusion test would therefore have to be preferred to a mutation test. A consequence, however, is that as the exclusion test does not provide any information about the mother, PGD can be used in cases where the mother may not have the disease. In order to respect the right not to know, embryos are therefore created and destroyed whilst that is *possibly* unnecessary.

The National Health Council considers the use of the exclusion test acceptable, but the Minister does not follow this advice. Both tests are unacceptable to the Minister because:

> The central point is that in all cases IVF and PGD are carried out, even if they are not at all necessary because both parents are at this point genetically healthy. In a number of cases an expensive treatment that is taxing on the woman is carried out whilst it is unnecessary. A child is created by means of unnecessary technical procedures, which can result in higher morbidity and mortality. Also, a number of surplus embryos are created without any necessity.[101]

As a consequence, parents who do not wish to be informed of their genetic status will not be able to use PGD. They can only decide whether or not to have children, ie take a chance on the health of their future children. If we measure this outcome against an ethics based on human rights, it will not pass. Also, it is very questionable whether a utilitarian review can offer support to the formulated policy. Certainly, the costs of the burden on the woman and the risks to the child through a possibly unnecessary PGD treatment are all considered, but the utilitarian will *also* value the parents' wish not to be informed, and the risks associated with possible Huntington sufferers becoming parents without a PGD treatment.

The Minister's Opinion rather testifies to an ethic to which dignitarians have no problem subscribing. The limitation is not imposed to protect any articulated human right, nor because of high costs or small benefits, but to prevent the 'unnecessary' creation and destruction of embryos. This last concern enjoys a higher priority than the right not to know, even though it is articulated in several international bioethical instruments.

3. Defects with Variable Expression or Incomplete Penetrance

The third question posed by the Minister to the National Health Council concerned PGD for disorders with variable expression or incomplete penetrance. In the case of variable expression, a genetic mutation causes a disease that may occur in a serious or less serious form. In the case of incomplete penetrance, some

[101] *Supra* note 93, at 6.

of the people who have the mutation stay healthy, and others develop the disease. The National Health Council formulated a subtle answer: depending on the individual circumstances of the case, PGD could be indicated. The Minister does not follow this advice:

> At this moment it appears that in these situations the selection through PGD of an embryo with regard to future health depends, to a certain degree, on chance. Even with embryos that are affected it is possible that in the course of their lives they develop no disease or—in the case of variable expression—only develop the disorder to a lesser degree. A PGD treatment appears to offer some certainty, but in the light of the knowledge currently available regarding such conditions, that is only a false certainty. That too, for me, is a reason to take a reserved stance.[102]

Again, a review based on autonomy will yield a different outcome. After all, from a human rights perspective the parents' wish will be respected, provided it is preceded by a prior informed consent, and the rights of third parties are not harmed. The assessment of whether PGD should be pursued although no scientific certainty exists as to how often and to what extent the mutation will lead to the disease, can then be made by the parents. In these cases, PGD does not infringe anyone's rights, so it should be permitted. For a utilitarian review more information would be needed concerning the individual circumstances of the case, which is quite similar to the viewpoint of the National Health Council. A utilitarian would not, in any case, advocate a general ban on PGD for disorders with a variable expression or incomplete penetrance.

The manner in which the Minister employs the precautionary principle on this issue is striking. The principle becomes operational because there is scientific uncertainty about the risk linked to disorders with variable expression or incomplete penetrance. The precautionary principle empowers the government to take protective measures. In the case at hand, a ban on PGD for disorders with variable expression or incomplete penetrance protects embryos that otherwise might be unnecessarily destroyed. As a result of the ban, the likelihood that children are born with a genetic defect increases. It is difficult to reach any other conclusion than that the precautionary principle is employed here to pursue a dignitarian agenda.

4. HLA-Typing for the Benefit of Donor Creation

The fourth question put to the National Health Council was whether a PGD treatment should be permitted to determine whether a future child could function as a donor for a seriously ill older sibling. The question here is whether PGD may be used in order to create saviour siblings. Based on the current

102 *Ibid*, at 7.

legislation, the Planning Decision, it is *not* permitted to use PGD to have the future child serve as a donor to another child.[103] The National Health Council advised that the Planning Decision should be altered such that PGD would be permitted for the purpose of donor creation.

Even more so than the preceding three questions, the subject of saviour siblings goes to the heart of an ethic based on human dignity. If a child is born with the intention of being used to aid another child, this means that the newborn child is being instrumentalized. In her decision-making, the Minister attaches central importance to the question of whether instrumentalization occurs in the case of saviour siblings:

> The question whether this is a matter of instrumentalization is pivotal. If the child to be conceived is primarily to function as a donor, I deem PGD not permissible for this reason.[104]

The danger of instrumentalization is not only taken seriously by dignitarians, but also by supporters of a human rights perspective. To a dignitarian, however, any form of instrumentalization will be reason to prohibit PGD. From a human rights perspective, that critical point is not reached until the phenomenon of saviour siblings affects the viability of a community based on respect for human rights. That would be the case if, by choices made at the embryonic stage, citizens were to lose their feelings of individual autonomy and responsibility, as a result of which the *collective* responsibility would also be threatened.[105] Still, the Opinion contains clues to suggest that the Minister's ethical framework is predominantly dignitarian. The primary evidence for this claim lies in the following passage:

> But there is—with a view to this situation—another point that I find even more essential. Is it permissible, with regard to donors, to create a child through all sorts of technical procedures like IVF and PGD, if conception could have been achieved naturally and without technical interference with the embryo? In other words, is it justified to perform all sorts of technical procedures—of which it is not yet certain whether these do not cause any harm in the long-term—in the very beginning of existence with the intention of producing a donor for a third party?[106]

The Minister rejects PGD for conceiving saviour siblings, not only on the basis of the argument of instrumentalization, but also because it requires a technology that necessitates the creation and destruction of embryos. This argument is sound to a dignitarian, but by no means convincing from a perspective based on human rights or utility. Does this argument close the door for a PGD indication for HLA-typing for the purpose of donor creation? As the Minister thinks the interests of the child to be conceived should be decisive, the answer to that

[103] See the appendix to Art 3 of the Planning Decision Clinical Genetic Research and Heredity Advice (*Planningsbesluit klinisch Genetisch Onderzoek en Erfelijkheidsadevisering*), 3.4: 'PGD for reasons of having the future child serve as a donor for another child cannot be deemed acceptable and should therefore be excluded in the protocol.'

[104] *Supra* note 93, at 8.

[105] Brownsword, *supra* note 78, at 561.

[106] *Supra* note 93, at 8.

question is in the negative. If PGD also benefits the health of the child to be conceived, it may be used. The Minister considers it permissible 'that in this procedure the embryo's suitability as a donor is also reviewed'.[107] If I am not mistaken, this means that at-risk parents can have another child for the sole purpose of creating a donor for a previously born child, as long as they do not tell the attending physician. Hence a slip of the tongue during the first consultation can (if the attending physician abides by the law) have far-reaching consequences for the sick child. This pragmatic solution to the problem of instrumentalization does not offer a solution if PGD is not *also* in the interest of the child to be conceived. Concretely, this translates into a ban on using PGD if the elder child has a non-hereditary condition, and IVF and PGD are employed solely for the creation of a donor and without these techniques benefiting the intended donor child itself. This last ban possibly is compatible with a human rights perspective, given the importance of protecting constructive notions of human dignity. Utilitarians will not understand this ban quite so well, and indeed might consider it unacceptable.

5. Additional Sex Selection

In the Netherlands, the Embryo Act currently prohibits carrying out procedures with the aim of selecting the sex of a future child. An exception to this ban exists if sex selection is carried out to prevent a serious sex-linked disorder. The Planning Decision follows suit. The National Health Council, however, has argued that the Planning Decision should be amended so as to make sex selection possible on non-medical grounds. The reasoning of the Council is that if PGD is used to prevent serious disorders while, at the same time, revealing the sex of the embryos, then there is no reason to deny parents that choice.

The Minister does not follow the advice of the National Health Council on this matter, and deems sex selection on non-medical grounds unacceptable.[108] She bases her stance on the argument that to permit additional selection on sex would mean that:

> in equal cases, a distinction is made solely on the basis of sex, namely on the basis of preference of the future parents for a boy or a girl. No justification can be found in legislation for this non-medical distinction. To allow such preferences would cause equal cases to be treated unequally.[109]

Objections to sex selection are usually, as in this case, based on the argument that it will lead to discrimination. Even though this is unarguably the case on a basic level (as one sex is preferred to the other), the point from a human rights perspective is whether such discrimination is unlawful or unjust. That is only the

[107] *Ibid.* [108] *Ibid*, at 9.
[109] See also Harris, 'No Sex Selection Please, We're British' (2005) 31 JME 286.

case if the discrimination harms others. It is not immediately clear how implanting, for instance, a girl would harm an existing boy. Especially if the selection is the result of a desire for a balanced family make-up, selection cannot be deemed disrespectful to the existing boy. Even if it is about a simple preference, the question of whether it is a matter of discrimination can only be answered by comparing the treatment of the boy and the later-born girl. From this perspective, objections based on discrimination do not yield a convincing argument against sex selection.[110] By now it will be clear that also from a utilitarian view, a ban on sex selection is by no means obvious. A consideration that may explain the ban on sex selection is contained in the Minister's final remark on this matter:

> Additional sex selection is a form of instrumentalization of the child and must remain prohibited.[111]

E. The Opinion Critically Reviewed

On the basis of this analysis, the ethical stance of the Minister at first glance appears overly inclusive. It contains elements of each corner of the bioethical triangle: utilitarian, human rights, and human dignity. As discussed, each of these three separate perspectives leads to different and often opposing answers to the question of whether PGD should be allowed for a specific indication. That makes it hard, if not impossible, to predict to which choices this plurality will concretely lead. Deliberation and reflection become extremely complex, and if an important place is given to human dignity as constraint, maybe even impossible. As a consequence, it will be hard to legitimize the policy pursued by the Minister. After all, the starting point can lead to opposing results. Another possible consequence is that ultimately the policy will please no one.

More striking yet is the fact that a closer analysis of the different questions regarding the availability of PGD yields a much more univocal picture of the ethical starting point on which the Opinion is founded. It is primarily human dignity, usually in a form that inclines towards dignitarianism, which leads to the conclusion that the current legislation should not be amended to make PGD available in the cases discussed. However, by not clearly articulating this view, by hiding it between references to utility, risk, and informed consent, an open and transparent debate is rendered more difficult. A clear-cut statement that human dignity is the starting point for the formulated policy, which means that the protection of a ten-celled embryo can be a reason for prohibiting PGD for the conception of a saviour sibling, would help such a debate. It would of course provoke questions as to why in one case human dignity leads to a prohibition whereas in another case it does not. Yet those questions oblige authorities to

[110] Brownsword, *supra* note 78, at 561. [111] *Supra* note 93, at 9.

rationalize their choices, which will only benefit a deliberative democracy. It has been argued that in our globalized world an ethic founded on respect for human rights should be preferred to one based on utility or dignitarianism. That this choice is of great importance is shown by the fact that the answers to the five questions would probably have been different from the final decisions of the Minister.

6. Conclusions and Final Considerations

In the opening section of this essay, I sketched a human civilization that is in a critical condition. This widely supported view feeds the enthusiasm for biotechnology and human genetics, and even eugenics. At the same time, however, that same technology gives rise to fear and aversion, which frustrates both technological development and its regulation. The main objective of this essay was to outline a framework that makes regulatory choices possible that pass a cosmopolitan review of legitimacy. To this end, PGD served as a case study.

The technology involved in PGD and its applications were briefly explained in Part 2. Relative to IVF, PGD involves few additional clinical risks. However, PGD *does* raise new ethical questions, among which is the possibility that PGD could be at the base of a neo-eugenic project. The label 'eugenics', however, is no reason to prohibit the technology *solely for that reason*. Part 3 provided a case study within the overall case study. It explored the manner in which the WTO guards the cosmopolitan legitimacy of measures that control risks involved with GMOs. I explored how the legitimacy dilemma posed by the judicial review of democratically promulgated policy has been handled. The answer lies in proceduralizing and formalizing the review as much as possible. For GMOs, this leads to an important role for 'science' and limitations on the precautionary principle.

The next part of the chapter was based on Brownsword's views, specifically, his mapping out of the bioethical battleground that surrounds human genetics. The three prevailing schools within the ethical plurality were briefly characterized and then assessed in terms of their suitability as a basis for a cosmopolitan legitimacy review. The arguments for deciding upon a human rights review were found to be numerous and, in my opinion, convincing. My argument differed from Brownsword's with regard to the possible role of the precautionary principle. Although agreeing with him, and many others, that human dignity is a constitutive requirement of a community based on respect for human rights, which must therefore be protected, I suggested that precaution could turn out to be an enemy of human dignity as empowerment. This was illustrated by means of several policy documents from international practice. Unlike the regulation of GMOs, where a limited role for the precautionary principle is desirable and possible, no value whatsoever should be attached to the principle within a regulatory framework for human genetics based on respect for human rights. The procedural instrument of informed consent has to be protected from the parasitic qualities of the precautionary principle. The need to protect human dignity can be better fulfilled by means of interpreting and articulating new international principles.

In Part 5, the final part of the essay, the Dutch Health Minister's Opinion regarding the availability of PGD treatment was reviewed against these principles. First and foremost, it was concluded that the ethical foundations of the Opinion, measured by formal criteria such as transparency and consistency, were anything but robust. This puts pressure on the internal and cosmopolitan legitimacy of the policy.

Finally, if we were to approach the question of access to PGD from a human rights perspective, a more liberal policy than currently exists in the Netherlands would result.

4

Constitutional Patriotism and the Right to Privacy:

A Comparison of the European Court of Justice and the European Court of Human Rights

*Francesca Bignami**

In October 1970, the heads of state of what was known at the time as the European Economic Community proclaimed: '[A] united Europe must be founded upon the common heritage of respect for the liberty and the rights of men.'[1] Decades later, at the celebration of the fiftieth anniversary of the European Union, the heads of state of today's European Union echoed that earlier proclamation:

> In the European Union, we are turning our common ideals into reality: for us, the individual is paramount. His dignity is inviolable. His rights are inalienable. Women and men enjoy equal rights.
>
> We are striving for peace and freedom, for democracy and the rule of law, for mutual respect and shared responsibility, for prosperity and security, for tolerance and participation, for justice and solidarity.[2]

This vision of Europe, one resting not simply upon economic convenience but upon a common tradition of rights and democracy, is a persistent theme in the history of European integration. With the legitimacy crisis of the 1990s, it has also become a pressing theme of European integration. Rights, according to some, might serve as the building blocks of a common European identity.

* Many thanks to Mattias Kumm, Peter Lindseth, Thérèse Murphy, and Ulf Öberg for their comments.

[1] European Community, Foreign Ministers of the Six European Community Countries, 'First Report of the Foreign Ministers to the Heads of State and Government of the Member States of the European Community', Part One, para 5 (27 October 1970), reprinted in C. Hill and K.E. Smith (eds) *European Foreign Policy: Key Documents* (2000) 76.

[2] Declaration on the occasion of the 50th anniversary of the signature of the Treaties of Rome. Berlin, 25 March 2007.

Europe might not have a single cultural, linguistic, ethnic, or religious tradition, as did the nation states of the nineteenth century. But it does, according to constitutional theorists, possess a liberal democratic tradition of self-government and individual rights. Based on this commonality, Europeans might recognize that they inhabit a single society. They might recognize themselves as rights-bearing and duty-owing members of the very same human community.

Constitutional patriotism, as this form of identity has been called, holds tremendous promise.[3] Yet because constitutional patriotism serves to bind a community together, it carries some of the dangers associated with the other, historically dominant form of community—the nation state. Nineteenth and twentieth-century nationalism was belligerent, intolerant, and monolithic. Why should we think that twenty-first-century constitutional patriotism will be any different? In other words, in the European search for solidarity, the core principles of liberal democracy might very well be undermined. The practice of constitutional patriotism might very well betray its normative aspirations of tolerance and inclusiveness.

Although the dark side of constitutional patriotism has been remarked by many, there have been few attempts to scrutinize Europe's emerging constitutional culture for signs of intolerance. Such illiberalism can be manifested in many ways. Foreigners might be unreasonably excluded through immigration policies. Ethnic and religious minorities might not be treated fairly. Or diversity among the national communities that constitute the European Union might be suppressed in the interest of greater unity.

This chapter examines attitudes towards national diversity in one piece of the emerging European constitution—the right to privacy. There is a thick constitutional culture of privacy in Europe. The familiar debate of how to balance the right to privacy against freedom of expression, the market, and public security can be heard in many places: before the Court of Justice and the European Court of Human Rights, in the European Parliament and the European Council, and before Europe's numerous data privacy ombudsmen. And in these places, the less familiar problem of tolerance of diversity within Europe, among Europe's different national communities, also arises. These communities all recognize a private sphere that sometimes deserves to be protected and that sometimes must give way to the demands of others. Yet these same communities also differ in their definition of the private sphere and the circumstances under which this private sphere must be sacrificed for the sake of other values such as freedom of expression. Thus the right to privacy poses squarely the dilemma of constitutional patriotism: can a right that is common to all of Europe nonetheless acknowledge and accommodate national diversity?

[3] J.-W. Müller, *Constitutional Patriotism* (2007); Müller and Scheppele, 'Constitutional Patriotism: An Introduction' (2008) 6 Intl J Constitutional L 67; Müller, 'A General Theory of Constitutional Patriotism' (2008) 6 Intl J Constitutional L 72.

The rapid development of a Europe-wide constitutional culture of privacy is due, in no small measure, to the emergence of new technologies. Because of widely available, cheap digital technologies, citizens and government officers of different European nations can communicate instantaneously. This facilitates immensely a breathtaking range of social relations, market transactions, and government policies across national borders. Yet the very same technologies put enormous pressure on the right to privacy. Put simply, today, spying is easy for everyone. A standard, pocket-size camera can take photographs, record conversations, and make video clips. A common flash drive can store vast data sets containing many gigabytes worth of electronic information. With a simple Internet connection, these digital files can be shared with hundreds of thousands of people throughout the world. These digital files can be combined with other files, taken from other sources, to create more complete and intrusive personal profiles. How can the law protect privacy in the dramatically different technological environment of today? Now that our images and personal habits can be captured so easily can we really claim to expect privacy and to expect that the law will defend our privacy? This context of rapid technological innovation—and differing national responses to innovation—is essential to understanding the European right to privacy.

This chapter proceeds as follows. The first section presents the theory of constitutional patriotism and the related concept of constitutional tolerance. The second section moves to the right to privacy. It outlines the legal framework for the right to privacy, with special attention to data protection because of the historically important role of this policy area in the European Union. The third section explores the challenge of diversity for European constitutionalism with two recent judgments involving a similar issue of privacy law but decided by two different European courts: *Criminal Proceedings against Bodil Lindqvist*, decided by the European Court of Justice (ECJ), and *Princess of Monaco*, decided by the European Court of Human Rights (ECtHR). Both cases took on the relationship between privacy and freedom of expression, but while the ECJ deferred to the national court's resolution of the matter, the ECtHR decided the question and did so contrary to the national court's ruling.

Based on the theory of constitutional patriotism, the last section argues in favour of the deferential approach taken by the ECJ. Although the doctrinal tools available to the two courts differ, the same deferential result could be accomplished by the ECtHR through greater resort to its doctrine of margin of appreciation. In doing so, the ECtHR would recognize, as the Court of Justice has done, that the answer to many difficult questions of constitutional law can only be found in the thick institutional and cultural setting of national constitutional orders. The European constitutional order, still, is too thin to settle many of the heated conflicts among rights-holders that emerge routinely at the national level.

This favourable assessment of the ECtHR's margin of appreciation is somewhat controversial. The ECtHR's resort to the doctrine has been criticized by many human rights scholars as the abandonment of rights, morality, and reason in the face of powerful, oppressive states.[4] Deference to the Member States, according to the critics, represents capitulation to repressive human rights practices. Other commentators, however, have defended the margin of appreciation, both from a principled, moral theory of rights[5] and from a pragmatic position related to the limits of international law.[6] This analysis adds to the earlier defences of the margin of appreciation by revealing the constitutional nature of ECtHR adjudication. Given the frenetic pace of European integration and the ECtHR's role in affirming rights, the ECtHR should be regarded as not simply an international human rights tribunal, but also a constitutional court. The ECtHR, like the ECJ, is entrusted with a critical element of Europe's constitutional self-identity—fundamental rights. In other words, the ECtHR is an institutional guarantor of European constitutional patriotism. In adjudicating rights, therefore, the ECtHR should be conscious of the importance of fundamental rights in their particular geographical and historical instantiations to national self-identities and local solidarities. And before resolving any particular rights conflict—like the clash between privacy and speech in *Princess of Monaco* —the ECtHR should consider whether indeed the best constitutional outcome for Europe would instead be tolerance of national diversity.

1. Constitutional Patriotism

The idea of a constitution as not simply a mode of organizing power within a society but as a source of allegiance that enables the formation of that society is commonly associated with German political theory. As Jan-Werner Müller explains, the circumstances of post-war Germany placed the Basic Law and the Constitutional Court at the heart of German political culture.[7] The Constitutional Court has been immensely successful at resolving highly contentious political disputes. Today, therefore, the Constitutional Court is one of the most highly respected institutions in German political life. At the same time, the Basic Law emerged as the focal point in the struggle to defend the German state against the twin threats of neo-fascism and communism. German citizens were urged,

[4] Letsas, 'Two Concepts of the Margin of Appreciation' (2006) 26 OJLS 705, at 731–2; Benvenisti, 'Margin of Appreciation, Consensus, and Universal Standards' (1999) 31 NYU J Intl L & Pol 843, at 847.

[5] See eg Carozza, 'Subsidiarity as a Structural Principle of International Human Rights Law' (2003) 97 American J Intl L 38, at 61–3.

[6] See eg Helfer and Slaughter, 'Toward a Theory of Effective Supranational Adjudication' (1997) 107 Yale LJ 273, at 315–17.

[7] Müller, 'A "Thick" Constitutional Patriotism for the EU? On Morality, Memory and Militancy', in E.O. Eriksen, C. Joerge, and F. Rödl (eds), *Law and Democracy in the Post-National Union* (2006), at 375.

based on their allegiance to the Basic Law's democratic values, to reject political extremism and come to the defence of German democracy.

In the view of many thinkers, a constitution for the European Union would engender a similar form of allegiance, not to any particular national order, but to a Europe-wide political community. Probably the best-known exponent of European constitutional patriotism is the renowned political philosopher Jürgen Habermas. According to Habermas, the democratic deliberation facilitated by the adoption of a constitution would promote identification with the European project.[8] This form of constitutionalism would engender genuinely post-national citizenship. That is, individuals would develop a sense of belonging to a Europe-wide community based on their participation in a democratic process, not based on their blood ties or cultural background.

As many have pointed out, deliberative democracy is a very demanding form of rule by the people.[9] Citizens are called upon to constantly re-examine and revise their values and interests, in communication with the other citizens that inhabit the public sphere. For the normative ideal to be achieved, four key criteria must be satisfied. (1) Deliberation must be inclusive; all citizens should have the opportunity to participate in the deliberation. (2) The citizens who participate in the deliberation must have equality of voice. They must all be heard and considered. (3) In the deliberation, deliberative as opposed to strategic rationality must prevail. Agreement must come from genuine consensus on the political principle under discussion. In other words, decisions must be made through appeal to commonly accepted reasons—what Habermas calls validity claims—not through bargaining among competing interests. Such bargaining is considered harmful to democracy because certain interests, inevitably, are more powerful than others and therefore dominate the bargaining process. Moreover, even if power could be assumed away, collective decision-making based on compromises among particularistic interests is not believed capable of obtaining the common good. (4) Last, deliberation must continue indefinitely until consensus is reached. Given the limits on the time and resources that can be dedicated to deliberation, consensus on every item on the public agenda is unattainable. Therefore, in the absence of consensus, a decision must be taken temporarily, following a commonly accepted decision rule.

Not only is deliberative democracy an extremely ambitious form of democracy, it is also highly universalistic. The society that self-governs under this theory could very well be the world, not simply a particular nation or region of the world. Unlike deliberative democracy in American political theory, rooted in the republican tradition, Habermas's starting point is man's universal capacity for language. The basic similarities underpinning language are the grounds for

[8] See Habermas, 'Why Europe needs a constitution' in E.O. Eriksen, J.E. Fossum, and A.J. Menéndez (eds), *Developing a Constitution for Europe* (2004), 19–34 at 27–8.

[9] See eg Knight and Johnson, 'Aggregation and Deliberation' (1994) 22 Political Theory 277.

claiming that all individuals share the same 'communicative rationality' through which they can reach agreement on matters of common concern.[10] An individual's capacity to recognize others as equals in the deliberative enterprise is rooted in reason, not in geography or history. Thus it should be possible to agree that the essentials of deliberative democracy be extended to all.

For these reasons, the version of constitutional patriotism expounded by Habermas has given rise to deep scepticism.[11] The type of reason and universalism required by Habermas's theory is considered to be too far removed from the realities of human nature for deliberative democracy to be achievable. But the idea of constitutional patriotism has had enormous purchase in Europe—the possibility of European citizenship based on the liberal ideals contained in a constitution and experienced as part of a constitutional culture. This idea spurred the long process that culminated in the Constitution for Europe. Of course, that Constitution failed in popular referenda held in France and the Netherlands. But the possibility of rooting European citizenship in a set of commonly held ideals about the right and good mode of conducting life in a human community has persisted.[12]

The liberal values found in most modern constitutions can only be considered virtuous. Yet building allegiance to a community based on such values is problematic. The act of subscribing to a single set of beliefs—even virtuous ones—carries the danger of suppressing alternative beliefs. In other words, constitutional patriots might not tolerate dissent from their version of the right and good. This intolerance might not be shown only to those extremists bent on destroying liberal values, such as neo-Nazi and militant Communist parties. It might also silence legitimate disagreement about the correct interpretation and application of critical elements of the liberal democratic canon. Jans-Werner Müller incisively frames the critique:

> [C]ivic nationalism—under which one is tempted to subsume constitutional patriotism—is still nationalism, and not automatically less fraught with danger in contrast with cultural nationalism. In particular, it still aims at homogeneity among citizens in a way that would have been recognizable to antiliberals like Carl Schmitt as indispensable for democracy, and that more or less directly endangers values such as inclusiveness, individualism, and diversity.[13]

The irony of intolerant constitutional patriotism is especially acute in the European Union. There, hostility towards diversity not only goes against the

[10] Gehring, 'Communicative Rationality in European Governance? Interests and Communicative Action in Functionally Differentiated Single Market Regulation', in E.O. Eriksen, C. Joerges, and J. Neyer (eds), *European Governance, Deliberation and the Quest for Democratisation* (2003), 57 at 69.

[11] See eg McCormick, 'Habermas, Supranational Democracy, and the European Constitution' (2006) 2 European Constitutional L Rev 398.

[12] Kumm, 'Why Europeans Will Not Embrace Constitutional Patriotism' (2008) 6 Intl J Constitutional L 117.

[13] Müller, 'Three Objections to Constitutional Patriotism' (2007) 14 Constellations 195, at 199.

grain of the values at the core of liberal democracy but also does violence to the constitutional system already in place. Among legal scholars, Joseph Weiler has put forward the most exhaustive analysis of the dangers of constitutional patriotism for the European Union.[14] The most fundamental attribute of the European constitutional system to have emerged over the years, Weiler argues, is constitutional tolerance.[15] In accepting the authority of the pronouncements of the European Court of Justice, national courts allow the views of other political communities to shape the constitutional fabric of their own communities. As Weiler puts it, European law is 'accepted as an autonomous voluntary act, endlessly renewed on each occasion, of subordination, in the discrete areas governed by Europe, to a norm which is the aggregate expression of other wills, other political identities, other political communities.'[16] Likewise, the European Court of Justice defers to the views of national courts in fashioning European law. From this relationship of mutual deference, Weiler draws the conclusion that a written European constitution, adopted by a constitutional convention, is undesirable. Whether such a constitution would indeed undermine constitutional tolerance, as Weiler argues, is debatable. But the important point is that the European constitutional order is intrinsically pluralist, and any effort to impose a hegemonic belief system in the interest of constitutional patriotism would be antithetical to the character of Europe. In fashioning a constitutional identity for Europe, therefore, it is imperative that the courts and other law-making institutions permit legitimate disagreement over the content of that identity. It is essential to preserve constitutional tolerance.

2. The Right to Privacy

The right to privacy is a perfect vehicle for exploring the tension between constitutional patriotism and constitutional tolerance. The right to privacy has long been an element of the European constitutional system. Yet it is an element on which national systems display considerable variation. Privacy is a claim that routinely clashes with other claims: my right to keep certain matters to myself conflicts with the right of others to learn about me for the purpose of public debate, market transactions, public security, and more. How a constitutional system mediates between the right of secrecy and the right to know is a defining feature of that society and one that varies tremendously among political communities. How such pluralism has been handled in Europe can offer guidance on the more general problem of reconciling a single set of—European—constitutional ideals with national diversity.

[14] Weiler, 'In Defence of the Status Quo: Europe's Constitutional Sonderweg', in J.H.H. Weiler and M. Wind (eds), *European Constitutionalism beyond the State* (2003), 7–23.
[15] *Ibid.* [16] *Ibid*, at 21.

The European right to privacy is the product of two distinct but overlapping political systems: the European Union and the Council of Europe. Historically, the Council of Europe came first. The Council of Europe system also has the broadest coverage: the application of the right to privacy in the Council of Europe system was never limited to certain domains of state activity as it was in the European Union, with its focus on the market. But today, especially in the domain of data privacy, the European Union is the more important system. Compared to the Council of Europe, it has a denser framework of laws and administrative instruments with a tighter enforcement system of regulators and courts.

The two political systems overlap because of the influence that the first-in-time has been able to exert over the second-in-time. As is well known, the European Court of Justice looks to the European Convention on Human Rights (ECHR) and the jurisprudence of the European Court of Human Rights to establish new rights. Furthermore, the European Union's legislation on data privacy draws on the earlier Council of Europe Convention in that area.

A. Council of Europe

1. European Convention on Human Rights

The right to private life is protected under Article 8 of the European Convention on Human Rights.

Article 8—Right to respect for private and family life

1. Everyone has the right to respect for his private and family life, his home and his correspondence.
2. There shall be no interference by a public authority with the exercise of this right except such as is in accordance with the law and is necessary in a democratic society in the interests of national security, public safety or the economic well-being of the country, for the prevention of disorder or crime, for the protection of health or morals, or for the protection of the rights and freedoms of others.

Initially, the European Court of Human Rights interpreted the right of privacy to apply only against the government and only in those situations in which the space or the type of fact fit squarely with commonplace understandings of what counted as private.[17] As with other Convention rights, however, the Court soon acknowledged that the right to privacy applied not just against the government but against other individuals, on the theory that states should be held responsible for all rights violations occurring within their territories. The Court also gradually expanded those categories of information deemed to be private and therefore protected under the right to private life. Today, any information about a person, no matter how banal or how widely known, is considered private as long as it is

[17] I. Cameron, *National Security and the European Convention on Human Rights* (2000), 70–100.

systematically stored and collected.[18] Likewise, even behaviour in a public place is considered private if it is recorded and disseminated in a manner that could not reasonably have been anticipated by the individual concerned.[19]

Under the case law of the Court, an interference with the right to private life must satisfy three conditions to be considered legal. First, if the processing is done by a public authority or for a public purpose, it must be authorized by a law, accessible to the public, with precise enough provisions to curb arbitrary government action and to put citizens on notice of possible incursions into their private sphere.[20] Secondly, the purpose of the interference must be legitimate. Namely, the purpose must be related to one of the categories listed in Article 8. It must be 'in the interests of national security, public safety or the economic well-being of the country, for the prevention of disorder or crime, for the protection of health or morals, or for the protection of the rights and freedoms of others.'[21]

Thirdly, the interference with private life must be proportional. The proportionality inquiry generally consists of three distinct steps. Is there evidence that government action can achieve the stated purpose? Is the government action necessary for accomplishing the stated purpose or would alternative means accomplish the same purpose with a lesser burden on the privacy right? And even though there might be no alternative means for accomplishing the same purpose, is the burden on the right to privacy nonetheless intolerable, requiring the law to be withdrawn? The burden of justification under the proportionality test lies with the government and varies tremendously, depending on the gravity of the interference with the right to private life and the public interest being pursued: the more significant the interference, the higher the burden on the government; the more important the public purpose, the lower the burden on the government.[22]

2. Convention for the Protection of Individuals with Regard to Automatic Processing of Personal Data

Long before the European Court of Human Rights interpreted the right to privacy to apply broadly to any type of information about a person, a specific set of rules relating to personal data was negotiated by the members of the Council of Europe. The Convention for the Protection of Individuals with Regard to Automatic Processing of Personal Data was adopted in 1980 and was opened for ratification in January 1981.[23] As of May 2007, it has been ratified by

[18] *Segerstedt-Wiberg and Others v Sweden*, Application no 62332/00, 6 June 2006, paras 71–2.

[19] *Von Hannover v Germany*, Application no 59320/00, 24 June 2004, para 50.

[20] *Amann v Switzerland*, Application no 27798/95, 16 February 2000, para 50.

[21] ECHR, Art 8.

[22] See Opinion of Advocate General, *European Parliament v Council and v Commission*, Cases 317/04 and 318/04, paras 228–30 (2006).

[23] 28 January 1981, Europ TS No 108. See C.J. Bennett and C.D. Raab, *The Governance of Privacy: Policy Instruments in Global Perspective* (2003) 72.

thirty-eight of the forty-six members of the Council of Europe and it has been signed but not ratified by six more Member States. The rules contained in the Convention apply to all those engaged in personal data processing—market actors, the media, government agencies, and others.

For purposes of understanding the legal rules contained in the Convention, it is useful to divide sequentially personal data processing into different phases: collection, storage, use, and dissemination to third parties. Under the Convention, personal data must be 'obtained and processed fairly and lawfully' (Article 5a). Generally speaking, this is interpreted to mean either that individuals must consent to the collection and use of their information or that a piece of legislation must authorize personal data processing, specifying the public reasons that necessitate personal data processing. The Convention restricts the amount and type of personal information that may be collected: it must be gathered for 'specific and legitimate purposes' (Article 5b), and it must be 'adequate, relevant and not excessive in relation to the purposes for which they are stored' (Article 5c). Certain categories of personal information, if collected, must be treated with special care: 'Personal data revealing racial origin, political opinions or religious or other beliefs, as well as personal data concerning health or sexual life, may not be processed automatically unless domestic law provides appropriate safeguards. The same shall apply to personal data relating to criminal convictions' (Article 6).

Moving past collection to the storage of information, data protection actors must guarantee personal data is 'accurate and, where necessary, kept up to date' (Article 5d). Safeguards must be put into place, to protect the data from theft and other forms of security breaches: '[A]ppropriate security measures shall be taken for the protection of personal data stored in automated data files against accidental or unauthorised destruction or accidental loss as well as against unauthorised access, alteration or dissemination' (Article 7).

As for use and dissemination, personal data may be used only for those purposes originally contemplated and only by those organizations for whom the data was originally intended. It may be put to other uses and shared with other organizations only if doing so is necessary to fulfil the original purposes of data collection (Article 5b). Lastly, the time during which personal information may be retained is limited: personal information is to be 'preserved in a form which permits identification of the data subjects for no longer than is required for the purpose for which those data are stored' (Article 5e).

The Convention also pays some attention to remedies. It provides that individuals should have a right of access to their personal information. This right guarantees that individuals will know what information about them is held, by which institutions. Moreover, once individuals have obtained access to their information, they have an opportunity to correct any incorrect information. Finally, they have a right to demand that unlawfully processed information—say, information retained for longer than necessary—be deleted.

The basic theory behind these rules is that the less personal the data processing, the better. The objectionable nature of personal data processing can only be comprehended in light of the fundamental rights origins of data privacy. Each time information about an individual is collected, personal autonomy is put at risk. Even though liberal societies disagree as to what is private and what is public, they all acknowledge that, in certain spaces and over certain matters, individuals are entitled to privacy.[24] If they so choose, individuals may bar access by others to their private sphere. This entitlement is rooted in the duty of respect for others. Although certain types of personal information might appear to be far removed from this private sphere of control and autonomy, once such information is combined, manipulated, and disseminated, it might very well be revealing of the private sphere.

Of course, there are other rationales for data privacy. Personal information in the hands of powerful actors can be used for illiberal ends like the suppression of political dissent and discrimination based on religion, race, or ethnic origins. If personal information is wrong, it can lead to all sorts of disastrous consequences for the individuals involved. They might be unfairly detained or wrongly denied a welfare benefit. And if personal information is stolen, individuals can be robbed of their assets or fraudulently charged with onerous debts. Yet, at the root of data privacy is the principle of autonomy. Even in a world in which, thanks to technology, acquiring knowledge about others is virtually effortless, personal autonomy must be respected.

Data protection rules and the fundamental right to privacy are not only related in political theory but also in legal doctrine. Most of the rules reviewed earlier can be conceived as guidance for the investigation required under fundamental rights law. Since every instance in which personal data is collected constitutes an intrusion into private life, personal data must be collected and stored for specific and legitimate purposes and may not be used in a way incompatible with those purposes.[25] To ensure that personal data processing can accomplish the government's purpose—as is required under proportionality—all such data must be 'adequate' and 'relevant' to that purpose.[26] Likewise, such data must be 'accurate and, where necessary, kept up to date.'[27]

Not only must data processing be able to achieve the government purpose, but it must be necessary to achieving that purpose. Hence the requirement that the amount of data processed and the period during which it is stored be no more than necessary to accomplish the purpose.[28] Since the processing of certain types of personal data—revealing racial origin, political opinions, or religious or other beliefs; concerning health or sexual life; or relating to criminal convictions—constitutes a particularly grave interference with privacy, special precautions must

[24] See Benn, 'Privacy, Freedom, and Respect for Persons', in J.R. Pennock and J.W. Chapman (eds), *Nomos XII: Privacy* (1971) 1; A. Margalit, *The Decent Society* (N. Goldblum trans, 1996), 203–4.

[25] Convention 108, Art 5b. [26] *Ibid*, Art 5c. [27] *Ibid*, Art 5d.

[28] *Ibid*, Arts 5c and 5e.

be taken. Finally, as a special safeguard for the privacy right, individuals should have the right to check their personal data to make sure that it is accurate and that, in all other respects, it is being processed in accordance with the law. Most of these limitations on data collection and processing can be conceived as less rights-burdensome means of accomplishing the same government purpose. Each time a personal data processing system is put into place, it must contain safeguards and restrictions that ensure that the privacy interest in personal data is compromised as little as possible. Last, under the Convention, exceptions are permitted in certain classes of cases, but only if the resulting burden on the privacy right is proportionate to the aim being pursued by permitting the exception.[29]

B. The European Union

1. The Charter of Fundamental Rights

Fundamental rights have long been part of the legal and political discourse of the European Union. Until recently, however, such rights were imported into the EU legal order from national constitutions and the European Convention on Human Rights. As the Court of Justice has never tired of repeating:

> It must also be stated that fundamental rights form an integral part of the general principles of law whose observance the Court ensures. For that purpose, the Court draws inspiration from the constitutional traditions common to the Member States and from the guidelines supplied by international instruments for the protection of human rights on which the Member States have collaborated or to which they are signatories. In that regard, the ECHR has special significance.[30]

In 2000, the European Union finally obtained its own statement of fundamental rights with the Charter of Fundamental Rights. The Charter includes a number of rights that reflect the circumstances of modern-day society. The right to privacy is one of them. The Charter dedicates two articles to privacy. The first, Article 7, copies the following language from the European Convention on Human Rights:

> Everyone has the right to respect for his or her private and family life, home and communications.

The second, however, innovates by specifically guaranteeing privacy in the age of electronic technology and the World Wide Web:

> **Article 8** *Protection of personal data*
>
> 1. Everyone has the right to the protection of personal data concerning him or her.
> 2. Such data must be processed fairly for specified purposes and on the basis of the consent of the person concerned or some other legitimate basis laid down by law.

[29] *Ibid*, Art 9.
[30] Case C-305/05, *Ordre des barreaux francophone et germanophone*, 2007 ECR para 29.

Everyone has the right of access to data which has been collected concerning him or her, and the right to have it rectified.

3. Compliance with these rules shall be subject to control by an independent authority.

Much of the substance of this right of personal data protection should be familiar from the earlier discussion of the Council of Europe Convention on Automatic Processing of Personal Data (Convention 108). The requirement of an independent authority is also familiar to data protection advocates. Independent authorities, established to oversee compliance with data protection rules, have been a fixture of national data protection schemes since the early 1970s. By the mid-1990s, consensus on the desirability of an independent authority had emerged at the European level, first in the EU Directive discussed below and then in a special protocol to Convention 108.[31]

2. *The Data Protection Directive*

In 1995, the Data Protection Directive was adopted by the European Union.[32] Today, all Member States have implementation legislation in place. The Directive builds on Convention 108 in a number of respects. Although the Directive only sets down rules for activities that come under the market-oriented First Pillar, those rules are far more detailed than those contained in Convention 108. The Directive begins with a restatement of the Convention's core provisions on lawfulness, specific purposes, necessity, accuracy, and limited data retention (Article 6). Like the Convention, the Directive recognizes certain categories of personal information that should receive special protection (Article 8), gives individuals a right of access to their personal information (Article 12), and imposes a duty to adopt security measures to protect personal data (Article 17). These provisions, however, are far more detailed and onerous than the spare principles set down in the Convention. The Directive also elaborates on lawfulness by setting out those circumstances under which collection and use of personal data is permitted (Article 7).

Running throughout the Directive is the proportionality principle familiar from the Council of Europe context. Government actors and private firms that seek to collect and process personal information must demonstrate that their operations satisfy the proportionality principle—that the burden imposed on the right to personal data protection is proportionate to the aim being pursued through their data processing operations (Article 7(f)). Any exceptions to the data

[31] Additional Protocol to the Convention for the Protection of Individuals with regard to Automatic Processing of Personal Data regarding supervisory authorities and transborder data flows, Strasbourg, 8 November 2001.

[32] Council and European Parliament Directive 95/46/EC on the Protection of Individuals with Regard to the Processing of Personal Data and on the Free Movement of Such Data [1995] OJ L281/31.

protection guarantees contained in the Directive must be proportionate to the aims being pursued in taking advantage of the exception (Articles 11, 12, 13, 26). To the extent that other rights—in particular freedom of expression—are permitted to interfere with the right to personal data protection, that interference must be proportional (Articles 9, 13).

The Directive creates three new sets of rights and duties with respect to the Convention. First, at the time that personal information is collected or transmitted to third parties, individuals must be informed of the identity of the data processor, the purposes of the data processing, and a number of other aspects of the data processing operation (Articles 10, 11). Secondly, in a provision drawn directly from French data protection law, an individual is entitled to object to automatic data processing with potentially adverse effects for that individual (Article 15). Thus, for instance, an individual denied credit by a financial institution based entirely on a computerized review of that individual's application may request that the application be reviewed by an employee of the bank. Thirdly, the transfer of personal data to third countries is prohibited unless those countries' data protection schemes are deemed adequate or one of a number of other conditions are met (Articles 25, 26).

Perhaps most importantly, the Directive has the enforcement apparatus that the Convention lacks. It requires that all Member States establish an independent data protection authority (Article 28). These authorities must have the powers necessary to oversee and enforce compliance with data protection law. The Directive anticipates that under certain circumstances firms and government agencies that process personal data will be required to notify the data protection authority of their operations and, occasionally, obtain prior authorization (Articles 18–20). The Directive also requires that national data protection authorities have the power to inspect data users so as to be able to detect data protection violations (Article 28). Data protection authorities must have remedial powers, such as the power to prohibit data processing operations that go contrary to the rules. And they must have the power to sanction those firms and government agencies that break the rules (Article 28).

On the European level, as opposed to the national level, two separate committees have been established to enforce the Directive. One committee is responsible for regulating third-country transfers, in line with privacy principles. The other—known as the Article 29 Working Party—has broad-ranging jurisdiction to promulgate opinions on most data protection issues that arise under the Data Protection Directive. Although the Working Party's opinions are not formally binding, they are generally followed by other EU institutions and national authorities; as a consequence, the Working Party's legal interpretations are treated as if they were binding by most regulatory authorities. In addition to this regulatory apparatus, an EU-level data protection authority has been established. The European Data Protection Supervisor, as this authority is known, is responsible for enforcing data protection rules against EU institutions.

3. Additional EU Data Protection Legislation

The Directive is the European Union's most comprehensive piece of data protection legislation. It applies, however, only to activities that fall under the First Pillar, namely market-based activities. A separate set of data protection instruments cover cooperation on immigration and cooperation on criminal justice under the Third Pillar. Because their essentials largely replicate the terms of Convention 108 and the Directive, only the briefest of overviews is necessary here.

To effectively cooperate on immigration and criminal justice, a number of EU databases have been established. These databases are covered by separate sets of data protection provisions. When national authorities exchange immigration and crime-related information through the Schengen Information System, they are bound by the data protection guarantees in the Schengen Convention.[33] The terms of the Europol Convention apply to the exchange of crime-related information through the Europol Information System.[34] When national prosecutors and courts exchange information through the Eurojust Information System, they are bound by the data protection provisions of the Decision establishing that information system.[35] And the data protection provisions of the basic regulation creating the Customs Information System apply when national customs officers exchange information related to smuggling and other offences through that computer system.[36] Currently under negotiation is a Third Pillar Framework Decision that will replace this complicated, largely overlapping, set of laws with one single legislative scheme.[37] The Framework Decision will also significantly extend the coverage of EU data protection guarantees: when national police, prosecutors, and the courts cooperate to combat crime under the Third Pillar and, in the process, exchange personal information on criminal suspects, they will be bound by the terms of the Framework Decision.

Today, therefore, the only domain of European social life that is still untouched by EU data protection law is intelligence-gathering by national spy

[33] Title IV of the Convention of 1990 implementing the Schengen Agreement of 14 June 1985 on the gradual abolition of checks at the common borders and integrated into the framework of the European Union pursuant to the Protocol annexed to the Treaty on European Union and the Treaty establishing the European Community, 2000 OJ L239/19.

[34] Convention Based on Article K.3 of the Treaty on European Union, on the Establishment of a European Police Office, 1995 OJ C316/2.

[35] Decision 2002/187/JHA of 28 February 2002 setting up Eurojust with a view to reinforcing the fight against serious crime, 2002 OJ L63/1.

[36] Council Regulation (EC) No 515/97 of 13 March 1997 on mutual assistance between the administrative authorities of the Member States and cooperation between the latter and the Commission, 1997 OJ L82/1.

[37] Note from Presidency to delegations on Proposal for a Council Framework Decision on the protection of personal data processed in the framework of police and judicial cooperation in criminal matters, Council of the European Union, Doc No 7315/07, Brussels, 13 March 2007.

agencies. National spy agencies like Germany's Office for the Protection of the Constitution and Federal Intelligence Service generally do not come within the reach of EU law at all. These agencies have been expressly excluded from the ambit of the European Union's Third Pillar. To the extent that spy agencies do come within the jurisdiction of the European Union, they would have to be regulated under the Second Pillar; at present, there are no proposals to do so.

3. The European Constitutional Culture of Privacy

In a series of cases the European Court of Human Rights and the European Court of Justice have grappled with the meaning of these privacy provisions. They have determined the scope of the private sphere protected by European law and the nature of the safeguards afforded under European law. The courts have also decided under which circumstances it might be permissible to intrude upon privacy, to further the legitimate purposes of other members of society. The right to speak, to be safe, and to transact in the marketplace can all come into conflict with the right to privacy. These moral and legal quandaries have become especially difficult with the massive diffusion of new information technologies. Information technologies hold extraordinary promise for human creativity and social communication yet they also pose an immense threat to privacy. Put simply, with the widespread availability of new information technologies, the temptation to spy—on one's neighbours or one's citizens—is extraordinary.

This section examines how the courts have addressed these issues in the unique context of Europe's plural legal order. The focus is on two cases involving the balance between the right to privacy and the right to speech. The first, decided by the European Court of Justice, considered the interference with the right to privacy caused by a personal web page, available throughout the world via the Internet. Although the Court recognized that privacy and speech were protected in the European constitutional order, it left the task of balancing the two values to the national court and the national constitutional order. The second case, decided by the European Court of Human Rights, reached a very different conclusion. That case involved the interference with a celebrity's right to privacy caused by the publication of paparazzi photographs in a weekly magazine. In its judgment, the Court not only recognized the importance of the right to expression and the right to privacy in the European legal order but also decided in favour of the celebrity's right to privacy. The reasoning in these two cases both demonstrates the tension between European constitutionalism and long-standing national legal orders, and reveals the jurisprudential techniques developed by Europe's courts to accommodate national diversity.

A. *Criminal Proceedings against Bodil Lindqvist*

The European Court of Justice had occasion to consider at length the Data Protection Directive in the case of *Criminal Proceedings against Bodil Lindqvist*.[38] Lindqvist volunteered as a catechist in her local parish in Sweden. At the same time, she took a computer class in which she learned how to create websites. She then created her own personal web page, on which she posted information designed to assist her fellow parishioners preparing for catechism. The website also included information on herself and eighteen colleagues from her parish, all identified by name. The information on her colleagues covered a variety of topics: their telephone numbers; their jobs and hobbies; their family circumstances; and, in the case of one colleague, the fact that she had injured her foot and therefore was on half-time medical leave from her job. Lindqvist posted this information without first obtaining the consent of her fellow catechists, although when she learned that some of them objected to the dissemination of their information, she promptly removed it from the website.

Under the Swedish law implementing the Directive, all personal data processors had to register their operations with the Swedish data protection authority. They also had to obtain authorization from the data protection authority if they planned to process sensitive data such as health information or to transfer personal data abroad, to a country outside the European Union. Lindqvist was unaware of this requirement when she first created her website, but subsequently she was informed of the requirement by an acquaintance. She therefore paid a visit to her local police station to inquire about her legal duties, at which point the police referred the matter to the public prosecutor. Lindqvist was prosecuted and convicted for various breaches of the Swedish law. Her punishment was a criminal fine of 4,000 SEK.

On appeal, the Swedish court referred a number of questions regarding the proper interpretation of the Directive—and therefore also the Swedish implementing law—to the European Court of Justice. The Court easily found that uploading personal data onto Lindqvist's website constituted 'personal data processing' covered by the Directive. With a bit more difficulty, the Court determined that Lindqvist's website did not qualify for any of the exceptions to the personal data processing covered by the Directive. Lindqvist invoked two exceptions allowed under the Directive: one for activities which fall 'outside the scope of Community law' and one for data processing 'by a natural person in the course of a purely personal or household activity.' The Court decided that her web page did not qualify for the latter exception because it was accessible to the entire universe of Web users; it was not confined to the home or to the personal realm. The Court went on to interpret the exception for activities 'outside the scope of Community law' as limited to activities falling under the Second and

[38] Case C-101/01, [2003] ECR I-12971.

Third Pillars. The Court therefore held that this exception too was unavailable to her: even though Lindqvist's website did not constitute a classic exercise of one of the four market freedoms—it was religious and charitable in nature—neither did it constitute an exercise of Second or Third Pillar powers. As a result, the Swedish registration system for personal data applied to personal websites, not only to Web operators like large commercial vendors and Internet providers.

It is worthwhile dwelling for a moment on the extraordinary reach of the Data Protection Directive. In narrowly interpreting the exception for activities falling outside Community law, the Court adopted an expansive reading of the Directive. The provision at issue reads, in full:

> This Directive shall not apply to the processing of personal data . . . in the course of an activity which falls outside the scope of Community law, such as those provided for by Titles V and VI of the Treaty on European Union [Second and Third Pillars] and in any case to processing operations concerning public security, defence, State security (including the economic well-being of the State when the processing operation relates to State security matters) and the activities of the State in areas of criminal law.

In giving meaning to the phrase 'outside the scope of Community law', the Court relied heavily on the reference to the Second and Third Pillars that immediately followed, as well as on the core subject areas of Second and Third Pillar listed thereafter. It found, using the canon of construction of *ejusdem generis*, that only activities of the same nature as public security, defence, state security, and criminal law were to be considered 'outside the scope of Community law.' In doing so, the Court departed from the opinion of the Advocate General, who had advised that activities should be considered within the scope of Community law only if they entailed the exercise of one of the four market freedoms or if they were regulated under a specific Community law, enacted under one of the Community's competences. The Court's interpretation also departed from the plain meaning of the Directive: the use of the phrase 'such as' suggested that the Second and Third Pillars and the types of operations typical of those pillars should be taken as an illustrative, not exhaustive, list of activities falling outside Community law. It was clear that the Court's interpretation was driven not so much by the text as by the spectre of future litigation. If the Court were to hold otherwise and find that the Directive applied to only market-related activities, how would future courts distinguish between market-related and other activities?

Setting aside the question of whether this interpretation was right, it accentuates the tension between a European law drafted in the early 1990s, before information technology had become a pervasive feature of daily life for the vast majority of citizens, and the application of that same law today. The drafters of the Directive could not have anticipated the incredible democratization of information technologies that has occurred over the past ten years. Because of this democratization, privacy principles today are relevant not only to big economic and government operators with the capacity to gather large amounts of personal information—banks, telecommunications providers, government welfare agencies. With digital technology and the Internet, everyone can gather and disseminate large amounts of personal information, and therefore everyone is a potential infringer of the right to data privacy. But even though privacy principles should certainly guide individuals like Lindqvist who benefit from the tremendous potential of information technologies, it is doubtful that the same regulatory scheme that applies to large commercial and government operators is also appropriate for individual citizens.

The Court did, however, limit the Directive's scope in one way. It found that the mere act of making personal data available to individuals in third countries through the World Wide Web did not constitute a transfer of that personal data to third countries. Although the Directive nowhere defines the term transfer, the Court found that this interpretation was necessary to prevent a specific set of duties—applicable only to third-country transfers—from becoming a general set of duties—applicable to all operations using the World Wide Web. In the Swedish case, this interpretation relieved individuals like Lindqvist of the duty to obtain regulatory authorization before posting personal information on the Web, and it relieved the Swedish data protection authority of the burden of reviewing and approving applications from thousands of individuals like Lindqvist.

Of greatest interest to the theme of constitutional patriotism was Lindqvist's challenge to her prosecution based on her right to freedom of expression. She claimed that both the requirements of the Directive as well as the application of the Directive in her specific case breached the right to freedom of expression. In her view, the terms of the Directive were not precise enough to satisfy the fundamental rights requirement that any interference with the right to expression be based on a law specific enough to place citizens on notice of the extent and nature of the government interference. (The freedom of expression analysis is identical to the three-step privacy analysis reviewed earlier: any interference with the right must be authorized by law, must seek to accomplish a legitimate purpose, and must be proportionate to the burden placed on the right.) Lindqvist also argued that the Directive failed to satisfy the proportionality requirement: the burden on the right to expression placed by the Directive's various privacy safeguards was disproportionate. And according to Lindqvist, even if the Directive as a whole were to be found lawful, in her particular case the burden on her freedom of expression should be considered disproportionate: the interference with her colleagues' right to privacy was minimal, given that most of the personal information on the web page was already available publicly or was trivial. Under such circumstances, her right to free expression should prevail, not the other way around.

The Court rejected this challenge but, at the same time, directed the Swedish court to take into account Lindqvist's right to freedom of expression. The Court acknowledged that some of the Directive's terms were vague and allowed for considerable choice at the national level when the Directive was implemented into national law. The Court dismissed these concerns, however, as intrinsic to the policy area: a far-reaching regulatory scheme seeking to cover data protection throughout the economy was necessarily broadly drafted. In addition, the Court found that the Directive took adequate account of freedom of expression by requiring that, in the application of the Directive, the right to freedom of expression always be balanced against the right to data protection as well as the other rights and interests advanced in the Directive. The Court concluded by directing the Swedish court to conduct the balancing exercise:

> Thus, it is, rather, at the stage of the application at national level of the legislation implementing Directive 95/46 in individual cases that a balance must be found between the rights and interests involved.
>
> In that context, fundamental rights have a particular importance, as demonstrated by the case in the main proceedings, in which, in essence, Mrs Lindqvist's freedom of expression in her work preparing people for Communion and her freedom to carry out activities contributing to religious life have to be weighed against the protection of the private life of the individuals about whom Mrs Lindqvist has placed data on her internet site.
>
> Consequently, it is for the authorities and courts of the Member States not only to interpret their national law in a manner consistent with Directive 95/46 but also to make sure they do not rely on an interpretation of it which would be in conflict with the fundamental rights protected by the Community legal order or with the other general principles of Community law, such as inter alia the principle of proportionality.

By sending the fundamental rights question back to the Swedish court, the Court of Justice not only avoided a difficult issue. It allowed for considerable national discretion in deciding the matter. The fundamental rights balance between privacy and speech is by no means a precise science and, by holding that the Swedish court was responsible for the balancing, the Court acknowledged that the outcome would be distinctive to Swedish law. In other words, the Swedish court was required as a participant in the European constitutional order to take into account both privacy and freedom of expression, but the accommodation of these two fundamental values was to be distinctive to Sweden's constitutional order. Of course, in the future, the Court of Justice might choose to give a European solution to the privacy-speech conundrum. Yet for the time being, the Court has decided to recognize that national constitutional cultures are the better site for the resolution of such thorny issues.

This constitutional settlement is reminiscent of many other areas of Community law. In deciding on the public policy exceptions to the fundamental market freedoms, the Court of Justice has always been reluctant to interfere with national legal orders in those cases in which the exception cannot be reduced to science but rather appears to have long-standing moral or religious roots. In a series of cases starting in the mid-1980s, the Court of Justice was called upon by the British courts to decide on the question of whether local rules prohibiting commerce on Sundays were compatible with the duty to allow for free movement of goods under the EC Treaty. In *Torfaen Borough*, the Court found that such rules could, in theory, reduce trade in goods among Member States and therefore, to use the doctrinal language of European law, be 'caught' by Article 28 of the EC Treaty.[39] The Court also found, however, that the prohibition on Sunday trading might be justified as a legitimate social policy. This social policy has such obvious Christian origins that there was no need for the Court be explicit; Sundays, in the Christian tradition, should be devoted to God, not to the pursuit of worldly, commercial activities. But rather than conduct the proportionality analysis itself, the Court sent the question back to the British court. (As with the right to privacy, before a public policy reason can trump the right to free movement of goods, the proportionality principle must be satisfied: the restriction on the free movement of goods must be necessary for satisfying the public policy end and the burden on free movement of goods must be proportionate to the end being pursued.) Thus the Court recognized that the British court was best placed to address the inordinately difficult challenge of reconciling religion and the life habits that accompany religion with the demands of contemporary economic life and membership in a free trade area. In essence, the Court acknowledged that this was a matter of the British—not European—constitutional order and that only the British court could conceivably resolve the dilemma.

In fact, not even the British courts were up to the challenge. After a number of conflicting rulings in the British lower courts, the matter went to the House of Lords. The House of Lords, rather than decide the issue, referred another question to the Court of Justice: what criteria should it use to decide the proportionality of regulatory restrictions on Sunday trading? Only reluctantly, therefore, did the Court of Justice settle this thorny problem. And rather than answer directly the question put to it by the House of Lords and elaborate on the appropriate proportionality criteria, the Court simply declared that the Sunday trading rules were proportionate.[40] By giving a preliminary ruling in the sparest of terms, the European law on the correct compromise between old mores and new commerce was left as vague as possible.

Another illustration of the Court's deferential approach to national constitutionalism is the *Omega* case.[41] The contending values in that case were free movement of services and human dignity. Omega owned and operated a laserdrome in Germany. Although Omega was a German company, it operated under a franchise agreement with a British firm, and therefore it was covered by the free movement of services provisions of the EC Treaty. Soon after Omega opened its

[39] Case 145/88, *Torfaen BC v B & Q plc* [1989] ECR 3851, para 14. The Court notoriously reversed itself on this point in Cases C-267 and 268/91, *Criminal Proceedings against Keck & Mithoard* [1993] ECR I-6097. That point, however, is not at issue in this discussion.

[40] Case C-169/91, *Stoke-on-Trent* [1992] ECR I-6635.

[41] Case C-36/02, *Omega Speilhallen* [2004] ECR I-9609.

doors to the public, it was ordered by the local authorities to cease part of its operations—games involving shooting laser beams at human targets. The justification for the order was the offence to human dignity perpetrated by such laser games. According to the German authorities, 'the games which took place in Omega's establishment constituted a danger to public order, since the acts of simulated homicide and the trivialisation of violence thereby engendered were contrary to fundamental values prevailing in public opinion.'[42]

Omega appealed to the German administrative courts. The courts found that, indeed, such games constituted an offence to human dignity and therefore had been banned legitimately. However, the highest administrative court referred to the Court of Justice the question of the compatibility of the German right to human dignity with the European guarantee of free movement of services: Could Germany take advantage of the public policy exception to the principle of free movement of services based on a right that, at least in the specifics, was not common to all Member States, ie a right to human dignity that included a prohibition on laser games?

The Court of Justice answered in the affirmative. It found that respect for human dignity was as much a part of European law as it was of German law and therefore the protection of human dignity qualified as a legitimate reason for restricting free movement. The means or 'system of protection' in favour of human dignity in one Member State did not have to mirror the system of protection in other Member States for it to be considered legitimate. Of course, consistent with the Court's earlier case law, a rights-based restriction on free movement, like any other restriction, had to satisfy the proportionality test. The Court of Justice then concluded by reviewing the German court's holding and finding that it had applied these European principles in a permissible fashion.

Different from *Lindqvist* and *Torfaen Borough*, the Court of Justice in *Omega* did not accommodate national diversity by sending the task of balancing constitutional values back to the national court. Rather, accommodation was accomplished by resort to another judicial technique: the Court reviewed the findings of the German authorities under the deferential standard of 'margin of discretion'. The Court explained this standard of review as follows: '[T]he specific circumstances which may justify recourse to the concept of public policy may vary from one country to another and from one era to another. The competent national authorities must therefore be allowed a margin of discretion within the limits imposed by the Treaty.'[43] The Court scrupulously applied this standard to the facts of the case. In examining the rationale for the German ban and the proportionality of the ban, the Court deferred extensively to the reasons given by the local administrative authority and the referring court. And in reaching the conclusion that the ban was permissible, the Court took care not to substitute its judgment for that of the German authorities. Instead of concluding that the ban

[42] *Ibid*, at para 7. [43] *Ibid*, para 31.

was justified, the Court found that the ban could not be regarded as 'unjustifiably undermining the freedom to provide services.'[44] Although, on the face of it, this word choice might appear trivial, it is indeed important. It signifies that, under European law, the authority to decide on the balance between the market and the right to human dignity rests primarily with the German courts; the European Court of Justice may interfere only in those extreme cases in which that national authority appears to have been abused. It is for the German courts to determine whether a rights-based restriction on the market is justified and for the European Court of Justice to guarantee that this national determination is not unjustified.

These cases demonstrate that on tough constitutional questions, like the balance between privacy and expression or between human dignity and market rights, the Court of Justice gives precedence to national constitutionalism. By sending constitutional questions back to national courts and employing the deferential review standard of margin of discretion, the Court avoids the pitfalls of constitutional patriotism. It affirms a set of principles common to the entire European Union yet also defers to national constitutional values. In other words, it tolerates constitutional diversity.

Another way to understand these judicial techniques of constitutional tolerance is by reference to the sociology of the European centre. In cases like *Lindqvist* and *Omega*, the Court rightly recognizes that the public sphere of Europe is still too thin to provide an answer to most of the difficult questions of constitutional law. In the language of Rawls, in these areas, there are no terms of fair cooperation to which Europe's gradually emerging society of citizens could agree. To take the privacy example, although European citizens might agree that individuals should be entitled to speak about others and, at the same time, to lead part of their lives in private, the deeper question of how to resolve conflicting claims to speech and privacy has no answer. If the Court of Justice were to attempt to resolve the matter, it would be deciding entirely at random. Of course, courts like the Court of Justice contribute to making a constitutional culture. Yet, they do not do so in a social void. They do so against the background of a thick set of institutions and social practices that serve to enable collective life in a community. There is no objective indicator of the point at which the institutional and cultural fabric of a new society is thick enough that a court like the European Court of Justice can begin giving meaningful answers to constitutional questions like the privacy-speech balance. But if the Court prematurely attempts to resolve such questions, it risks at best, irrelevance, at worst, delegitimation. At best, the Court's judgment will be obeyed but will be entirely marginal to future debates on that constitutional question. At worst, the lack of social foundation for the Court's judgment will justify disobedience in that case or in future cases.

[44] *Ibid*, para 40.

B. *Von Hannover v Germany (Princess of Monaco)*

Because of its history, the European Court of Human Rights (ECtHR) has had many more opportunities than the European Court of Justice (ECJ) to address the right to privacy—and the inevitable conflicts that arise between the right to privacy and the right to free expression. In doing so, the ECtHR has been far less shy than the ECJ about reversing national decisions. It has shown considerably less deference towards national courts and has intervened actively in domestic decision-making, sometimes on the side of privacy, sometimes on the side of speech.

The ECtHR's doctrinal tools for negotiating the relationship between national and supranational constitutionalism are slightly different from those of the ECJ. The system of jurisdiction established under the European Convention on Human Rights does not contemplate preliminary references. Rather, individuals must exhaust their legal remedies at the national level, obtaining a final national judgment, before they may apply to the ECtHR for relief. One consequence of this system of jurisdiction is that the ECtHR cannot send legal issues back to national courts for final resolution. In its judgments, the ECtHR must decide once and for all whether there has been a breach of the Convention. The judicial dialogue made possible in the European Union by virtue of the preliminary reference system has no analogue in the European Convention on Human Rights.

The ECtHR, however, employs a deference doctrine similar to the margin of discretion that was afforded to Germany in the *Omega* case: Member States, in balancing fundamental rights against other fundamental rights or public interest objectives, benefit from a 'margin of appreciation'.[45] The concept does not appear in the European Convention on Human Rights but rather is a creation of the ECtHR. It was first articulated by the ECtHR in the 1970s in a series of cases involving the public emergency derogation to human rights.[46] Later, the concept was extended to a number of other human rights areas, including interferences with the right to privacy under Article 8 and with the right to free expression under Article 10. The deference given to Member States under the doctrine of margin of appreciation can save an interference with a Convention right at two distinct phases of the legal analysis: in assessing the importance of the public policy or competing individual right that serves as the government justification

[45] Indeed, on their face, the doctrines of margin of discretion and margin of appreciation are identical. The linguistic difference is an artefact of different English translations from the same, French term of art '*marge d'appréciation*'. For a comparison of the ECJ's case law with the ECtHR's doctrine of margin of appreciation, see Sweeney, 'A "Margin of Appreciation" in the Internal Market: Lessons from the European Court of Human Rights' (2007) 34 Legal Issues of Economic Integration 27, at 28–36.

[46] Cameron, *supra* note 17, at 28; H.C. Yourow, *The Margin of Appreciation Doctrine in the Dynamics of European Human Rights Jurisprudence* (1996).

for the interference with the right, and in assessing the proportionality of the interference with the right.

The rationale for affording Member States discretion in making a determination on fundamental rights is similar to the one explored earlier in the context of the ECJ: national authorities are tightly integrated into their national social and cultural settings and are therefore in a better position than the ECtHR to give specific effect to the highly abstract commitments of the Convention. As the ECtHR stated in an early case on freedom of expression, 'by reason of their direct and continuous contact with the vital forces of their countries, State authorities are in principle in a better position than the international judge to give an opinion on the exact extent of these requirements [of free expression and the conditions under which a State's interference with free expression is justified]'.[47]

Most commentators agree that as the ECtHR has matured as a court, it has become tougher on defending Member States and more reluctant to defer to them using the doctrine of margin of appreciation.[48] This is certainly evident in the domains of privacy and freedom of expression.[49] And this reluctance to tolerate national constitutional diversity is dramatically at odds with the approach taken by the ECJ, analysed in the preceding section.

The recent case of *Von Hannover v Germany (Princess of Monaco)* illustrates the different approach taken by the ECtHR in resolving conflicting claims to privacy and freedom of expression.[50] At issue in the case were three separate German judgments, involving three different series of photographs of the Princess and her family. These photographs, taken without her consent, were published in a number of different German weekly magazines. At the time of publication of each series of photographs, the Princess of Monaco brought suit before the German courts to enjoin their publication. The litigation concerning the first series of photographs was the most exhaustive: it gave rise to extensively reasoned opinions by both the Federal Court of Justice (the highest court for the system of civil and criminal justice) and the Federal Constitutional Court. The subsequent litigation involving the two other series of photographs was decided based on this precedent and therefore, from a legal reasoning perspective, holds less interest.

The first series of photographs captured the Princess in a variety of situations. She was shown with Vincent Lindon, the French actor and her boyfriend at the time, in a secluded restaurant courtyard. She appeared together with her children in various public places, such as the market. And she was shown alone, horseback riding and shopping. Both the German court of first instance and the German

[47] *Handyside v The United Kingdom*, Application no 5493/72, 7 December 1976, para 49.

[48] See Sweeney, *supra* note 45, at 42; Yourow, *supra* note 46, at 56.

[49] In other areas, however, the movement away from the margin of appreciation is less clear. Indeed, according to some commentators, the ECtHR continues to rely heavily on the doctrine to decide a wide variety of cases. See Benvenisti, *supra* note 4, at 846; Carozza, *supra* note 5, at 57.

[50] Application no 59320/00, 24 June 2004, para 50.

court of appeals took the conventional view of privacy and held against the Princess. They found that as an 'absolute figure of contemporary history' the Princess could not protest the publication of her photographs. Because of their importance to contemporary debate, such figures could only claim a right to privacy when in the home, not when in public spaces. The highest court—the Federal Court of Justice—was more sympathetic to the Princess. It found that a concept of privacy limited to the home, even in the case of public figures like the Princess, was too narrow. Such figures should be able to prevent the public from prying even outside the home, in other spaces of 'seclusion'. In the Court's view, the restaurant courtyard in which the Princess was photographed eating with Vincent Lindon was precisely such a space of seclusion. The Court, therefore, enjoined the publication of the restaurant photographs but permitted publication of the remaining photographs since they involved entirely public places.

That, however, was not the end of the matter. The Princess of Monaco filed a constitutional complaint with the Federal Constitutional Court. In one respect, the Constitutional Court expanded her right to privacy. It found that children's privacy should receive special constitutional protection. Children, the Constitutional Court reasoned, were especially vulnerable to the autonomy harms caused by privacy violations because their personalities were still developing. Unless a public figure like the Princess of Monaco intentionally took centre stage at a public event with her children, the interest of the press and the public in knowing about the children must give way to the children's privacy. The Constitutional Court therefore enjoined the publication of all those photographs in which the Princess's children appeared. Remaining in the public eye were only those photographs in which she appeared horseback riding and shopping.

Based on this reasoning, the German courts rejected two later lawsuits brought by the Princess, seeking to enjoin the publication of two other sets of photographs. The first series of photographs showed the Princess enjoying different sports together with her husband, Prince Ernst August von Hannover. The second showed her stumbling at the pool of a private country club. Both series of photographs, according to the German courts, involved places fully visible to the public eye. They did not objectively demonstrate an effort, by the Princess, to create a place of seclusion. Nor were they taken of her together with her children. Therefore, before the German courts, the public interest in the photographs of the Princess prevailed.

The Princess had better luck before the European Court of Human Rights. The Court found that the publication of all the photographs at issue in the three German proceedings violated the Princess's right to private life. The Court first elaborated on the value of the right to private life. In doing so, it sought to identify, in abstract terms, the values underpinning the right to private life, and, in concrete terms, the circumstances under which the Court would recognize the right:

> [P]rivate life, in the Court's view, includes a person's physical and psychological integrity; the guarantee afforded by Article 8 of the Convention is primarily intended to ensure the development, without outside interference, of the personality of each individual in his relations with other human beings . . . There is therefore a zone of interaction of a person with others, even in a public context, which may fall within the scope of 'private life'.

How would the Court recognize that zone which fell within the scope of 'private life'? The Court said that it would consider 'whether the material thus obtained was envisaged for a limited use or was likely to be made available to the general public.' Since the photographs showed the Princess in her daily life, not engaged in public functions, the Court found that the photographs belonged to her private life.

The Court then turned to the competing right of freedom of expression. It emphasized the importance of free expression and the press to a democratic society: the press had a duty to impart 'information and ideas on all matters of public interest.' Yet, at the same time, the Court underscored the duties and responsibilities of the press: 'it must not overstep certain bounds, in particular in respect of the rights of others.'

The balance between the two rights, however, was struck in favour of the Princess and the right to privacy. The Court reached this conclusion based on the low value it attached to the speech involved in the case. The Court employed an implicit hierarchy of speech. At the bottom of this hierarchy, the Court placed the photographs of the Princess and declared them unworthy of protection, at least when in conflict with the right to privacy. The Court drew three distinctions. First, according to the Court, words were superior to images because words generally communicated ideas whereas images could convey very personal information. Secondly, expression concerning politicians was considered more valuable than expression concerning private individuals. In the view of the Court, the Princess was a private individual because she did not exercise official functions on behalf of the State of Monaco. The third distinction employed by the Court was between information on an individual's public life and information on that same individual's private life. Only in special circumstances did the public have the right to learn of the latter. These photographs combined the worst of all categories: as photographs of a private figure and of the private life of that figure, they deserved no protection. Thus the Court held that Germany, in permitting publication of the photographs, had violated the right to privacy under the European Convention on Human Rights. In reaching this decision, the Court made only passing reference to Germany's margin of appreciation on the rights question.

The purpose of this chapter is not to assess the correctness of the Court's balance between privacy and expression in the *Princess of Monaco* case. This consideration of the Court's reasoning is simply designed to show that there is ample room for disagreement on the matter. It would not be unreasonable to attach greater value to the photographs than was done in the case and, consequently, to find that the right to know about the Princess's activities trumped the Princess's right to privacy. This, in essence, is the conclusion that was reached by the German courts. And yet the European Court of Human Rights failed to take heed of Germany's constitutional settlement.

4. Tolerant and Intolerant Constitutional Patriotism

The contrast between the case law of the two courts is striking. While the ECJ deferred to national law on the privacy-speech question, the ECtHR handled the matter directly under European law. In this instance, given the circumstances of the two cases, the ECJ followed the better path. The *Lindqvist* judgment both affirmed a European commitment to privacy and free expression *and* made room for diverse moral orderings of public life at the national level. In the future, the ECtHR would do well to make heavier use of the legal doctrine of margin of appreciation so as to negotiate a similar compromise between European universalism and national particularism.

Certainly there are reasons for European courts to intervene directly in privacy cases. A national court's resolution of the right claims might be so clearly one-sided—in favor of the government's interference with privacy or a competing rights claim—that the decision constitutes a breach of European law. Furthermore, as a matter of institution-building, supranational courts must sometimes intervene. If supranational courts always hold for defending states, against complainants, they risk appearing to be mere puppets of those states, not independent lawgivers. But neither of these reasons was present in the *Princess of Monaco* case. The German courts carefully considered the Princess's claims and, in the course of the litigation, progressively limited access to the Princess's private life. The remaining right to know—the photographs of the Princess in non-secluded places by herself or in the company of other adults—had a plausible relationship to public debate. The Princess of Monaco is a member of a ruling royal family. She does not govern herself, but she does benefit directly from a system of public authority that, like all other such systems, should be subject to scrutiny, debate, and criticism. Germany's resolution of the privacy-speech conflict in favour of the press was not so biased as to constitute a breach of European law. Neither was the ECtHR at risk of appearing to be a tool of powerful sovereign states. Set against the Court's long track record of deciding cases against defending states, a judgment in favour of Germany would not have had the appearance of capitulation to national interest.

Instead, in the *Princess of Monaco* case, the countervailing consideration of tolerance for constitutional pluralism should have prevailed. The German courts were expounding a national view of the right and good organization of public debate. That view was consistent with the constitutional law of certain other national systems and at odds with others. In the *Princess of Monaco* litigation itself, the lower courts acknowledged that, had French instead of German law applied, the case would have come down in favour of the Princess.[51] But the German decision in favour of the press was in line with the law in other European countries. For instance, under English law, the press would also have prevailed.[52] Given this diversity and the absence, still, of a thick European public sphere, the ECtHR should have deferred to Germany's constitutional settlement.

This, of course, was the outcome of the *Lindqvist* case. In that case, the ECJ deferred to national constitutionalism by deliberately not deciding the privacy-speech conflict. Instead, the ECJ sent the legal question back to the Swedish court. This move was made possible by the preliminary reference system, a system of jurisdiction unavailable to the ECtHR. But the ECtHR's long-standing doctrine of margin of appreciation can serve a similar end. Based on the margin of appreciation enjoyed by Member States, the ECtHR could have deferred to the privacy-speech balance struck by the German courts.

Many scholars take a dim view of the doctrine of margin of appreciation. The ECtHR's resort to the doctrine is cast as the abandonment of rights, morality, and reason in the face of powerful, oppressive states.[53] Yet, as demonstrated by this chapter's review of the ECJ's use of analogous doctrines, the margin of appreciation can serve good purposes too. It is not just a face-saving device for a supranational court opposed by powerful Member States. With deference doctrines, courts can construct a supranational constitutional order and, at the same time, recognize and respect national constitutionalism. Deference doctrines like the margin of appreciation can be used to safeguard constitutional pluralism even as a single, European constitution is being built.

To understand the contribution that such doctrines can make to constitutional pluralism, it is important to appreciate the difference between ordinary appellate review and review by one court of another court's determination of law based on a deferential standard of review. In an ordinary judicial system, higher courts generally do not give any special consideration to the determinations of law made by lower courts. In other words, if a determination of law is appealed to a higher court, the higher court will decide the legal question *de novo*.[54] To take an example familiar from the earlier discussion of the *Princess of Monaco* case, even

[51] See Judgment of the Federal Court of Justice, BGHZ 131, 132, 19 December 1995, translated in B. S. Markesinis and H. Unberath, *The German Law of Torts: A Comparative Treatise* (4th edn, 2002), 449.

[52] See Neill, 'Privacy: A Challenge for the Next Century' in B.S. Markesinis (ed), *Protecting Privacy* (1999), 1; *Naomi Campbell v MGN Limited* [2004] UKHL 22, 6 May 2004.

[53] See Letsas, *supra* note 4, at 731–2; Benvenisti, *supra* note 4, at 847.

[54] This discussion does not address appellate review of questions of fact. On questions of fact, higher courts generally defer to lower courts because lower courts are better equipped to establish the facts in a particular case. Moreover, a question of fact is, by definition, limited to the particular case before the court and therefore a lower court's determination will not reverberate throughout the legal system.

though the German lower courts had already decided that a public figure like the Princess of Monaco could not assert a privacy right outside the home, the German Federal Court of Justice and then the Constitutional Court each decided the question anew. This is because the courts in a single judicial system all share in the same judicial power and the courts at the apex of the system stand in a hierarchical relationship to the courts at the bottom of the system. In the interest of coming to the best resolution of a question of law—questions which, by definition, have implications for an entire legal system—a court of last resort is bound to give fresh consideration to the various interpretive possibilities.

By contrast, deferential standards of review carry an acknowledgment of the independent power of the government body under review. Such standards are employed routinely by national courts tasked with reviewing parliamentary laws and executive decisions.[55] The judicial branch recognizes that, under traditional separation of powers principles, the legislative and executive branches are vested with independent lawmaking and administrative powers. Courts are not supposed to legislate or to execute, just to adjudicate. In other words, judicial intervention is warranted only if the decision of the legislature or the executive appears to be so misguided as to violate a command set down in the constitution or, in the case of an administrative decision, the applicable legislation. Judicial deference is the doctrinal recognition of this system of separate powers.

Deference in supranational judicial systems signifies the same type of institutional relationship but between courts rather than branches of government. Each court—European and national—is vested with independent powers. National courts are authorized to interpret their national constitutions, the two European courts to interpret the emerging European constitutional order. Deferential review by the European courts represents an acknowledgment of the extensive set of moral commitments set down in national constitutions and the privileged role of national courts as interpreters of those moral commitments. Only if national constitutional authority is abused, should the European courts intervene. In other words, only if national adjudication violates the minimum standards set down in Europe's still-fledgling constitution, should the European courts decide.

This argument in favour of the ECtHR's resort to the margin of appreciation does not take lightly the human rights critique of the doctrine. Human rights, undoubtedly, contain a universal, moral core.[56] And because human rights inhere in individuals as human beings, not as citizens of one nation or another, they should not vary depending on geography. They should give rise to the same

[55] See R.J. Pierce Jr, 1 *Administrative Law Treatise* 137 (4th edn, 2002).

[56] The debate on universalism versus diversity in the human rights arena and the implications for the ECtHR's margin of appreciation is extensive. For the universalist position and a critical view of the doctrine of margin of appreciation, see Benvenisti, *supra* note 4, at 847; Letsas, *supra* note 4, at 731–2.

treatment everywhere. Yet defending a universal set of human rights is not incompatible with recognizing constitutional diversity. As the European Court of Justice did in the *Lindqvist* case, it is possible both to affirm a common, European privacy right and to hold that certain, morally repugnant interferences with the right will constitute a human rights violation but that otherwise, depending on the national context, the privacy right might be trumped by competing rights or public policies. The recognition of such pluralism, of course, narrows the reach of human rights law. Yet if Europe truly values its different traditions of social organization and political contestation, it does not seem that it could be otherwise.[57]

In the human rights world, the doctrine of the margin of appreciation has been the subject not only of critique but also praise. The praise comes, essentially, in two varieties: the first casts the doctrine as advancing certain principles that are part and parcel of the moral architecture of human rights[58] whereas the second treats the doctrine in pragmatic terms, as a necessary judicial tool in an international realm of sovereign states reluctant to accept the dictates of international institutions.[59] The constitutional patriotism analysis in this article contributes to this literature by advancing a separate reason for the normative desirability of the margin of appreciation: because the ECtHR defends a Europe-wide constitution it should be sensitive to the imperatives of tolerance that are part of that constitution.

In essence, this analysis treats the ECtHR as a constitutional court, not simply an international human rights tribunal. The rights that are defended by the Court are not simply universal moral standards but individual entitlements that arise because of citizenship in a particular social community and political order—Europe. The ECtHR's stature as a constitutional court is not a matter of explicit design but rather is a by-product of the intertwined nature of rights in the European Union and the Council of Europe and the incremental, yet spectacular, success of European integration. This chapter's earlier description of the many norms and institutions that guarantee a European right to privacy is but one example of the overlapping and self-reinforcing nature of rights in Europe today. The European Union's court—the ECJ—is widely recognized as a constitutional court. When the ECJ interprets the treaties and establishes fundamental principles of law, it gives shape to public power and individual rights throughout the European Union. It makes very little sense to deny this same status to the ECtHR when both the positive instruments and the judicial precedent that serve

57 Weiler, 'Human Rights, Constitutionalism and Integration: Iconography and Fetishism', in E. O. Eriksen, J.E. Fossum, and A.J. Menéndez (eds), *supra* note 8, 59–68, at 66.

58 Carozza defends the margin of appreciation based on subsidiarity. Carozza, *supra* note 5, at 46–50, 56–63. Mahoney argues that the doctrine is based on subsidiarity, democracy, and cultural diversity. Mahoney, 'Marvelous Richness of Diversity or Individious Cultural Relativism?' (1999) 19 Human Rights LJ 1, at 2–3.

59 See eg Helfer and Slaughter, *supra* note 6, at 282–91.

as the basis for their decisions are inextricably interwoven. Thus, under the theory of constitutional patriotism, both courts should be self-conscious of the importance of fundamental rights in their particular geographical and historical instantiations to national self-identities. And before reaching a European determination on any particular rights dilemma—like the privacy versus speech conflict in *Lindqvist* and *Princess of Monaco*—both courts should consider whether such a determination is truly warranted or whether, indeed, the best constitutional outcome is tolerance of national diversity.

5. Conclusion

The emerging European right to privacy reveals both the identity-creating potential and the diversity-suppressing dangers of constitutional patriotism. The right has spread to virtually every corner of European governance. Under the European Convention on Human Rights, privacy has been construed broadly to protect individuals against most types of observation—even in public places—and to protect individuals against the collection of most types of personal information—even if that information is considered banal or widely available. In addition to the European Convention on Human Rights, the Council of Europe has Convention 108. Convention 108 sets down a series of guarantees specific to *data* privacy that have been enacted domestically throughout Europe.

In the European Union, the Charter of Fundamental Rights recognizes not simply a general right to privacy but also a right to the protection of personal data. This more specific privacy right is also guaranteed throughout the common market by the European Union's detailed Data Protection Directive. And when national police, customs officers, and courts cooperate on law enforcement and immigration policy, they are covered by a similar set of EU data protection laws.

Already the European Court of Human Rights and the European Court of Justice have been called upon to interpret this legal framework for privacy. These cases raise hard questions of constitutional law since individual claims to privacy routinely come into conflict with the right of others to know—to engage in public debate, transact in the marketplace, or pursue public ends. Just by examining one subset of this case law—the privacy-speech balance—it is possible to catch a glimpse of the community-building possibilities of a shared, European commitment to rights. In both the *Lindqvist* case and the *Princess of Monaco* case, the ECJ and the ECtHR proudly declared Europe's respect for free expression and privacy. And they explained the importance of these rights to the—European—understanding of the morally good ordering of life in a human community. Quite obviously, for the ECJ and the ECtHR, liberal values constitute a source of European unity.

But the ringing judicial affirmation of liberal European rights in the *Lindqvist* and *Princess of Monaco* cases also carried illiberal undertones. Because with a court's declaration of rights comes the imperative to protect rights and decide cases, regardless of another court's resolution of the very same rights claim. In an intrinsically pluralist political order like Europe, this admittedly noble commitment to rights can become the illiberal imposition of a uniform, hegemonic constitutional identity. Thus sometimes the impulse to defer to national courts on difficult questions of constitutional law, such as the privacy–speech balance in the *Lindqvist* and *Princess of Monaco* cases, can be the normatively superior outcome. In essence, the ECtHR and the ECJ are not like any ordinary set of constitutional courts. Their constitutional order is founded upon historically rooted, radical diversity. It is a nascent constitutional order, lacking the social and institutional infrastructure that would enable a genuinely European solution to many of the constitutional conflicts that emerge daily before national courts. Deference, for the European courts, is an essential instrument of tolerant—and patriotic—constitutionalism.

5

New Technologies, the Precautionary Principle, and Public Participation

Laurence Boisson de Chazournes

The issue of scientific and technological innovation is not new for the legal arena. Ever since the industrial revolution, scientific progress has been so fast that the law has often been late in responding to new technological challenges. Indeed, as Weeramantry has pointed out, 'inventions are piling up at a rate which far outpaces the possibilities of legal control. The law is not ready with the appropriate armory of new concepts to challenge these new weapons of intrusion.'[1] This is particularly true for 'new technologies', the progress of which, especially in recent years, is faster and more difficult to regulate compared to 'traditional technologies'. In the context of the present contribution, the emphasis will be on legal responses to risks posed by new technologies. It is accepted however that new technologies may also bring benefits and pose new opportunities.[2]

There is no widely accepted definition of 'new technologies'. In the relevant literature (not only that of the legal kind), new technologies seem to be addressed in two distinct ways. In a stricter sense, the expression only concerns technologies in the domains of communication and information. This definition is, we could say, a 'static' one, because it only covers certain categories of technologies. In a larger sense, the expression 'new technologies' is used as a synonym of 'high technology activities', encompassing technologies in the domains of biology, space, communications, computer, and nuclear activities. This definition relies on the novelty of the technological means at stake, meaning that a 'new technology' of today will no longer be 'new' in the future, and that, more generally, all technologies are 'new' when they are first introduced into society. The meaning of 'new' is thus subject to evolution and has to be continuously updated. It is this latter definition of 'new technologies' which is most commonly found in legal writings.

[1] C.G. Weeramantry, *Justice without Frontiers: Protecting Human Rights in the Age of Technology* (1997), at 72.

[2] See further the chapter by Roger Brownsword in this volume.

Science makes action possible, namely in the form of new technologies, but often science is unable to predict the collateral effects of actions with sufficient certainty. Human creativity does not seem to have limits. As it has been said, '*nos pouvoirs excèdent notre savoir*'.[3] For international law, this predicament first appeared with respect to nuclear activities and then in the environmental area and the field of biotechnology. International law has to face contrasting values: on the one hand the interests of humanity and of future generations as served by environmental protection, respect for human rights, and sovereignty over natural resources; on the other hand, freedom of scientific research, free trade, and the protection of intellectual property rights.[4] The International Court of Justice (ICJ) has pointed out this dialectic of values in the international legal order:

> Throughout the ages, mankind has, for economic and other reasons, constantly interfered with nature. In the past, this was often done without consideration of the effects upon the environment. Owing to new scientific insights and to a growing awareness of the risks for mankind—for present and future generations—of pursuit of such interventions at an unconsidered and unabated pace, new norms and standards have been developed, set forth in a great number of instruments during the last two decades. Such new norms have to be taken into consideration, and such new standards given proper weight, not only when States contemplate new activities but also when continuing with activities begun in the past.[5]

The risks posed by new technologies have caused a paradigm shift in the international legal order. Indeed, these risks require a new mode of both thinking and of action to prevent their realization and their potentially irreversible consequences. Traditional international law is familiar with the issue of risk (risk to international peace and security, risks in International Humanitarian Law, risk to human rights, etc) and has long since developed means of risk avoidance and prevention. However, this perception of risk as defined by the International Court of Justice is mostly based on a reactive approach. That is, it is dependent on the certainty of the imminent occurrence of the risks. For instance, efforts to prevent a threat to peace and security often occur when clear evidence of the threat is established, including the beginning of hostilities or the movement of troops.

The management of risks deriving from new technologies, however, requires a new approach based on *uncertainty* and *future events*. New legal tools, mechanisms, rules, and principles are thus needed to face the specific challenges posed by those risks. These innovative legal concepts go beyond the simple prevention or

[3] Y. Lambert-Faivre, *Droit du dommage corporel systèmes d'indemnisation* (3ème édition, 2000), at 614.

[4] Laghmani, 'Le droit international face aux nouvelles technologies: rapport introductif', in R. Ben Achour and S. Laghmani (eds), *Le droit international face aux nouvelles technologies* (2002) 23 ff.

[5] Gabcikovo-Nagymaros Project (Hungary/Slovakia) (1997), ICI Reports (1997), at para 140.

reduction of *present* and *well-known* risks foreseen in traditional international law to encompass a new rationale, ie that of precaution.

The present contribution will focus, first, on the definition of the notion of precaution and its constitutive elements, and will analyse the manner in which the precautionary approach brings new concerns and new paradigms into the international legal order. Precaution's interwoven links with public participation will then be developed in Part 2, particularly as they involve the notions of risk assessment, risk management, and risk communication. The chapter will conclude in Part 3 with remarks on the societal dimension of precaution.

1. Risks and International Law: The Virtues of Precaution

Precaution starts from the premise that the absence of scientific certainty should not be used as a pretext to postpone decisions if there is a risk of serious or irreversible damage to the environment or to public health. Precaution is a key concept in international sustainable development law and is increasingly penetrating the area of human rights and health. A commonly given definition of the precautionary principle/approach[6] is the one contained in the 1992 Rio Declaration on Environment and Development. It reads as follows:

> In order to protect the environment, the precautionary approach shall be widely applied by States according to their capabilities. Where there are threats of serious or irreversible damage, lack of full scientific certainty shall not be used as a reason for postponing cost-effective measures to prevent environmental degradation.[7]

The Rio Declaration on Environment and Development played a pioneering role in the process of crystallizing the concept of precaution in international law. Precaution tends to defy standard or classical legal assumptions. It is in essence a meta-legal principle, allowing legal provisions to incorporate considerations beyond those resulting from strictly positive law.[8] While these characteristics endow the precautionary principle with a certain originality, they also determine the complexity of its analysis.

A. Precaution and International Law

Law has long been a reflection of Cartesianism and positivism, dominant schools of thought of modern times. New challenges confronted by humanity as a whole

[6] In this chapter, the terms 'the precautionary principle', 'the precautionary approach', and 'precaution' will be used interchangeably.

[7] Principle 15 of the Rio Declaration on Environment and Development, UN Doc. A/CONF.151/26, vol I, annex I, 1992.

[8] J. Carbonnier, *Flexible Droit* (1998), at 66; Desideri, 'La précaution en droit privé' (2000) 15 Recueil le Dalloz 238.

and trepidation sparked by technological progress have led to a questioning of legal Cartesianism and of what has been described as 'Euclidian law'.[9] This revolution did not take place directly within the legal system, but resulted from the remodelling of both philosophical and scientific thought. On the philosophical level, 'the heuristics of fear', developed by Hans Jonas,[10] pioneered the process of reflection on our interaction with the world, with science and with humanity. Major catastrophes (Chernobyl) and uncertainties (the consequences of climate change) have led to the definition and assessment of other parameters of state action in the international sphere. In this sense, precaution is a tool jockeying to be included in the new understanding of human relations to the environment in the international legal order. This order uses precaution as a tool to 'juridicize' these new concerns. International law cannot ignore precaution.

Precaution favours the emergence of new paradigms that contribute to the development of a global risks governance regime. These paradigms are those of uncertainty, interdependence, and anticipation. Precaution thus influences the philosophy of law in its entirety.

1. Precaution and the New Paradigms of International Law

(a) Regulating uncertainty

The precautionary approach developed following concerns over the regulation of uncertain ecological phenomena. Traditionally, international law has been based on the dogmas of Cartesian rationality. As a result, only those phenomena characterized by a certain scientific certainty were addressed by international regulation.

When the Vienna Convention for the Protection of the Ozone Layer was adopted in 1985, the extent and the nature of the human impact on the ozone layer were largely shrouded in uncertainty.[11] In the same vein, scientific uncertainty was widely prevalent in 1992 when two of the key multilateral environmental agreements were opened for signature in Rio, namely the UN Framework Convention on Climate Change and the Convention on Biological Diversity.[12]

[9] Timsit, 'Pour une nouvelle définition de la norme' (1988) 36 Recueil le Dalloz 267.

[10] H. Jonas, *Le principe responsabilité: une ethique pour la civilisation technologique* (1993), at 13.

[11] See the Preamble of the 1985 Vienna Convention for the Protection of the Ozone Layer: 'Aware *also* of the need for further research and systematic observations to further develop scientific knowledge of the ozone layer and possible adverse effects resulting from its modification.', available at <http://www.unep.org/ozone/pdfs/viennaconvention2002.pdf> (last accessed 11 December 2007).

[12] See the Preamble of the 1992 United Nations Framework Convention on Climate Change: 'Noting that there are many uncertainties in predictions of climate change, particularly with regard to the timing, magnitude and regional patterns thereof.', (1992) 31 ILM, at 851. See also the Preamble of the 1992 Convention on Biological Diversity: 'Aware of the general lack of information and knowledge regarding biological diversity and of the urgent need to develop scientific, technical and institutional capacities to provide the basic understanding upon which to plan and implement appropriate measures', (1992) 31 ILM, at 822.

One can also refer to the Cartagena Protocol on Biosafety which was adopted in 2000 (hereinafter the 'Cartagena Protocol') at a time of major scientific uncertainty regarding the risks of genetically modified organisms for the environment and public health—uncertainty which in fact persisted after the Cartagena Protocol's entry into force in 2003.[13]

(b) International law and the challenge of interdependence

A precautionary approach emphasizes the interconnections and inter-sectoral aspects of the environmental problems to be regulated. It thus redefines the methods leading to the development of law at two levels. First, it excludes a compartmentalized approach towards regulation, acknowledging the intrinsic and extrinsic linkages among the phenomena to be controlled or managed. Lawmakers no longer limit themselves to the known impact of a human activity or of a new technology on certain elements of the environment. Rather, they also take into consideration the impact, even if uncertain, on other environmental components. For example, an effective and adequate regulation of the ozone layer cannot ignore desertification, climate change, and atmospheric pollution. These four elements are inherently linked. This new kind of complexity therefore calls, as its corollary, for a new and more comprehensive kind of legal approach. Indeed, the very concept of global environmental governance requires a holistic and systemic treatment of all its components.

Secondly, a precautionary approach also presumes a flexible and open-ended view of law. Traditionally, international law is based on a straight and focused form of legal reasoning that makes it difficult to consider emerging and interconnected issues. Precaution, on the other hand, as a technique of managing uncertainty is very different from any notion of a finite law. It is built upon a foundation that allows adjustment and openness. The procedural technique of developing a framework convention that is later complemented by the adoption of additional protocols addressing the progression of scientific knowledge facilitates this kind of 'legal openness'. The 1994 Protocol to the 1979 Convention on Long-range Transboundary Air Pollution regarding *newly introduced* reductions of sulphur emissions represents a typical example of this phenomenon.[14] That this new reduction is essentially based on precaution is clear from the terms of the Protocol's Preamble, which relevantly states '[r]esolved to take precautionary measures to anticipate, prevent or minimize emissions of air pollutants and mitigate their adverse effects'.

[13] See the Preamble of the 2000 Cartagena Protocol on Biosafety to the Convention on Biological Diversity: 'Aware of the rapid expansion of modern biotechnology and the growing public concern over its potential adverse effects on biological diversity, taking also into account risks to human health', (2000) 39 ILM, at 1027.

[14] The Protocol partially amends the 1985 Helsinki Protocol on the Reduction of Sulphur Emissions or their Transboundary Fluxes by at least 30%, available at <http://www.unece.org> (last accessed 11 December 2007).

(c) An anticipatory approach

By definition, precautionary measures attempt to control events that have not yet occurred and may never occur. Precaution focuses on the future and therefore constitutes a technique of legal anticipation with regard to global risks.[15] For instance, the Preamble of the Convention on Biological Diversity carries an explicitly anticipatory message: 'Noting that it is vital to *anticipate*, prevent and attack the causes of significant reduction or loss of biological diversity at source' (emphasis added). Precaution rejects a curative legal approach. Given that it is not characterized by a curative logic, precaution differs in many aspects from the polluter pays principle which is closely linked to compensation for environmental damage. Furthermore, precaution does not depend simply on a cause and effect relationship. Precaution goes beyond the principle of prevention which ties action to scientific certitude. This need for anticipation is addressed in legal systems through the precautionary principle.

B. Precaution: A Means for Law to Address New Concerns

Precaution is a tool for introducing new values into the international legal system, which essentially lead to new modes of thought (1) and of action (2).

1. The 'Juridicization' of a New Mode of Thought: Complexity

The new thought method to which international law refers when it incorporates a precautionary approach is that of complexity, or, to quote Edgar Morin, that of 'complex thought'.[16] What constitutes complexity? The idea of a complex and temporary order has replaced that of a determinism clear, simple, universal, and eternal.[17] Considering precaution as 'the law of complexity' highlights two of its substantive character traits: uncertainty and globality.[18] Uncertainty, because taking into account complexity comes precisely from the refusal of a simplistic approach. Precaution is therefore 'the law of uncertainty', hostile to the dogma of Cartesian rationality. International public action is called upon to create a new paradigm: the management of uncertainty.[19]

[15] See Mbengue, 'Towards a Precautionary International Law: About Uncertainty, Interdependence and Anticipation in Global Environmental Governance', 2 *IHDP Update, Newsletter of the International Human Dimensions Programme on Global Environmental Change* (2004) at 13.

[16] E. Morin, *Introduction à la pensée complexe* (1990).

[17] Milani, 'La complexité dans l'analyse du système-monde: l'environnement et les régulations mondiales' (2000) 46 Revue Droit et Société 425, at 426.

[18] Diebolt, 'La complexité comme paradigme pour concevoir une régulation juridique adéquate' (2000) 46 Revue Droit et Société 485.

[19] On the concept of scientific uncertainty, see *infra*.

Being the law of complexity, precaution is also the 'law of globality'.[20] Indeed, complexity stresses the transversality and interconnection between phenomena, especially in the environmental field. An effective and efficient management of these phenomena requires a global approach. In this context, precaution manifests itself as a fitting legal technique to manage the interrelation and interaction between various phenomena.

Precaution as a legal technique presents itself as an appropriate and adequate path for the administration of common goods and spaces. The uncertainty of the potential consequences of a state's actions on the well-being of other societies requires the elaboration of global strategies for the management of these common spaces. In the environmental field, in which precaution finds its principal application, the concrete issues at stake in terms of international relations often extend beyond the borders of states and regional groupings. Such is the case in relation to climate change, protection of biodiversity, deforestation processes, and use of shared natural resources, especially fresh water resources.

2. The 'Juridicization' of a New Mode of Action: Anticipation

In substance, precaution obeys a particular *ratione tempori* process. The legal temporality to which precaution refers is the future. In this, it is 'the law of anticipation'. The logic of precaution, centered essentially on the uncertainty of the effects of human activity, highlights the necessity to take into account the future and potential effects of this activity. Precaution anticipates in that it seeks to govern situations that have not yet arisen. Therefore it is not a body of law focused on the concept of reaction. Precaution as a legal technique therefore guarantees the right to anticipation in the international legal system. The dogmas of objectivism and realism have often led law in general and international law in particular to ignore a large portion of the legal temporality—the future—taking into account exclusively the past and the present.[21]

As a law of anticipation, precaution is *a fortiori* a manifestation of 'the law of prospective'. It even adds a touch of originality by raising the dimension of uncertainty to the status of a fundamental element of consideration in the elaboration of any public or private prospective policy.

As a result, the stages of thought and action in the precautionary approach are intimately linked. Indeed, it is difficult to conceive of anticipation without uncertainty, and uncertainty without anticipation. Similarly unlikely is the conception of uncertainty and anticipation without any reference to a global

[20] See Cameron and Abouchar, 'The Precautionary Principle: A Fundamental Principle of Law and Policy for the Protection of the Global Environment', (1991) 14 Boston College Intl and Comparative L Rev 1, at 17.

[21] Gurvitch, 'La multiplicité des temps sociaux', in G. Gurvitch, *La vocation actuelle de la sociologie*, vol II, (1963) 325. According to this author, '*la vie sociale s'écoule dans des temps multiples, toujours divergents, souvent contradictoires et dont l'unification relative, liée à une hiérarchisation souvent précaire, représente un problème pour toute société.*'

approach. The uncertainty of the spatial extent of the analysed phenomena requires global anticipation—that is, not ignoring the potential implications of a phenomenon in other spaces, or those of one system on others. Precaution calls for the re-examination of the notion of spatial sovereignty as sketched by traditional international law, and puts emphasis on a need for a special 'territorialization' in a global context.

Precaution is not a volatile notion in international law. On the contrary, international law, in order to guarantee legal security and predictability, seeks to identify criteria permitting the durable establishment of the legal content of precaution.

C. The Notion of Precaution: Identifying the Criteria of Definition

Precaution has four fundamental criteria, which define it as well as justify *ratione materiae* its application in given situations. These criteria endow it with a particular structure, an original basis compared to other principles or approaches of international law. Three of these criteria (risk, damage, and scientific uncertainty) justify the application of precaution *a priori*, whereas the last criterion (capacities) intervenes *a posteriori* to objectively determine its applicability, permitting the passage from 'application' to 'applicability', from 'desirability' to 'feasibility'.

1. The 'Risk' Criterion

(a) Meaning and content of the risk criterion

Risk is the very essence of precaution. Precaution's *raison d'être* originates in law's aspiration to assess and manage risk in our societies. Risk is a more or less foreseeable and contingent danger that can cause damage. It is therefore arbitrary in its essence. Volatility is its nature; its occurrence can be unforeseen, even unexpected. As long as there is any trace of doubt as to the occurrence of an event, there is risk.[22] In an attempt to legally and precisely qualify the risks targeted by precaution, it is useful to recall the typology of risks compiled by Nicolas de Sadeleer, who was inspired by lessons drawn from German thought.[23] According to him, there are three main categories of risks:

- 'Unacceptable' or 'definite' risks, for which the causal link between the event and the damage is scientifically proven, even if doubt remains as to the

[22] See Martin, 'Le concept de risque et la protection de l'environnement : évolution parallèle ou fertilisation croisée ?', in M. Prieur and C. Lambrechts (eds), *Les Hommes et l'environnement, en hommage à Alexandre Kiss,* Paris, Frison-Roche (1998) 451. See also Kiss, 'Droit et risque', in (1991) 36 Archives de Philosophie du Droit 49.

[23] N. de Sadeleer, *Les principes du pollueur-payeur, de prévention et de précaution (Essai sur la genèse et la portée juridique de quelques principes du droit de l'environnement)* (1999), at 175 ff. See also N. de Sadeleer, *Environmental Principles: From Political Slogans to Legal Rules* (2002), at 91–226.

time it will take for the damage to occur. These risks should be eliminated by the principle of prevention.

- 'Residual' risks, those inherent to normal human activity that must be tolerated (eg the risk implicit in driving a car or taking a plane). These risks, which 'rest on purely speculative considerations without any scientific foundation', do not need to be taken into account in the decision-making process.[24] In order to avoid situations which would be absurd for human activity, residual risks have to be excluded from the precautionary principle's range of application.
- 'Uncertain' risks, whose existence has not been established by science, but cannot be dismissed. These must be addressed by the precautionary principle.

The legal terminology used in international instruments in relation to risks varies. In some instruments, references are made to 'threat'; elsewhere, however, references are made to the idea of 'potential' nuisance or danger.

(b) The components of a precautionary risk assessment

The main difficulty with the risk criterion lies in its assessment. Challenges such as quantifying the probability of occurrence of risk and qualifying the risk itself arise. International law does not provide definite answers, but international practice provides indications of what is, in objective terms, the evaluation of risk.[25] The Cartagena Protocol for instance, provides for a risk assessment procedure in its Annex III. The objective of risk assessment under the Cartagena Protocol 'is to identify and evaluate the potential adverse effects of living modified organisms on the conservation and sustainable use of biological diversity in the likely potential receiving environment, taking also into account risks to human health.' The Cartagena Protocol was a pioneer in formulating principles of risk assessment as well as integrating precaution into the assessment of risks posed by new technologies such as biotechnology:

3. Risk assessment should be carried out in a scientifically sound and transparent manner and can take into account expert advice of, and guidelines developed by, relevant international organizations.
4. Lack of scientific knowledge or scientific consensus should not necessarily be interpreted as indicating a particular level of risk, an absence of risk, or an acceptable risk.

[24] N. de Sadeleer, *Environmental Principles: From Political Slogans to Legal Rules* (2002), at 158.

[25] As an example, it is interesting to point out that the WTO Appellate Body in the case *EC – Measures Concerning Meat and Meat Products (Hormones)* noted that the risk mentioned in Art 5, para 1 of the SPS Agreement was not only a 'risk ascertainable in a science laboratory', but also a 'risk in human societies as they actually exist, in other words, the actual potential for adverse effects on human health in the real world where people live and work and die', see *infra*. The Appellate Body thus affirmed that the scientific evidence which the European Communities referred to did not concern the type of hormone at stake; for this reason risk assessment was deemed insufficient. See the Report of the Appellate Body, paras 187 and 199–200, available at <http://www.wto.org>.

5. Risks associated with living modified organisms or products thereof, namely, processed materials that are of living modified organism origin, containing detectable novel combinations of replicable genetic material obtained through the use of modern biotechnology, should be considered in the context of the risks posed by the non-modified recipients or parental organisms in the likely potential receiving environment.
6. Risk assessment should be carried out on a case-by-case basis. The required information may vary in nature and level of detail from case to case, depending on the living modified organism concerned, its intended use and the likely potential receiving environment.[26]

The Cartagena Protocol goes further by identifying a methodology for the risk assessment of risks deriving from modern biotechnology. Annex III of the Protocol emphasizes the importance of precaution by making clear that 'The process of risk assessment may on the one hand give rise to a need for further information about specific subjects, which may be identified and requested during the assessment process, while on the other hand information on other subjects may not be relevant in some instances.' Risk assessment is based on the following steps:

1) An *identification* of any novel genotypic and phenotypic characteristics associated with the living modified organism that may have adverse effects on biological diversity in the likely potential receiving environment, taking also into account risks to human health.
2) An *evaluation of the likelihood of these adverse effects* being realized, taking into account the level and kind of exposure of the likely potential receiving environment to the living modified organism.
3) An *evaluation of the consequences* should these adverse effects be realized.
4) An *estimation of the overall risk* posed by the living modified organism based on the evaluation of the likelihood and consequences of the identified adverse effects being realized.
5) A recommendation as to *whether or not the risks are acceptable or manageable*, including, where necessary, identification of strategies to manage these risks.
6) *Where there is uncertainty regarding the level of risk*, it may be addressed by requesting further information on the specific issues of concern or by implementing appropriate risk management strategies and/or monitoring the living modified organism in the receiving environment.[27] (emphasis added)

It is important to note that these steps do not constitute a risk assessment in itself. They are only 'risk assessment techniques' to refer to the terminology used in the

[26] Annex III of the Cartagena Protocol on Biosafety, *supra* note 13.
[27] *Ibid.*

Agreement on Sanitary and Phytosanitary Measures (SPS Agreement).[28] The Panel in the *EC – Biotech* case recalled, for instance, the distinction to be made between a 'risk assessment' *per se* and 'risk assessment techniques' taken into account in a process of risk assessment:

To the extent the European Communities is arguing that there is no requirement to assess 'risks', we disagree. By its own terms, Article 5.1 requires [WTO] Members to base their SPS measures on an appropriate assessment 'of the risks to human, animal or plant life or health.' The immediate context of Article 5.1 confirms this view. Article 5 is captioned '*Assessment of Risk* and Determination of the Appropriate Level of Sanitary or Phytosanitary Protection.' Moreover, Articles 5.2 and 5.3 of the *SPS Agreement* specify relevant factors to be taken into account by Members 'in the assessment of risks' and 'in assessing the risk to animal or plant life or health.' Finally, Article 5.7 of the *SPS Agreement* provides that in circumstances where relevant scientific evidence is insufficient and a Member has adopted a provisional SPS measure based on available pertinent information, it must seek to obtain the additional information necessary for a more objective assessment of 'risk'... As the European Communities points out, Article 5.1 provides that Members must base their SPS measures on an appropriate assessment of risks, 'taking into account risk assessment techniques developed by relevant international organizations.' *In our view, the phrase 'taking into account risk assessment techniques developed by relevant international organizations' does not address the issue of whether risks are to be assessed, but rather how risks are to be assessed. This is clear from the reference to 'techniques' of risk assessment. Contrary to the European Communities, we therefore do not consider that the phrase in question supports the view that no assessment of risks is required.* To the contrary, the phrase in question would, in our view, be unnecessary if there were no requirement to assess risks.[29] (emphasis added)

The components of a precautionary risk assessment have been specifically analysed in a case brought before the Court of First Instance of the European Community, ie the *Pfizer Animal Health v Council of the Europe Union* case.[30] Pfizer submitted that the contested European regulation should be annulled, since the European Community institutions had made errors in the assessment

[28] For a comparison of the provisions on risk assessment and precaution in the Cartagena Protocol on Biosafety and the Agreement on Sanitary and Phytosanitary Measures (SPS Agreement), see Oliva, 'The Cartagena Protocol on Biosafety and the Agreement on Sanitary and Phytosanitary Measures: What will Decisions Regarding GMOs have to be Based on?' (2002) 13 Int'l Legal Perspectives 22. The SPS Agreement is available at <http://www.wto.org>. See, for instance, Art 5.1 of the SPS Agreement: 'Members shall ensure that their sanitary or phytosanitary measures are based on an assessment, as appropriate to the circumstances, of the risks to human, animal or plant life or health, taking into account risk assessment techniques developed by the relevant international organizations.' See *infra* the notion of risk analysis under the *Codex Alimentarius* Procedural Manual.

[29] *European Communities – Measures Affecting the Approval and Marketing of Biotech Products*, Reports of the Panel, WT/DS291/R, WT/DS292/R, WT/DS293/R, 29 September 2006, paras 7.3021–7.3022.

[30] *Pfizer Animal Health v Council of the Europe Union*, Judgment of the Court of First Instance, 11 September 2002, Case T-13/99, available at <http://www.curia.europa.eu> (last accessed 11 December 2007).

and management of the risks to human health associated with the use of virginiamycin as a growth promoter, as well as in their application of the precautionary principle.

Let us recall that the Preamble to the contested regulation shows that the European Council, in adopting the measure, took the view that the use of virginiamycin as an additive in feedingstuffs involved a risk to human health and that, accordingly, it was necessary to withdraw the authorizations to use the product.[31] Both parties to the dispute admitted that, at the time of the adoption of the contested regulation, neither the reality nor the seriousness of the risk had been scientifically proven. It was against that background that the Council relied on the precautionary principle as justification for adopting the regulation.

Pfizer did not dispute that, in principle, the Community institutions may take preventive measures if, following a risk assessment, it is found that the use of an antibiotic, such as virginiamycin, as a growth promoter in animals involves a risk of a transfer of antimicrobial resistance from animals to humans and, consequently, of a reduction in the effectiveness of certain medicinal products used in human medicine for the treatment of dangerous infections. However, Pfizer maintained that the Community institutions did not correctly assess that risk, and argued essentially that these institutions had made their decision for reasons of political expediency without a proper scientific basis. For Pfizer:

> the Community institutions must show that the risk, although it has not actually become a reality, is nevertheless probable. The existence of a very remote risk should be allowed given the concrete positive elements arising from the use of the product concerned. In any event, the Community institutions cannot legitimately apply a test which Pfizer describes as a zero risk test. Such a test is inappropriate since it is impossible to satisfy. It amounts essentially to requiring *probatio diabolica* from the industry, something which is recognized as unlawful in all the legal systems of the Member States... It is never possible to prove conclusively that a chemical or pharmaceutical compound or anything created by modern technology represents a zero risk to public health now or that it will do so in the future. To apply such a test would quickly lead to the paralysis of technological development and innovation.[32]

The Council of the European Union considered for its part that the contested regulation was adopted on the basis of an adequate assessment of the scientific knowledge available at the time of its adoption. The Council confirmed that any such measure withdrawing authorization cannot be based on a test described as zero risk. However, the fact that the competent authorities had, at a given time, considered that a particular additive meets the conditions for authorization and have therefore authorized it does not imply that the manufacturer is freed from

[31] Council Regulation (EC) No 2821/98 of 17 December 1998 amending, as regards withdrawal of the authorization of certain antibiotics, Directive 70/524/EEC concerning additives in feedingstuffs, available at <http://www.legaltext.ee/text/en/T40644.htm> (last accessed 13 December 2007).

[32] *Pfizer Animal Health v Council of the Europe Union*, *supra* note 30, para 130.

the onus of proving that its product continues to meet such conditions. In the view of the Council:

Scientific knowledge and the risks to human health associated with use of a particular product evolve. Consequently, when faced with new scientific evidence that the use of an additive poses a hazard to public health and that the hazard has reached alarming proportions since the additive was first authorized, the Community institutions are fully entitled to require the manufacturer in question to demonstrate that its product continues not to represent a risk to human health.[33]

The Court of First Instance, in a very interesting and detailed dictum, explained the purpose of a risk assessment in cases where the precautionary principle finds application:

It is appropriate to bear in mind that, as the Court of Justice and the Court of First Instance have held, where there is scientific uncertainty as to the existence or extent of risks to human health, the Community institutions may, by reason of the precautionary principle, take protective measures without having to wait until the reality and seriousness of those risks become fully apparent... It follows, first, that as a result of the precautionary principle, as enshrined in Article 130r(2) of the Treaty, the Community institutions were entitled to take a preventive measure regarding the use of virginiamycin as an additive in feedingstuffs, even though, owing to existing scientific uncertainty, the reality and the seriousness of the risks to human health associated with that use were not yet fully apparent. *A fortiori*, the Community institutions were not required, for the purpose of taking preventive action, to wait for the adverse effects of the use of the product as a growth promoter to materialize... Thus, in a situation in which the precautionary principle is applied, which by definition coincides with a situation in which there is scientific uncertainty, a risk assessment cannot be required to provide the Community institutions with conclusive scientific evidence of the reality of the risk and the seriousness of the potential adverse effects were that risk to become a reality... However, it is also clear from the case-law that a preventive measure cannot properly be based on a purely hypothetical approach to the risk, founded on mere conjecture which has not been scientifically verified.[34]

And the Court of First Instance went on to conclude that:

The precautionary principle can therefore apply only in situations in which there is a risk, notably to human health, which, although it is not founded on mere hypotheses that have not been scientifically confirmed, has not yet been fully demonstrated. In such a situation, risk thus constitutes a function of the probability that use of a product or a procedure will adversely affect the interests safeguarded by the legal order. Hazard (danger) is, in this context, commonly used in a broader sense and describes any product or procedure capable of having an adverse effect on human health (see in that regard, at an international level, the provisional communication from the Codex Alimentarius Commission of the Food and Agriculture Organization of the United Nations and the World Health

[33] *Ibid*, para 134.
[34] *Ibid*, paras 139–43.

Organization, CX 2/20, CL 1996/21-GEN, June 1996). Consequently, in a case such as this, the purpose of a risk assessment is to assess the degree of probability of a certain product or procedure having adverse effects on human health and the seriousness of any such adverse effects.[35]

2. *The 'Damage' Criterion*

Once the decision-maker has an idea of the likelihood of occurrence of the suspected risk, he will naturally reflect on the possibilities of shielding himself from it. Must this risk be reduced or even eliminated regardless of the importance or severity of damage it could provoke? Or on the contrary, is intervention required only if what is at stake is worth the effort? His attitude is obviously subject to variations according to the probability of occurrence and especially to the importance of damage.

The terminology used to denote damage in international instruments varies. Reference can be made directly to the concept of damage.[36] Some instruments refer to the concept of 'impact'.[37] In spite of these terminological variations, the formulation of precaution in international instruments embodies an original, if not special, concept of damage. The latter is usually bound to a threshold of severity, which limits the application of the precautionary principle. This threshold usually invokes concepts such as 'severity' and 'irreversibility'. The 'damage' criterion is an important element of the definition of precaution, even though it is intrinsically linked to the risk criterion.

3. *The 'Scientific Uncertainty' Criterion*

(a) Content and meaning of the criterion

Whenever precaution is formulated in an international instrument, the criterion of scientific uncertainty is bound to it. The element of uncertainty is a *sine qua non* condition to the application and even to the legitimacy of the precautionary principle.[38] Indeed, the latter differs from the principle of prevention precisely in its reference to the aforementioned element. The 'preventive model' is forced to depend constantly on science and its expertise, the only way to allow for a certain

35 *Ibid*, paras 146–8.

36 See Principle 15 of the Rio Declaration on Environment and Development: 'Where there are threats of serious or irreversible damage.' The Rio Declaration is also available at <http://www.unep.org>. See also the United Nations Framework Convention on Climate Change: 'Where there are threats of serious or irreversible damage, lack of full scientific certainty should not be used as a reason for postponing such measures' (available at <http://www.unfccc.int>).

37 See the Convention on the Protection and Use of Transboundary Watercourses and International Lakes: 'The precautionary principle, by virtue of which action to avoid the potential transboundary impact of the release of hazardous substances shall not be postponed' (available at <http://www.unece.org>).

38 Remond-Gouilloud, 'Le risque de l'incertain: la responsabilité face aux avancées de la science', in (1993) X(4) La Vie des Sciences 341, at 355.

objectivization of the risks taken. Yet if prevention's strength is found in its reference to scientific knowledge, this also reveals its limitations. If risk is known, preventive measures can be taken in full knowledge of the situation. Prevention is only possible for known phenomena. It is difficult to prevent what is unfamiliar, and even more difficult to do so with the unknown.[39]

To denote scientific uncertainty, international instruments refer to 'the absence of complete scientific certainty',[40] 'the absence of absolute scientific certainty',[41] or even 'uncertain, unreliable, or inadequate data' and 'the lack of adequate scientific data'.[42] However, a precautionary measure must be anchored to a minimum level of knowledge, a basis of scientific data presenting a certain consistency.[43]

(b) Scientific uncertainty versus predictability in risk analysis: the case of the WTO

Risk and, more especially, uncertain risk constitute an important foundation of precaution. The application of precautionary measures needs to be applied to the whole procedure of risk analysis, which is based on the cumulative combination of risk assessment, risk management, and risk communication. Precaution can be implemented through these three distinct but often overlapping and interacting phases. However, in areas such as international trade law, the precautionary implementation of these concepts may be confronted by the issue of predictability.

Multilateral environmental agreements (MEAs) with potential trade implications such as the Cartagena Protocol on Biosafety or the Stockholm Convention on Persistent Organic Pollutants (POPs)[44] enunciate rights and obligations dealing with trade. These agreements imply *de jure* and *de facto* that a party to these MEAs who is also a WTO member may be in a situation whereby it has to impose import restrictions regarding products which pose a scientifically uncertain risk for the environment and for public health. There might thus be a conflict between precautionary policies and measures and rules of the global trade regime, and especially of the WTO.

[39] Rehbinder, 'The Precautionary Principle in an International Perspective', in *Miljorettens Grundsporgsmaal* (1994) 95.

[40] See the Declaration on Sustainable Development of the United Nations Economic Commission for Europe, Bergen, 16 May 1990, available at <http://www.unece.org/leginstr/cover.htm>.

[41] See the Convention for the Protection of the Marine Environment and the Coastal Region of the Mediterranean, available at <http://www.unep.ch/regionalseas>; see also the Rio Declaration on Environment and Development, available at <http://www.unep.org>.

[42] See the Convention on the Conservation and Management of Straddling Fish Stocks and Highly Migratory Fish Stocks, available at <http://www.un.org/Depts/los>.

[43] See for instance the Convention for the Protection of the Marine Environment of the North-East Atlantic, available at <http://ec.europa.eu/world/agreements>, according to which the precautionary principle should be applied when there are 'reasonable grounds for concern that substances or energy introduced into the marine environment' may cause hazards.

[44] Stockholm Convention on Persistent Organic Pollutants (POPs) (2001) 40 ILM 532.

The WTO has built much of its credibility as the central arbiter and steward of the multilateral trade system on its claim of predictability. This has led to a quite specific understanding of the notion of risk in the trade community. The weight given to the concepts of 'scientifically identified risk' or 'identifiable risk' by the Panel in the *Hormones* case is illustrative of such a tendency.[45] This being said, one must note that the Appellate Body has contextualized and elaborated on the term 'scientifically identified risk' used by the above-mentioned panel:

It is not clear in what sense the Panel uses the term 'scientifically identified risk'. The Panel also frequently uses the term 'identifiable risk', and does not define this term either. The Panel might arguably have used the terms 'scientifically identified risk' and 'identifiable risk' simply to refer to an ascertainable risk: if a risk is not ascertainable, how does a Member ever know or demonstrate that it exists? In one part of its Reports, the Panel opposes a requirement of an 'identifiable risk' to the uncertainty that theoretically always remains since science can *never* provide *absolute* certainty that a given substance will not *ever* have adverse health effects. We agree with the Panel that this theoretical uncertainty is not the kind of risk which, under Article 5.1, is to be assessed. In another part of its Reports, however, the Panel appeared to be using the term 'scientifically identified risk' to prescribe implicitly that a certain *magnitude* or threshold level of risk be demonstrated in a risk assessment if an SPS measure based thereon is to be regarded as consistent with Article 5.1. To the extent that the Panel purported to require a risk assessment to establish a minimum magnitude of risk, we must note that imposition of such a quantitative requirement finds no basis in the SPS Agreement. A panel is authorized only to determine whether a given SPS measure is 'based on' a risk assessment. As will be elaborated below, this means that a panel has to determine whether an SPS measure is sufficiently supported or reasonably warranted by the risk assessment.[46]

The WTO Appellate Body has abstained so far from pronouncing itself on the legal status of the precautionary principle and has refused, as a result, to recognize its prevalence over rights and obligations contained in the WTO Agreements—and particularly the Agreement on Sanitary and Phytosanitary Measures. Nevertheless, it has considered it as a principle contained in the *corpus juris* in force within the WTO, for example in Article 5.7 of the SPS Agreement.[47] As such, precaution appears as a right under the WTO SPS Agreement. Indeed, the Panel in the *EC – Biotech* case recognized that:

[45] *EC – Measures Concerning Meat and Meat Products*, Report of the Panel, 18 August 1997, doc *WT/DS26/R/USA*, para 8.257.

[46] *EC – Measures Concerning Meat and Meat Products*, Report of the Appellate Body, 16 January 1998, doc WT/DS26/AB/R, WT/DS48/AB/R, para 186.

[47] Article 5.7 reads as follows: 'In cases where relevant scientific evidence is insufficient, a Member may provisionally adopt sanitary or phytosanitary measures on the basis of available pertinent information, including that from the relevant international organizations as well as from sanitary or phytosanitary measures applied by other Members. In such circumstances, Members shall seek to obtain the additional information necessary for a more objective assessment of risk and review the sanitary or phytosanitary measure accordingly within a reasonable period of time.' On this article, see Mbengue, *supra* note 15.

Article 5.7 should be characterized as a *right* and not an exception from a general obligation. The view that Article 5.7 is not an exception in the nature of an affirmative defence is also consistent with the statement by the Appellate Body in *Japan – Agricultural Products II* that 'Article 5.7 operates as a *qualified* exemption from the obligation under Article 2.2 not to maintain SPS measures without sufficient scientific evidence.' Had the Appellate Body been of the view that Article 5.7 operates as an exception under which an importing Member could justify an inconsistency with an applicable obligation, it would, in our view, have been more natural and appropriate to use the term 'exception' rather than the term 'exemption', as the term 'exemption' connotes freedom from, and hence inapplicability of, an obligation.[48]

Other elements militate in favour of an application of precaution in the treatment of risk by the WTO.[49] The Panel in the *Asbestos* case admitted that it is not possible to require a level of absolute certainty from a Member who wishes to invoke Article XX of the GATT:

to make the adoption of health measures concerning a definite risk depend upon establishing with certainty a risk . . . would have the effect of preventing any possibility of legislating in the field of public health.[50]

The interpretation of the scope of risk assessment by the WTO dispute settlement bodies has also given space to the acceptance of a precautionary treatment of risk in the WTO framework. The case *Australia – Measures Affecting Importation of Salmon* provided an opportunity for the Appellate Body to explain that 'the "risk" evaluated in a risk assessment must be an ascertainable risk . . . This does not mean, however, that a Member cannot determine its own appropriate level of protection to be "zero risk".'[51] Finally, the evaluation of risk on which a measure is based can include non-quantifiable data of a factual or qualitative nature and is not exclusively limited to purely quantitative scientific data. This interpretation was given in the *Hormones* case by the Appellate Body of the WTO which rejected the initial interpretation of the Panel, according to which the evaluation of risk would have to be quantitative and establish a minimum level of risk.[52] These elements relating to the precautionary approach

48 *European Communities – Measures Affecting the Approval and Marketing of Biotech Products*, *supra* note 29, paras 7.2969 and 7.2973.

49 See Marceau, 'Le principe de précaution dans la jurisprudence de l'OMC – Leçon inaugurale', University of Geneva, Faculty of Law, (2005) 2(3) EcoLomic Policy and Law 1, available at: <http://www.ecolomics-international.org/ecolomic_policy_and_law.htm> (last accessed 11 December 2007).

50 *European Communities – Measures Affecting Asbestos and Asbestos-containing Products*, Report of the Panel, 18 September 2000, WT/DS135/R, para 8.221.

51 *Australia – Measures Affecting the Importation of Salmon*, Report of the Appellate Body, 20 October 1998, WT/DS18/AB/R, para 126.

52 *EC – Measures Concerning Meat and Meat Products*, Report of the Appellate Body, 16 January 1998, doc WT/DS26/AB/R, WT/DS48/AB/R, para 186.

show that, despite the importance of predictability in the multilateral trading system, there is room for precautionary concerns.[53]

This recognition of precautionary concerns, however, has so far had a limited impact—as can be seen from the evolution of WTO case law since the *Hormones* case. Indeed, the criterion of scientific uncertainty which is at the core of precaution does not *per se* trigger the application of SPS Article 5.7. This has been shown particularly in the *Japan – Apples* case in which the Appellate Body ruled that:

> The application of Article 5.7 is triggered not by the existence of *scientific uncertainty*, but rather by the *insufficiency of scientific evidence*. The text of Article 5.7 is clear: it refers to 'cases where relevant scientific evidence is insufficient', not to 'scientific uncertainty'. The two concepts are not interchangeable. Therefore, we are unable to endorse Japan's approach of interpreting Article 5.7 through the prism of 'scientific uncertainty'.[54] (emphasis added)

4. *The 'Different Capacities' Criterion*

The risk, damage, and scientific uncertainty criteria would *a priori* justify the application of the precautionary principle. But a certain number of international instruments stipulate that measures of precaution apply *a posteriori* according to the concerned states' capacities to deal with the problem. Acknowledging capacity permits one to relate the precautionary principle to a proportional approach in light of a state's economic, social, and technological means. Precautionary measures do not aim to paralyze human activity. A measure of rationality and reason must guide the application of precaution. It is obvious that states at different stages of development cannot be submitted to the same requirements. *A fortiori*, precautionary measures can therefore be expected to vary from one state to another.

2. The Precautionary Principle and Public Participation

The precautionary principle does not stand alone, but is interwoven with other international law norms and principles. This is the case in regard to the principle of public participation. The link between public participation and the

[53] See Mbengue and Thomas, 'The Precautionary Principle: Torn between Biodiversity, Environment-related Food Safety and the WTO' (2005) 5(1/2) Intl J of Global Env Issues 36.

[54] *Japan – Measures Affecting the Importation of Apples*, Report of the Appellate Body, 26 November 2003, WT/DS245/AB/R, para 184. However, some commentators consider that, while uncertainty does not in itself trigger Art 5.7, it constitutes an essential element in determining whether or not the quality of the scientific evidence is sufficient for an adequate assessment of risks under the SPS Agreement. See eg Center for International Environmental Law *et al*, *Amicus Brief to the EC – Biotech Case*, see *supra* note 29, June 2004. For the text of Art 5.7 of the SPS Agreement see *supra* note 47.

precautionary principle is justified by the fact that the determination of a tolerable risk level generally requires the involvement of the public in one way or another.[55] The principle of public participation is stated in Principle 10 of the Rio Declaration on Environment and Development:

> Environmental issues are best handled with participation of all concerned citizens, at the relevant level. At the national level, each individual shall have appropriate access to information concerning the environment that is held by public authorities, including information on hazardous materials and activities in their communities, and the opportunity to participate in decision-making processes. States shall facilitate and encourage public awareness and participation by making information widely available. Effective access to judicial and administrative proceedings, including redress and remedy, shall be provided.[56]

This principle benefits from the dynamism implicit in the precautionary principle. The management of the uncertainty linked to human activity must not become solely the prerogative of public decision-makers. The precautionary principle upsets traditional decision-making processes by requiring increased transparency. Precaution considered as the symbiosis of technical, scientific, social, economic, cultural, political, and legal norms involves a plurality of agents. To do so, the implementation of the precautionary principle must give rise to an effective and efficient application of the principle of public participation. The state cannot be the only agent responsible for the evaluation of whether precaution should be applied in a particular situation. Scientists, companies, NGOs, local populations, and other concerned agents must participate in the decision-making process. In addition, adequate information as a corollary to participation guarantees a transparent decision-making process. As said by Marie-Angèle Hermitte, the precautionary principle is the bearer of 'a subversive capacity for the whole of the legal order and leads the way to democratic renewal'.[57]

A. Access to Information, Public Participation, and the Notion of Risk in the Context of the Cartagena Protocol on Biosafety and of the *Codex Alimentarius*

As emphasized earlier, the precautionary principle has an impact on the way public decisions are made. Another point to be highlighted is that the precautionary principle does not dictate a particular outcome and requires an ongoing debate between science and politics. The notion of risk analysis encapsulates the

[55] Boutillon, 'The Precautionary Principle: Development of an International Standard', (2002) 23 Michigan J of Intl L 429, at 440.
[56] Principle 10 of the Rio Declaration on Environment and Development, *supra* note 7.
[57] Cited by Laronche, 'Principe de précaution et renouveau démocratique', *Le Monde* (13 March 2001).

evolutive interaction between science and politics. There is a requirement of a constant re-evaluation of the risk, which means that precautionary measures should constantly be refined.

The outcome of a risk analysis should be the choice of an acceptable level of risk for a specific community. This choice of an acceptable level of risk requires a social debate. Such an understanding of the values underpinning a risk analysis process is reflected *en filigrane* in the dictum of the Appellate Body in the *Hormones* case:

> 'a *scientific* process aimed at establishing the *scientific* basis for the sanitary measure a Member intends to take.' To the extent that the Panel intended to refer to a process characterized by systematic, disciplined and objective enquiry and analysis, that is, a mode of studying and sorting out facts and opinions, the Panel's statement is unexceptionable. However, to the extent that the Panel purports to exclude from the scope of a risk assessment in the sense of Article 5.1 [of the SPS Agreement], all matters not susceptible of quantitative analysis by the empirical or experimental laboratory methods commonly associated with the physical sciences, we believe that the Panel is in error. Some of the kinds of factors listed in Article 5.2 [of the SPS Agreement] such as 'relevant processes and production methods' and 'relevant inspection, sampling and testing methods' are not necessarily or wholly susceptible of investigation according to laboratory methods of, for example, biochemistry or pharmacology. Furthermore, there is nothing to indicate that the listing of factors that may be taken into account in a risk assessment of Article 5.2 was intended to be a closed list. *It is essential to bear in mind that the risk that is to be evaluated in a risk assessment under Article 5.1 is not only risk ascertainable in a science laboratory operating under strictly controlled conditions, but also risk in human societies as they actually exist, in other words, the actual potential for adverse effects on human health in the real world where people live and work and die.*[58] (emphasis added)

There are different ways of stimulating a public debate; there are no prefixed rules on this matter. However, what can be stated is that access to information and participation in the relevant decision-making processes are two important pillars for a meaningful social debate to take place and have been identified as such under international law. Their satisfaction relates to the legitimacy of a decision-making process and its sound implementation.

The precautionary principle does not prescribe a totally restrictive approach, nor does it allow for a totally permissive approach to the regulation of new technologies. It requires an approach somewhere in between: tempering the permissive approach through the taking into account of risks which are uncertain. This permits a sound decision to be made based on adequate risk assessment.

Political entities are responsible for making adequate decisions in terms of risk assessment and risk management, as was clearly recognized by the Appellate Body in the *Hormones* case:

> We do not believe that a risk assessment has to come to a monolithic conclusion that coincides with the scientific conclusion or view implicit in the SPS measure. The risk

[58] *EC – Measures Concerning Meat and Meat Products (Hormones), supra* note 25, para 187.

assessment could set out both the prevailing view representing the 'mainstream' of scientific opinion, as well as the opinions of scientists taking a divergent view. Article 5.1 does not require that the risk assessment must necessarily embody only the view of a majority of the relevant scientific community. In some cases, the very existence of divergent views presented by qualified scientists who have investigated the particular issue at hand may indicate a state of scientific uncertainty. Sometimes the divergence may indicate a roughly equal balance of scientific opinion, which may itself be a form of scientific uncertainty. In most cases, responsible and representative governments tend to base their legislative and administrative measures on 'mainstream' scientific opinion. In other cases, equally responsible and representative governments may act in good faith on the basis of what, at a given time, may be a divergent opinion coming from qualified and respected sources. By itself, this does not necessarily signal the absence of a reasonable relationship between the SPS measure and the risk assessment, especially where the risk involved is life-threatening in character and is perceived to constitute a clear and imminent threat to public health and safety. Determination of the presence or absence of that relationship can only be done on a case-to-case basis, after account is taken of all considerations rationally bearing upon the issue of potential adverse health effects.[59]

The public should be able to express its viewpoint at both the risk assessment and risk management stages. This is foreseen through the concept of risk communication, a crucial tool for providing information about risks. In this regard, it is noteworthy that the Appellate Body in the *Asbestos* case implicitly acknowledged the role of the viewpoint of the consumers:

We consider it likely that the presence of a *known* carcinogen in one of the products would have an influence on *consumers' tastes and habits* regarding that product. We believe this to be true irrespective of whether the consumer of the cement-*based* products is a commercial party, such as a construction company, or is an individual, for instance, a do-it-yourself ('DIY') enthusiast or someone who owns or lives or works in a building. This influence may well vary, but the possibility of such an influence should not be overlooked by a panel when considering the 'likeness' of products containing chrysotile asbestos- . . . We consider it likely that the presence of a *known* carcinogen in one of the products will have an influence on consumers' tastes and habits regarding that product. It may be, for instance, that, although cement-based products containing chrysotile asbestos fibres are capable of performing the same functions as other cement-based products, consumers are, to a greater or lesser extent, not willing to use products containing chrysotile asbestos fibres because of the health risks associated with them.[60] (emphasis added)

It is also important to note the way the concepts of risk assessment, risk management, and risk communication have found a place within the context of the Cartagena Protocol on Biosafety. As noted earlier, this instrument is a good example of the operationalization of the precautionary principle, specifically in the area of biotechnology. The Protocol deals with the exchanges of genetically

[59] *Ibid*, para 194.

[60] *European Communities – Measures Affecting Asbestos and Asbestos-containing Products*, Report of the Appellate Body, WT/DS135/AB/R, 12 March 2001, paras 130 and 145.

modified organisms (GMOs), considered as products that pose certain potential risks. It establishes strict procedures aimed at regulating the international trade of GMOs. Annex III of the said instrument provides for an assessment of the scientific uncertainties with respect to a GMO product.[61]

The Protocol contains provisions on access to information. Article 20 provides for an 'Advance Informed Agreement' (AIA) procedure. The most rigorous procedures are reserved for GMOs that are to be introduced intentionally into the environment. These include GMOs that are destined to grow and that have the potential to pass their modified genes on to succeeding generations. Under the AIA procedure, the exporter starts by giving the government of the importing country detailed written information, including a description of the organism, in advance of the shipment. The competent national authorities of the importing country acknowledge receipt of this information within 90 days and then explicitly authorize the shipment within 270 days or state their reasons for rejecting it.

The AIA procedure enables recipient countries to assess risks that may be associated with a GMO before agreeing to import it. The AIA procedure does not apply to GMOs in transit through a country, GMOs destined for contained use (in a scientific laboratory for example), or to GMOs to be directly used as food or animal feed or for processing. A country may, however, under its domestic regulatory framework, and consistently with the objective of the Protocol, decide to subject such GMOs to risk assessment and other requirements.

Article 21, paragraph 1 deals with confidential information and provides that 'the Party of import shall permit the notifier to identify information in the context of an AIA procedure submitted under the procedures of this Protocol or required by the Party of import as part of the advance informed agreement procedure of the Protocol that is to be treated as confidential. Justification shall be given in such cases upon request.' This means that the recipient country may, if a justification exists, keep certain information about GMOs confidential. This means *a contrario* that there is information which is not confidential.

Article 23 of the Protocol deals specifically with the issue of public awareness and participation, encouraging states to involve non-state actors in national decisions on GMOs. It calls for cooperation on promoting public awareness of safe transfer, handling, and use of GMOs. It also underlines the need for education, thus contributing to transparency and informed decision-making.

Article 26 deals with socio-economic considerations. Not only should the Parties to the Protocol consult the public, in accordance to Article 23, but they may take into account 'socio-economic considerations arising from the impact of living modified organisms on the conservation and sustainable use of biological

[61] See *supra* note 13.

diversity, especially with regard to the value of biological diversity to indigenous and local communities.' Moreover, Article 26, para 2 encourages the Parties to 'cooperate on research and information exchange on any socio-economic impacts of living modified organisms, especially on indigenous and local communities.'

Another instrument which clarifies the notion of risk analysis is the *Codex Alimentarius*.[62] The *Codex Almentarius* Procedural Manual[63] provides for an understanding of what is meant by risk analysis or risk regulation. To this end, three distinct processes are envisaged: (1) risk assessment, concerning the scientific understanding of the potential risk; (2) risk management, regarding the taking into consideration of the various alternatives for regulators and policy-makers in order to reduce or eliminate the risk; and (3) risk communication, dealing with the exchange of information between the most important stakeholders. The Manual defines each of the components of a risk analysis:

> ***Risk Assessment:*** A scientifically based process consisting of the following steps: (i) hazard identification, (ii) hazard characterization, (iii) exposure assessment, and (iv) risk characterization.
> ***Risk Management:*** The process, distinct from risk assessment, of weighing policy alternatives, in consultation with all interested parties, considering risk assessment and other factors relevant for the health protection of consumers and for the promotion of fair trade practices, and, if needed, selecting appropriate prevention and control options.
> ***Risk Communication:*** The interactive exchange of information and opinions throughout the risk analysis process concerning risk, risk-related factors and risk perceptions, among risk assessors, risk managers, consumers, industry, the academic community and other interested parties, including the explanation of risk assessment findings and the basis of risk management decisions.[64]

The notion of risk analysis includes *inter alia* uses, cost/benefit analysis, appropriateness of the measure, proportionality, preventive measures to avoid the uncertain risk, and risk communication, risk communication covering interactive exchanges among the most important stakeholders. These processes are not sequential, but rather iterative and interdependent and have been construed to require adoption in a mutually supportive way. At the level of risk management, the Procedural Manual refers to the notion of 'other legitimate factors relevant for the health protection of consumers and for the promotion of fair practices in food trade.'[65] This leaves room for the elaboration of international standards and guidelines reflecting those 'other legitimate factors'. The

[62] The text of the *Codex Alimentarius* is available at <http://www.codexalimentarius.net>.
[63] The various editions of the *Codex Alimentarius* Procedural Manual are available at <http://www.codexalimentarius.net/web/procedural_manual.jsp>.
[64] *Codex Alimentarius* Commission 17th Procedural Manual, 'Definitions for the Purposes of the Codex Alimentarius', at 42, available at: <ftp://ftp.fao.org/codex/Publications/ProcManuals/Manual_17e.pdf>.
[65] *Ibid.*

Codex Alimentarius Commission has developed 'Criteria for the Consideration of the Other Factors':

(i) when health and safety matters are concerned, the *Statements of Principle Concerning the Role of Science* and the *Statements of Principle Relating to the Role of Food Safety Risk Assessment* should be followed;
(ii) other legitimate factors relevant for health protection and fair trade practices may be identified in the risk management process, and risk managers should indicate how these factors affect the selection of risk management options and the development of standards, guidelines and related texts;
(iii) consideration of other factors should not affect the scientific basis of risk analysis; in this process, the separation between risk assessment and risk management should be respected, in order to ensure the scientific integrity of the risk assessment;
(iv) it should be recognized that some legitimate concerns of governments when establishing their national legislation are not generally applicable or relevant world-wide;
(v) only those other factors which can be accepted on a world-wide basis, or on a regional basis in the case of regional standards and related texts, should be taken into account in the framework of Codex;
(vi) the consideration of specific other factors in the development of risk management recommendations of the Codex Alimentarius Commission and its subsidiary bodies should be clearly documented, including the rationale for their integration, on a case-by-case basis;
(vii) the feasibility of risk management options due to the nature and particular constraints of the production or processing methods, transport and storage, especially in developing countries, may be considered; concerns related to economic interests and trade issues in general should be substantiated by quantifiable data;
(viii) the integration of other legitimate factors in risk management should not create unjustified barriers to trade; particular attention should be given to the impact on developing countries of the inclusion of such other factors.[66]

The concept of risk analysis, and in particular risk communication, requires further refinement. Two of its main facets are the right to access to information and the right to participate in the relevant decision-making processes as provided for by the Aarhus Convention. This Convention is analysed in the following section.

B. The Aarhus Convention and Genetically Modified Organisms (GMOs)

The 1998 Aarhus Convention on Access to Information, Public Participation in Decision-Making and Access to Justice in Environmental Matters[67] (hereinafter the 'Aarhus Convention') covers not only certain, but also uncertain risks

[66] *Ibid*, 'Appendix: General Decisions', at 159–60.
[67] The text of the Aarhus Convention is available at <http://www.unece.org/env/pp/>.

deriving from human activities and the use of technologies in the environmental area. The Aarhus Convention was adopted under the auspices of the United Nations Economic Commission for Europe, which means that its membership is significantly more limited than the previous instruments mentioned. That said, the links established between public participation and GMOs are important to assess, as they reveal some of the social contours of the precautionary principle.

1. The Pillars of the Aarhus Convention

The main goal of the Aarhus Convention, as enunciated in Article 1, is 'to contribute to the protection of the right of every person of present and future generations to live in an environment adequate to his or her health and well-being.' For this purpose, the Convention enunciates three substantive pillars: (1) the right of access to environmental information; (2) the right to public participation in environmental decision-making; and (3) the creation of administrative and judicial mechanisms to restore environmental damage. In the context of the present contribution only the first two pillars will be dealt with.

The first pillar—the right of access to information—is a fundamental element of a democratic society.[68] The right of access to information finds its expression in Article 4 of the Convention. Notably, environmental information may be requested from governmental entities without necessarily demonstrating the existence of an 'interest'.[69] What renders the Aarhus Convention particularly interesting (and acclaimed by a broad range of NGOs) is that the definitions of both 'environmental information' and 'governmental entities' in the Convention are broader than most national standards.[70] The Convention, however, provides also for a certain number of exceptions to the access to information regime. In particular, Member States have the discretionary power to refuse to disclose requested information, whenever the disclosure 'would adversely affect', in the state's view, interests covered by the exemptions. The exemptions are rather broad, including international relations, intellectual property rights, and the confidentiality of industrial and commercial interests.[71]

The second pillar of the Convention permits the public to participate in governmental decision-making processes concerning the issuance of permits for activities likely to have an impact on the environment (Article 6). The dynamic of Article 6 imposes upon the Parties the obligation to inform their citizens early in the decision-making process of any initiative falling under the list of activities contained in Annex I of the Convention. Furthermore, the public has the right,

[68] See Sand, 'Information Disclosure as an Instrument of Environmental Governance', (2003) 2 Heidelberg J of Intl L 487.

[69] Art 4, para 1 of the Convention.

[70] See Art 2, para 3 of the Convention.

[71] Art 4, para 4 of the Convention. See also Lee and Abbott, 'The Usual Suspects? Public Participation under the Aarhus Convention' (2003) 66 MLR 80, at 90.

in accordance with Article 6, para 7, to submit comments, information, analysis, or opinions considered as relevant for the concerned decision-making process.[72]

2. *The Special Status of GMOs in the Context of the Aarhus Convention*

The issue of GMOs has received specific treatment in the context of the Aarhus Convention. Negotiations were highly charged with respect to this matter. Article 6, para 11 of the Convention states that '[e]ach Party shall, within the framework of its national law, apply, to the extent feasible and appropriate, provisions of this article to decisions on whether to permit the deliberate release of genetically modified organisms into the environment.' For a number of NGOs, this provision was considered too weak due to its flexible language and they thus called for stronger mechanisms to guarantee access to information and public participation on this matter. The outcome took the form of an amendment to the Aarhus Convention which was adopted in 2005. It has not yet entered into force. It reads as follows:

> For the existing text of Article 6, paragraph 11, substitute:
> 11. Without prejudice to article 3, paragraph 5, the provisions of this article shall not apply to decisions on whether to permit the deliberate release into the environment and placing on the market of genetically modified organisms.
> After article 6, insert a new article reading:
> Article 6 bis
> Public Participation in Decisions on the Deliberate Release into the Environment and Placing on the Market of Genetically Modified Organisms
> 1. In accordance with the modalities laid down in annex I bis, each Party shall provide for early and effective information and public participation prior to making decisions on whether to permit the deliberate release into the environment and placing on the market of genetically modified organisms.
> 2. The requirements made by Parties in accordance with the provisions of paragraph 1 of this article should be complementary and mutually supportive to the provisions of their national biosafety framework, consistent with the objectives of the Cartagena Protocol on Biosafety.[73]

The contours of the access to information and public participation pillars with respect to GMOs have thus been specified and have given rise to a specific regime whereby, in comparison to other new technologies, more information and public participation are required. Given that GMOs have been dealt with through the Cartagena Protocol in a way that acknowledges scientific uncertainties and the

[72] McAllister, 'The Convention on Access to Information, Public Participation in Decision-Making, and Access to Justice in Environmental Matters' [1998] Colorado J of Intl Env Law and Policy 187, at 189 ff.

[73] The text of the amendment is available at <http://www.unece.org/env/pp/>.

precautionary principle, the question that can be posed is whether the precautionary principle calls for more stringent measures with respect to access to information and public participation. Another question, of course, is whether other new technologies deserve a similar regime.

C. Practice in Litigation Fora

It is interesting to note that inter-state arbitration and investment arbitration cases have shed light on the contours of the right to information and the right to public participation with respect to other new technologies. Although the Aarhus Convention was not directly relevant in these cases,[74] some of their features inform the concepts applied, including those of access to information and the right to public participation with respect to new technologies.

The issue of access to information by states with regard to new technologies has also become crucial in inter-state relationships. Two recent disputes between Ireland and the United Kingdom—*Access to Information under Article 9 of the OSPAR Convention*[75] and *Mox Plant*—are perfect illustrations of this trend. The MOX plant is designed for reprocessing and making mixed oxide (MOX) fuel from plutonium and uranium oxides. MOX technology, if not controlled, may give rise to risks of releases of radioactive materials, including in a liquid or gaseous form which would pose a significant threat to the environment.

In the *Mox Plant* case, Ireland argued that the United Kingdom's failure to consult and cooperate fully and effectively with Ireland constituted irreparable prejudice to Ireland's claimed right to such consultation and cooperation. Ireland alleged that the United Kingdom had failed 'to provide it, on appropriate terms, *with information relating to the operation of the MOX plant and of related facilities and of shipments to the plant, and generally to consult with Ireland on an intergovernmental basis, within the framework of Article 123 of the Convention [on the Law of the Sea], as a coriparian of the Irish Sea.*'[76] It was in this vein that the International Tribunal for the Law of the Sea (ITLOS) ordered Ireland and the United Kingdom to cooperate and to enter into consultations in order to '*exchange further information* with regard to possible consequences for the Irish Sea arising out of the commissioning of the MOX plant.'[77] (emphasis added)

[74] In the *Access to Information under Article 9 of the OSPAR Convention* case, Ireland relied on the Aarhus Convention but the Tribunal refused to apply it. See Permanent Court of Arbitration, *Dispute concerning Access to Information under Article 9 of the OSPAR Convention*, Final Award, 2 July 2003, para 104.

[75] *Ibid.*

[76] Arbitral Tribunal Constituted Pursuant to Art 287, and Art 1 of Annex VII, of the United Nations Convention on the Law of the Sea for the Dispute Concerning the MOX Plant, *International Movements of Radioactive Materials, and the Protection of the Marine Environment of the Irish Sea, The MOX Plant Case*, Order No 3, para 45.

[77] International Tribunal for the Law of the Sea, *The Mox Plant Case, Request for Provisional Measures*, Provisional Measure no 1, 3 December 2001.

The importance of public participation in the management of environmental and sanitary risks deriving from technologies was also at stake in the Methanex case.[78] This case was an investment dispute between Canada-based Methanex Corporation and the United States, based on the provisions in the North American Free Trade Agreement's (NAFTA) Chapter 11 on investment. Methanex is a major producer of methanol, a key component in MTBE (methyl tertiary butyl ether), which is used to increase oxygen content and act as an octane enhancer in unleaded gasoline. The company launched an international arbitration against the United States in response to a March 1999 order by the State of California to ban the use of MTBE by the end of 2002. California argued that banning MTBE was necessary because the additive was contaminating drinking water supplies, and was therefore posing a significant risk to human health and safety and the environment. Methanex argued in its original submission that the ineffective regulation and non-enforcement of domestic environmental laws, including the US Clean Water Act, were responsible for the presence of MTBE in California water supplies. The company argued that the planned ban was tantamount to an expropriation of the company's investment.

On 9 August 2005, the Methanex Tribunal established under UNCITRAL Arbitration Rules issued an award dismissing all claims against the United States. One reason for the Tribunal's dismissal of Methanex's claims was based on the scientific evidence produced by a Report of the University of California (hereinafter the 'UC Report'). In its Award, the Tribunal recalled that public hearings on the UC Report were held. At these hearings, the authors of the UC Report presented their findings, and government officials and members of the public (including MTBE and methanol producers) had an opportunity to ask questions and present oral testimony. The Tribunal pointed out that those testifying, amongst others, included persons affected by MTBE water contamination, as well as individuals associated with the chemical and oil industries, and considered that the testimony received at these public hearings indicated broad-based support for the finding by the University of California that MTBE usage in gasoline constituted a serious threat to California's drinking water, and that a ban on the use of MTBE in California RFG was warranted. Of the 109 persons who testified, 69 were in favour of banning, 23 were against, and 17 testified on other issues.[79]

Furthermore, the Tribunal noted that the 1999 Executive Order which prohibited the use of MTBE was stated as follows: '[t]he findings and recommendations of the U.C. report, *public testimony* and regulatory agencies are that, while MTBE has provided California with clean air benefits, because of leaking

[78] In the Matter of an International Arbitration under Chapter 11 of the NAFTA and the UNCITRAL Arbitration Rules between Methanex Corporation (Claimant/Investor) and United States of America (Respondent/NAFTA Party), Final Award of the Tribunal on Jurisdiction and Merits, 3 August 2005, available at <http://www.state.gov/documents/organization/51052.pdf> (last accessed 11 December 2007).

[79] *Ibid*, paras 17–18.

underground fuel storage tanks MTBE poses an environmental threat to groundwater and drinking water.'[80] In light of these elements, the Tribunal stated:

Having considered all the expert evidence adduced in these proceedings by both Disputing Parties, the Tribunal accepts the UC Report as reflecting a serious, objective and scientific approach to a complex problem in California. Whilst it is possible for other scientists and researchers to disagree in good faith with certain of its methodologies, analyses and conclusions, the fact of such disagreement, even if correct, does not warrant this Tribunal in treating the UC Report as part of a political sham by California. In particular, the UC Report was subjected at the time to *public hearings, testimony* and peer-review; and its emergence as a serious scientific work *from such an open and informed debate is the best evidence that it was not the product of a political sham engineered by California*, leading subsequently to the two measures impugned by Methanex in these arbitration proceedings . . . It is convenient here to summarise the principal findings of fact which the Tribunal has made in regard to the scientific issues relating to MTBE: (1) The California ban on the oxygenate MTBE began as a policy decision of the California Senate which, as expressed in the California Bill, was contingent on the scientific findings of the UC Report and which was to be implemented by California in the light of its public hearings, testimony and peer review.[81] (emphasis added)

The Tribunal added:

Legislation in democratic systems involves, by its nature, participation by a wide spectrum of private individuals and interest groups in addition to the members of the legislature and the executive, insofar as its endorsement is also necessary for a bill to become law. While there may be circumstances in which facts would support an inference that one 'invisible hand' was lurking behind and controlling a seemingly democratic process which had been elaborately contrived to conceal its machinations, it is clear beyond peradventure that the facts in the record do not warrant such an inference here.[82]

Finally, it should be noted that the development of the right of access to information and the right to public participation has given rise to the emergence of new social and legal concepts in the field of the management of risks. One of these concepts is 'community pressure' which was clearly recognized in the *Tecmed* case, by an ICSID Tribunal established under the aegis of the Convention on the Settlement of Investment Disputes between States and Nationals of other States:[83]

The authorities of the Municipality of Hermosillo were the direct target of '*community pressure*'. The Municipality was one of INE's interlocutors at the time of consideration of the Permit's renewal. In view of the pressure that questioned the Municipality's grant of the permit to use the land where the Landfill was operated, the Municipality rendered an opinion on March 31, 1998, which explained that at the time of granting such permit the current legal provisions were not applicable and that those provisions came into force subsequently, establishing a minimum distance between landfills and urban centers which

[80] *Ibid*, para 18. [81] *Ibid*, paras 101–2. [82] *Ibid*, para 46.
[83] 575 UNTS 159, adopted 18 March 1965, entered into force 14 October 1966 (also known as the 'ICSID Convention' or the 'Washington Convention').

the Landfill did not comply with. However, *the Municipality expressed its agreement with the community about the need to relocate Cytrar's hazardous waste landfill operation to a different site and its support to conduct an audit of operations to determine whether the Landfill's operation entailed any risks.* That same day, the Health Commission of the Municipality rendered an opinion confirming that, although Cytrar's operation at the Las Víboras site met *the legal requirements for functioning and there were no 'legal, ethical or logical arguments' to seek the closing of the Landfill,* all necessary efforts should be made to relocate Cytrar's operations.[84] (emphasis added)

D. The Issue of Democratic Governance of a Society Facing New Technologies and their Use

Democracy is essential for allowing free scientific and technological progress. However, if it is true that science and democracy often complement each other, it is also true that science can escape the traditional checks and balances of democratic governance, by virtue of both its complexity and the lack of accountability mechanisms over the scientific community.

The restraints imposed by democratic governments on any form of power capable of challenging the state's institutions are of vital importance in order to limit the dangers of uncontrolled technological growth. For the first time in its history, humanity possesses today scientific and technological means powerful enough to lead to its destruction. This is true not only by reference to nuclear weapons and other new military technologies, but also in the field of environmental pollution, deforestation, and desertification, where new means of production increase the degree of exploitation of the world environment. This is why it has been said that 'we must decide first what society wants and needs—then tailor our technology to help where it can',[85] in order to avoid the risk of being ruled by technology.[86]

The numerous uncertainties surrounding virtually all new technological activities constitute a threat to modern societies that can be faced through the recourse to a series of measures such as the accountability of science to the public and the obligation to reveal relevant information.[87] This also raises the issue of which type

[84] *Técnicas Medioambientales Tecmed, SA v United Mexican States* (Case No ARB(AF)/00/2), Award of May 29, 2003, para 109.

[85] Daddario, 'Technology Assessment: A Legislative View', (1967–68) 36 George Washington L Rev 1033, cited by Weeramantry, *supra* note 1, at 27.

[86] See further the chapter by Roger Brownsword in this volume.

[87] This requirement has to be further specified in the human rights field. In the case *Guerra v Italy* (1998), adjudicated by the European Court of Human Rights, the applicants lived approximately one kilometre from a 'high risk chemicals factory' which produced fertilizers and other chemicals, when a series of accidents occurred. The Court stated that these environmental aspects adversely affected the private and family life of the applicant, thus violating Art 8 of the European Convention on Human Rights. The Court however did not state that there was a violation of Art 10, providing *inter alia* for a right to receive information without interference by the public authorities. European Court of Human Rights, *Guerra and others v Italy*, Judgment of 19 February 1998, Reports 1998-I, Case No 116/1996/735/932. The Court restated an opinion already expressed in *Lopez Ostra v Spain*, Judgment of 9 December 1994, Reports 1994, Case No 41/1993/436/51520.

of democratic principles should be promoted, ie representative democracy and/or participatory democracy controlled by the people's representatives. The challenge is not to be underestimated: new technologies are often extremely complex, and this can constitute a serious obstacle if they have to be placed within the reach of legislators only.[88] Experts are gaining power but should not be left as the only masters of the rules of the game. This relates to the balance to be achieved between risk assessment and risk management, as explained by the Court of First Instance of the European Community in the above-mentioned *Pfizer* case:

A full risk assessment may require long and detailed scientific research. The case-law... shows that unless the precautionary principle is to be rendered nugatory, the fact that it is impossible to carry out a full scientific risk assessment does not prevent the competent public authority from taking preventive measures, at very short notice if necessary, when such measures appear essential given the level of risk to human health which the authority has deemed unacceptable for society. In such a situation, the competent public authority must therefore weigh up its obligations and decide either to wait until the results of more detailed scientific research become available or to act on the basis of the scientific information available. Where measures for the protection of human health are concerned, the outcome of that balancing exercise will depend, account being taken of the particular circumstances of each individual case, on the level of risk which the authority deems unacceptable for society. So, where experts carry out a scientific risk assessment, the competent public authority must be given sufficiently reliable and cogent information to allow it to understand the ramifications of the scientific question raised and decide upon a policy in full knowledge of the facts. Consequently, if it is not to adopt arbitrary measures, which cannot in any circumstances be rendered legitimate by the precautionary principle, the competent public authority must ensure that any measures that it takes, even preventive measures, are based on as thorough a scientific risk assessment as possible, account being taken of the particular circumstances of the case at issue. Notwithstanding the existing scientific uncertainty, the scientific risk assessment must enable the competent public authority to ascertain, on the basis of the best available scientific data and the most recent results of international research, whether matters have gone beyond the level of risk that it deems acceptable for society. That is the basis on which the authority must decide whether preventive measures are called for. Furthermore, a scientific risk assessment must also enable the competent authority to decide, in relation to risk management, which measures appear to it to be appropriate and necessary to prevent the risk from materialising.[89]

3. Concluding Remarks: Understanding Precaution through its Societal Dimension

'To be or not to be.' As is the case for any new regulative legal instrument, precaution is confronted with this existential query. In terms of positive law, courts and

[88] Weeramantry, *supra* note 1, at 26–32.

[89] *Pfizer Animal Health v Council of the Europe Union, supra* note 30, paras 160–3.

tribunals have thus far been reluctant to raise precaution in explicit terms to the rank of a principle of customary international law. As an example, the WTO Panel in the *EC – Biotech* case stated:

It appears to us from the Parties' arguments and other available materials that the legal debate over whether the precautionary principle constitutes a recognized principle of general or customary international law is still ongoing. Notably, there has, to date, been no authoritative decision by an international court or tribunal which recognizes the precautionary principle as a principle of general or customary international law. It is correct that provisions explicitly or implicitly applying the precautionary principle have been incorporated into numerous international conventions and declarations, although, for the most part, they are environmental conventions and declarations. Also, the principle has been referred to and applied by States at the domestic level, again mostly in domestic environmental law. On the other hand, there remain questions regarding the precise definition and content of the precautionary principle... Since the legal status of the precautionary principle remains unsettled, like the Appellate Body before us, we consider that prudence suggests that we not attempt to resolve this complex issue, particularly if it is not necessary to do so. Our analysis below makes clear that for the purposes of disposing of the legal claims before us, we need not take a position on whether or not the precautionary principle is a recognized principle of general or customary international law. Therefore, we refrain from expressing a view on this issue.[90]

Another test for the precautionary principle's viability in the international legal order is related to its intrinsic and extrinsic aptitude to play its role. Intrinsically, it must play its role as an original and singular legal technique if it is not to be overtaken by other legal techniques. Extrinsically, the precautionary principle must be able to live up to its societal dimension. If not, it risks being stripped of all normative legitimacy.

The precautionary principle triggers new reflection on the 'social contract'. It proves that all attempts at social constructivism (that is, management of society according to a final and preordained plan) are destined to failure, or at least to serious questioning. Precaution brings with it a new trend: that of complexity, and therefore the questioning of all absolute assumptions which have long been the foundation of modern society. It takes place within a post-modern context, and is the bearer of several disturbances in the legal order.

The first is the remodelling of the interaction between law and science. The relativity of Cartesian dogma due to the appearance of new problems and challenges for the whole of humanity leads to a new mode of handling scientific expertise by law. Indirectly, it also leads to a rebalancing of the interaction between legal politics and science. Precaution speaks of a bond profoundly upset: science would be sought for the suspicions and doubts it raises rather than for the knowledge it offers.[91] This is not a dilution of science, but a repositioning.

[90] *EC – Biotech* case, *supra* note 29, paras 7.88–7.89.

[91] N. de Sadeleer, *Les principes du pollueur-payeur*, *supra* note 23, at 196.

Indeed, the precautionary principle will create another mode of interaction between normative processes and scientific expertise. The latter favours a new approach to law tending towards normative processes which are consolidated and renewed by the results obtained through scientific expertise held on an ongoing basis. This expertise calls for a constant adaptation of decision-making processes. This can be done through the adoption of protocols and amendments to existing instruments.[92]

The second is the remodelling of the law/economics interaction, linked to the cost which precaution could entail for a particular society. Precautionary measures can initiate a social psychosis which would kill initiative and innovation. Societies cannot allow such vagaries. To re-equilibrate the interaction between law and economy, the precautionary principle must be considered as a substantive element of 'sustainable development'. In this perspective its objective is not to halt economic activity, but to emphasize the necessity of integrating today's requirements in terms of environmental protection. As an element of sustainable development, the precautionary principle must strive for a 'durable better-being of humanity'[93] in all its interpretations, not just its economic interpretation. Thereby, the precautionary principle can be the guarantor of activities which are 'economically viable, socially equitable and ecologically sustainable.'[94]

The precautionary principle also requires that one rethink the interaction between law and effectiveness. In other words, one must redefine effective mechanisms to implement and underwrite this new body of law, whose essence is 'soft'. The aim is to avoid veering into the arbitrary or excessive use of discretionary power in the application of the principle. Courts and tribunals are therefore asked to play a fundamental role in determining or instituting the contours of the precautionary principle. The objectivization of the elements of this principle must depend on a body which will arbitrate between the contradictory interests of society's different agents. Through this objectivization, the precautionary principle will permanently deserve its right of citizenship in international law.

Finally, the precautionary principle is an example of a particular interaction between national and international law. There is a 'contamination' of the

[92] *Ibid*, at 198. N. de Sadeleer says: 'At first glance the precautionary principle seems to occupy a paradoxical position at the interface between science and normative decision-making. On one hand, it would reaffirm the primacy of political decision-making in determining the contents and timing of preventive measures, thereby limiting the role of scientists. On the other hand, although arising from a lack of scientific information, precaution calls for ever increasing scientific knowledge, thus serving to reinforce the power of experts and consequently the dependence of decision-makers on science. Scientific expertise, initially rejected as insufficient, would thereafter be sought to balance the scope of anticipatory measures.' (*Environmental Principles*, *supra* note 23, at 200–1).

[93] Concept developed by the World Commission on Dams in *Dams and Development: A New Framework for Decision-Making*, Report of the World Commission on Dams (2000) 27.

[94] *Ibid*, at 27.

domestic sphere by the international sphere expressed by a 'phenomenon of favorable emulation between international and national dimensions.'[95] The crises of contaminated blood and mad cow disease which erupted in European countries proved the precautionary principle's applicability in the international sphere, and required greater cooperation between states. But they also demonstrated its applicability in the domestic sphere,[96] and confronted decision-makers with public opinion, which is increasingly exacting as to the measures that should be taken to manage the threats and risks of today.

[95] Boisson de Chazournes, 'La mise en œuvre du développement durable', in S. Maljean-Dubois and R. Mehdi (eds), *Les Nations Unies et la protection de l'environnement: La promotion d'un développement durable* (1999) 65.

[96] See eg 'Conseil d'Etat, 28 juillet 2000, Association Force Ouvrière Consommateurs et autres, no 212115', in (2001–02) 86 Droit de l'environnement 46. See also Tribunal de Grande Instance d'Orléans, *Jugement correctionnel sur les faucheurs d'OGM*, 9 décembre 2005, in Ecole Nationale de la Magistrature, *Dossier documentaire «Justice, environnement, développement durable, risques industriels»* (2006) 55; Supreme Court of India, *Godavarman Thirumalpad v Union of India*, 2002 (10 SCC 606)—the European Communities referred to this judgment of the Supreme Court of India in *European Communities – Measures Affecting the Approval and Marketing of Biotech Products*, *supra* note 29, para 7.79: 'the European Communities noted that the precautionary principle is one of the "salutary principles which govern the law of the environment" in India and has been applied by the Indian Supreme Court.'

6

The Texture of Reproductive Choice:

Law, Ethnography, and Reproductive Technologies

Thérèse Murphy

I have two aims in this essay. The first is to explain why, as regards reproductive technologies, reproductive choice seems more and more to end up as a 'bad argument'; the second is to make clear why I believe it would be a serious blow if this framing of reproductive choice led to a diminution in the 'speakability' of choice. I am well aware that in certain quarters my latter aim means that, straight off, this essay will be classified as 'pro-choice'. And that this, in turn, could mean that my core argument will remain unread, prejudged as stale or overworked and predictably partisan. As I shall try to make clear, however, the description 'pro-choice' is only accurate as regards this essay if that term can be understood in a somewhat unconventional way. That is, what follows is 'pro-choice' if it is possible for the term to have a meaning which escapes the 'for or against' framing that is all too familiar from abortion politics in countries like the US. Increasingly, of course, 'choice' also needs to avoid being damned by association with bloated, out-of-control consumerism. At this juncture, however, that general trend feels less of an impediment than the grip of the oppositional frame on reproductive choice.[1]

Part 1 of the essay will tackle the first aim. Specifically, it will survey a number of reasons why reproductive choice seems like a flawed argument when used in the context of assisted reproductive technologies (ARTs). Amongst other things, it will emphasize the problems created by the overlap between ARTs and abortion, and by the burgeoning linkage of genetic and reproductive technologies. In Part 1, I shall also broach the question of 'reproductive choice for whom?', focusing in particular on the issue of gender neutrality—that is, whether

[1] Indeed, it might be useful at least to consider whether a market version of consumer choice could facilitate increased public funding for ARTs. Specifically, what merit, if any, would there be in *requiring* individuals and couples to choose which embryo to implant, and then using the money saved as a result of the reduction in multiple pregnancies and births and their associated costs (especially as regards neo-natal care) to increase public funding for ARTs? See, relatedly, Franklin, *infra* note 8.

and to what extent there ought to be equal reproductive choice for women and men.

In Part 2, I shall try to correct for the harm that could be caused by the prevalence of particular, negative assumptions about reproductive choice. I shall explain why the 'speakability' of choice in the field of ARTs needs to be made secure; why, in other words, choice must not become 'unspeakable' in this arena, hounded out as an unfit or improper frame for reasoning and explanation. Part 2 will open with a summary of two recent ethnographic studies of ART users, one from the UK, the other from the US.[2] These studies show prospective or 'would-be' parents and ART professionals adopting a language of choice. Their particular importance, however, is that they show choice being articulated in a layered or intricate manner. In other words, they show ART professionals and would-be parents finding a depth and scope within choice to express the (extreme) complexity of their circumstances. To my mind, their adoption of choice, and specifically their manner of reasoning therewith, demonstrate how important it is that human rights lawyers engage with accounts of reproductive choice 'in action' in ARTs. To be more precise, they demonstrate the potential value or importance of *additional* accounts of reproductive choice 'in action', accounts that will augment the extant 'lawyerly' ones which examine reproductive choice in particular, discipline-specific ways in a particular set of official regulatory sites, notably courtrooms and legislatures. Ethnographic work, I shall argue, can take human rights thinking to new sites. And, in so doing, it might bring an all-important texture to human rights arguments for reproductive choice, and in that way help us to get out of both defensive and head-in-sand ways of thinking and explaining reproductive choice. I shall also argue that engagement with ethnographic work is likely to have secondary, more general benefits: namely, it will remind us not to neglect the experience of rights users (and non-users) and, more generally, it could prove to be a useful spur for sluggish imaginations.[3]

1. A Bad Argument?

I began by saying that, as regards ARTs, reproductive choice can feel like a bad argument. That, of course, raises the question: what is a 'bad' argument? One contender for this description would be an argument that tends to get choked by controversy: in other words, an argument that might not even leave the starting

[2] Respectively, S. Franklin and C. Roberts, *Born and Made: An Ethnography of Preimplantation Genetic Diagnosis* (2006); C. Thompson, *Making Parents: The Ontological Choreography of Reproductive Technologies* (2005).

[3] On the latter point, see H. Garfinkel, *Studies in Ethnomethodology* (1967), at 11. On lawyers' tendency to instrumentalize, that is, to view concepts, etc as 'means to ends', see Riles, 'Anthropology, Human Rights, and Legal Knowledge: Culture in the Iron Cage' (2006) 108 American Anthropologist 52.

blocks. Another contender would be an argument that does not seem to fit 'the facts'. In what follows, I test reproductive choice against both of these standards.[4]

A. Choked by Controversy

1. Reproductive Choice is the 'Abortion Argument'

To say that reproductive choice is the 'abortion argument' is to risk immediate counter-argument; moreover, it is easy to see why this is the case. For starters, claims grounded in reproductive choice have certainly not been limited to the abortion setting. They have, for instance, also been central to arguments for access to contraception, and they have been particularly strong and convincing in that context. In addition, a range of scholars, including Ronald Dworkin, John Harris, Emily Jackson, and John Robertson, have made appeals to reproductive or procreative autonomy in arguing for a wide-ranging right to self-expression in reproductive matters, including the use of ARTs.[5]

It is therefore fair to say that describing reproductive choice as the 'abortion argument' is not particularly accurate. What I was aiming to capture, however, was the way in which appeals to reproductive choice have an almost relentless association with abortion and the problems that can cause. That said, ARTs have had a far less furious history than abortion. As a general rule, they have been protected by what Charis Thompson describes as a 'rhetorical advantage': 'they can be cast in a robustly bipartisan manner through appeal on the one hand to "family-building" and on the other to "reproductive choice".'[6] Relatedly, during their short history, ARTs do not seem to have attracted the sort of organized opposition that has been a feature of the response to certain other new technologies—most notably, genetically modified crops and food, which have been viewed with high levels of scepticism,

[4] The fact that reproductive choice is often understood narrowly in terms of rights of access to ARTs (including public funding thereof) arguably has been an additional choke on the argument.

[5] See eg R. Dworkin, *Life's Dominion: An Argument about Abortion and Euthanasia* (1983) and also his *Freedom's Law* (1996); J. Harris, 'Rights and Reproductive Choice', in J. Harris and S. Holm (eds), *The Future of Human Reproduction: Ethics, Choice and Regulation* (1998) 5; E. Jackson, *Regulating Reproduction* (2001); J. Robertson, *Children of Choice: Freedom and the New Reproductive Technologies* (1994).

[6] *Supra* note 2, at 238. But pronatalism does come in different forms, not all of which support access to reproductive technologies: for example, pronatalism is strong in the Muslim Middle East, but for Sunni Muslims third-party donation of gametes or embryos is prohibited and there is also evidence of strong social prohibitions amongst both Sunnis and Shi'ites. However, the development of intracytoplasmic sperm injection (ICSI) to remedy male infertility has led to increased use of ARTs: see Inhorn, 'Gender, Health, and Globalization in the Middle East: Male Infertility, ICSI, and Men's Resistance', in I. Kickbusch *et al* (eds), *Globalization, Women, and Health in the 21st Century* (2005) 113.

especially by many Europeans.[7] The stigma of infertility, as well as the fact that infertility is not a life-threatening illness, seem also to have helped to keep ARTs largely private and privatized.[8]

But these differences are not set in stone. In the first place, the separation of reproductive technologies from abortion has never been fully secure, not least because *in vitro* fertilization (IVF) can raise issues of abortion law. And, as I go on to argue in the next section, the burgeoning overlap between reproductive and genetic technologies has placed the human embryo at centre stage, which has revived, and intensified, earlier concerns about reproductive science racing ahead and about the dangers of interfering with nature and redesigning human reproduction and heredity.

Looking first at the overlap between ARTs and abortion law, I want to make two points. First, IVF raises questions of abortion law because the implantation of several embryos in an attempt to secure a pregnancy can lead to a high multiple pregnancy, which in turn raises the question of reducing that pregnancy by 'selective reduction'.[9] The reduction of a high multiple pregnancy may be necessary in order to protect a woman's health and perhaps her life. It can also increase the chances of a full-term pregnancy and a live birth, and it reduces the risk of spontaneous abortion.[10] It is also the case that, in some jurisdictions, the law on abortion is such that it could impact on ARTs by creating problems with respect to the existence of supernumerary or 'surplus' embryos—that is, embryos created *in vitro* which prove unnecessary or unsuitable for reproductive purposes.

The overlap between abortion law and ARTs has surfaced in the sphere of international human rights law and practice. For example, in March 2004, the Inter-American Commission on Human Rights admitted a case concerning Costa Rica, which involves a challenge to a ruling of the Constitutional Chamber of the Constitutional Supreme Court of Costa Rica prohibiting the use of IVF. The decision of the Constitutional Chamber cites a series of international human rights instruments, including the American Declaration on Rights and Duties of

[7] The US, Canada, and Argentina initiated dispute settlement procedures at the WTO against the EC for delaying approvals of GM crops within its borders: see *European Communities – Measures Affecting the Approval and Marketing of Biotech Products*, Reports of the Panel, WT/DS291/R, WT/DS292/R, WT/DS293/R, 29 September 2006. For discussion of *EC Biotech*, see the chapters by Laurence Boisson de Chazournes and Han Somsen in this volume.

[8] Of course, whether this has been a good thing is debatable. In relation to the UK, see eg the point made by Sarah Franklin during her inaugural lecture 'The Reproductive Revolution: how far have we come?' at the LSE on 24 November 2005: '[t]he sharp rise in multiple births and attendant pathology was an expensive cost to impose from the largely private Assisted Conception market of the 1980s and 1990s into the already-over-burdened National Health Service.'

[9] A high multiple pregnancy can also follow from medically induced superovulation, followed by unassisted or natural conception.

[10] R.J. Cook, B.M. Dickens, and M.F. Fathalla, *Reproductive Health and Human Rights: Integrating Medicine, Ethics, and Law* (2003), at 307 suggest that '[t]he challenge is to consider multi-foetal pregnancy reduction not as abortion per se, which ends pregnancy, but as a means of preserving pregnancy against the danger of spontaneous abortion of all embryos or foetuses in utero.'

Man, the American Convention on Human Rights, the Convention on the Rights of the Child, and the International Covenant on Civil and Political Rights, in support of its conclusion that fertilized eggs have a right to life and that IVF techniques are therefore unlawful.[11] The Court's conclusion would seem to be based on a mistaken interpretation of these instruments.[12] But, for present purposes, the key point is that the Court's focus on the right to life directs attention to a continuum between protection of the life of the foetus and that of the fertilized ovum: in this way, it links IVF to abortion law and politics. And this sort of linkage could, in turn, be bolstered by the fact that both the Vatican and organizations on the US Christian Right who are active at the United Nations appear to see 'reproductive rights'—as embodied in the Programme of Action agreed at the International Conference on Population and Development held in Cairo in 1994—as a covert reference to abortion and, more generally, an attack on motherhood and the 'natural' family.[13]

My second point about the overlap between IVF and abortion law, and the problems this may cause for 'choice' as an argument in the field of ARTs, requires us to take a look at the situation in Britain. The Abortion Act 1967 requires that a woman's reasons for wanting to terminate her pregnancy must be deemed acceptable by two medical practitioners.[14] Putting that another way, the Act is grounded in the idea that 'the legitimacy of an abortion depends upon its purpose.'[15] This, in and of itself, is unremarkable: it is a relatively widespread regulatory approach to abortion. What is interesting however is that the focus on purpose has no parallel in the wider field of English law on patient self-determination. Indeed, in recent years the English courts have developed the principle of patient self-determination to protect an individual's freedom to make choices about her medical treatment even where those choices might widely be considered irrational or repugnant. In so doing, the courts have made it clear that the principle applies with full and equal force to a competent pregnant woman who refuses medically indicated treatment (for example, a Caesarean section).[16] The upshot is that there is now an apparent anomaly in English law: the

[11] *Ana Victoria Sanchez Villalobos and others v Costa Rica*, Decision No 2000-02306.

[12] On this point, see the amicus brief submitted to the Inter-American Commission by the Center for Reproductive Rights, an international non-governmental organization: Amicus Curiae In Support of Petition 12.361, available at <http://www.reproductiverights.org/pdf/crt_amicus_IVF.pdf> (accessed 31 March 2008).

[13] See generally D. Buss and D. Herman, *Globalizing Family Values: The Christian Rights in International Politics* (2003). Paragraph 7.16 of the Cairo Programme of Action, UN Doc A/CONF.171/13/Rev.1 (1995), urges states to include services for 'prevention and appropriate treatment of infertility' as part of health services. Paragraph 7.17 makes reference to IVF, saying that IVF techniques 'should be provided in accordance with appropriate ethical guidelines and medical standards.'

[14] The 1967 Act does not apply in Northern Ireland.

[15] Jackson, 'Abortion, Autonomy and Prenatal Diagnosis' (2000) 9 SLS 467, at 468.

[16] *St. George's Healthcare NHS Trust v S* [1998] 3 WLR 936 (CA).

legitimacy of abortion depends upon its purpose, even though other choices by a competent adult patient—including the decision to reject a medically indicated Caesarean section—are protected by the law irrespective of the patient's reasons (including where she gives no reason at all).[17] The significance of this for present purposes is that, if the overlap between abortion and ARTs becomes a matter of increased focus, then the legitimacy of using ARTs could become more bound up with questions of purpose.

The overlap with abortion is, however, not the only problem for ARTs and reproductive choice. As mentioned earlier, there is also a growing problem generated by the overlap between reproductive technologies and genetic ones. This overlap has the potential to change the dominant framing of ARTs: for example, the coupling of pre-implantation genetic diagnosis (PGD) with IVF takes us away from the strongly emergent picture of IVF as 'assisting nature' towards a host of concerns about science racing ahead, 'playing God' and 'mere, sheer choice'. I explore this point below.

2. *On Demand and By Design*

The increasing linkages between genetic and reproductive technologies could be seen as taking us 'back to the future'. That is to say, burgeoning references to designer babies and parental eugenics could be said to resonate with the early labelling of ARTs as *artificial* reproduction technologies. References to artificiality were later nudged aside by 'assistance', but emergent concerns about design and demand may present a deep challenge to that naturalizing move. These concerns reorient us away from 'miracle babies' and from the more recent, more low-key depiction of ARTs as a 'helping hand' to nature; the emphasis, instead, is on the question of the boundaries to assisting nature.[18] Put briefly, we seem, as Roger Brownsword has pointed out, increasingly to be preoccupied by questions of 'rogues and red lights'.[19]

Of course, it's not really 'back to the future'; there are new elements as well, including the fact that, since the birth in 1978 of Louise Brown, the world's first 'miracle baby', millions of IVF children have been born. In the UK alone, one child in every 80 is born as a result of an assisted conception treatment.[20] (Indeed, the use of ARTs by married heterosexual couples, where 'cure' of

[17] See Jackson, *supra* note 15, at 488 noting that because abortion is 'an exclusively female medical procedure, it might be worth drawing attention to the conspicuous absence of a robust concept of autonomy' in this setting.

[18] But see *infra* at 205 for discussion of ICSI, a technological development that could be seen as further 'naturalizing' IVF by limiting the need for third-party sperm.

[19] Brownsword, 'Red Lights and Rogues: Regulating Human Genetics', in H. Somsen (ed), *The Regulatory Challenge of Biotechnology: Human Genetics, Food, and Patents* (2007) 39.

[20] See the Human Fertilisation and Embryology Authority, *HFEA Guide 2007/8*, available at <http://www.hfea.gov.uk> (accessed 31 March 2008).

naturally occurring infertility rather than the 'customization' of babies is at issue, is now widely considered to be a brave and commendable act.[21]) But the number of IVF children also helps to make clear why the human embryo is generating increasing concern. A higher incidence of ART use can be used to flesh out a picture of a worldwide store of human embryos, surplus to treatment or discarded following techniques such as PGD, which will be allowed to perish or used to progress controversial types of research, including human reproductive and therapeutic cloning.[22] Clearly there are hard questions that do need to be asked in relation to what Sarah Franklin calls the 'IVF-stem cell interface': I do not in any way want to limit the space for these important questions but I am concerned about the fact that emphasis on 'design and demand' can generate potentially misleading depictions of the 'plight' of the human embryo.

A second and related new element is the rise of the 'dignitarian alliance', so-called because, first, it focuses on the need for constraints or prohibitions on biotechnology in order to safeguard human dignity and, second, it draws on a 'mixture of Kantian, Catholic and communitarian credos.'[23] Another new element is the increased concern with the migration of people and bodily material, and relatedly with markets and the race for profits and shareholder value in the biosciences. Concerns about babies on demand and by design may also be fuelled by more general concerns about the 'rise and rise' of patient autonomy. In the UK, for example, patient autonomy has been depicted as out of control and a threat to the health system, health professionals, and perhaps even patients themselves. It has also been seen as a threat to the regulatory capacity of states. Moreover, concerns such as these become more pointed in the face of the claim that patient autonomy may in fact be an 'illusory quest'. The philosopher Onora O'Neill, for example, has argued that '[r]obust conceptions of autonomy may seem a burden and even unachievable for [us]; mere choosing may be hard enough.... It is a considerable relief to [us] that [we] are not asked to muster much in the way of individual autonomy.'[24]

To sum up: in the domain of ARTs, reproductive choice can feel like a bad argument because the overlap between, on the one hand, ARTs and abortion, and, on the other, reproductive and genetic technologies, means that invocations of choice tend to lead to other debates, both old and new, which are characterized by intense and persistent controversy. The overlaps, in other words, can mean

[21] In the UK, widows are another group for whom ART use has been represented as both apposite and heroic: see *R v Human Fertilisation and Embryology Authority, ex parte Blood* [1997] 2 All ER 687.

[22] Cf the Warnock Committee's construction of the 'pre-embryo'.

[23] Brownsword, 'Stem Cells and Cloning: Where the Regulatory Consensus Fails' (2005) 39 New England Law Rev 535, at 546. See, also, the chapters by Han Somsen and Roger Brownsword in this volume.

[24] O. O'Neill, *Autonomy and Trust in Bioethics* (2003), at 38.

that one gets mired in the 'unarguable'.[25] The end result: choice gets a raw deal. But these overlaps are not the only problem facing reproductive choice: as noted earlier, there is also the difficulty that reproductive choice does not exactly fit 'the facts' of ARTs. In what follows, I explore this in more detail, looking first at the question 'reproductive choice for whom?'

B. Reproductive Choice Does Not Seem to Fit the 'Facts' of Assisted Reproductive Technologies

1. Reproductive Choice for Whom?

The question 'reproductive choice for whom?' leads in many directions. I deal explicitly with one of these below where I consider the politics, the practicality, and the appropriateness of seeing reproductive choice in gender-neutral terms. First, however, I want to take a brief look at some other filters on access to ARTs. The aim here is to show how these filters can bolster the conclusion that choice does not fit the facts of ARTs.

Child welfare is perhaps the most prominent or well-known filter on access. For example, the legislation in the UK, the Human Fertilisation and Embryology Act 1990 (HFEA 1990), declares that treatment services must not be provided unless account has been taken of the welfare of any child who may be born as a result.[26] And interestingly, in the US, although there has been no federal legislation on access to ARTs, Thompson found that 'most clinics operate with an implicit, albeit variously interpreted and in some cases residual from earlier practice, heterosexual coupled norm' and that, in discussions about access, the 'default frame of discussion' amongst clinic staff was usually 'the perceived needs of a future possible child.'[27] I say more about child welfare later in the essay, noting in particular that its status as an official filter is in flux. In what follows, I concentrate on two other filters that can affect access to ARTs: namely, clinic accountability and cost.

Clinic accountability cropped up repeatedly as a filter on access to ARTs in the US clinics in Thompson's study. The publication of success rates and the need to maintain and build clinic business:

[25] According to the UK's Nuffield Council on Bioethics, *Genetically Modified Crops: The Ethical and Social Issues* (1999), at 96, views that 'have something of an "unarguable" quality' are such that 'no amount of information, explanation or rationalisation would move a person with such views from their position.'

[26] HFEA 1990, s 13(5). At the time of writing, Parliament is debating a revised legislative scheme: see the Human Fertilisation and Embryology Bill, available at <http://services.parliament.uk/bills/2007-08/humanfertilisationandembryology.html> (accessed 16 July 2008).

[27] Thompson, *supra* note 2, at 86. But she goes on to note that by the late 1990s, there were 'fewer references to the appropriateness of couples based on their assumed fitness or the high-income homes they would provide future children and more references to the right, need, or right time to reproduce and the corresponding privacy that such a decision merited. There was a subtle if partial shift from children's rights to adults' reproductive rights and a corresponding shift of ethical frame from social interventionism to privacy' (p 89).

dictated against treating patients for whom the procedures were unlikely to succeed. And pregnancy statistics that would attract other patients to the clinic would also be considerably worse if large numbers of such patients were treated. Similarly, as cycles and not patient courses of treatment became the new standard of accountability, the more diagnostics that could be done prior to embarking on a cycle of treatment, the better.[28]

It appears therefore that the question 'reproductive choice for whom?' is now strongly inflected by scientific indicators on who would make an appropriate patient, and that these indicators tend to operate alongside longer-standing notions of appropriateness grounded in child welfare and, in places, displace them.

Cost is another key filter. Treatment is expensive (in the UK, as of 2008, the typical cost of one cycle of IVF at a private clinic is from £4,000 to £8,000) and access is often contingent upon the ability to be self-funding.[29] Where access involves travel to another jurisdiction (to avoid waiting lists, or perhaps to avoid or sidestep the legal rules governing ARTs that operate in one's own country), it may be more expensive again. This cost filter brings us face to face with a long-standing feminist concern about the language of choice: namely, being 'for choice' is no guarantee of feminist outcomes, not least because the 'formal existence of choice within liberalism is valued regardless of its actualisability.'[30] The point can be illustrated by reference to *Webster v Reproductive Health Services*,[31] wherein the US Supreme Court upheld the constitutionality of a law prohibiting the use of public facilities and publicly employed staff in abortions. The Supreme Court did not accept that its opinion would limit choice: in its view, in curtailing public funds, the state was leaving pregnant women 'with the same choices as if the state had chosen not to operate any hospitals at all.'[32]

Reproductive tourism has aggravated feminist concerns about being 'for choice'. Specifically, it has sharpened the need to think about infertility and ARTs as part of a global politics of reproduction. 'Stratified reproduction', a term coined by Faye Ginsburg and Rayna Rapp, is now widely used to capture the inequalities surrounding fertility and infertility.[33] Indeed, the much-feted Cairo Programme of Action, which was welcomed as a key step towards a more complex approach to population, has also faced the criticism that 'at the level of

[28] *Ibid*, at 90.

[29] Thompson, *ibid*, at 88, notes that in her study of US clinics '[f]or some infertility physicians... having an elite clientele was an active and valued element of their working culture, whereas for some lab technicians [at these clinics] it was a moral issue and contravened a responsibility of the possessors of knowledge to surrender that knowledge without prejudice.'

[30] Vakulenko, 'Liberalism, Civilisation and the (Non-)Oxymoronic Limits of Tolerance' (2008) 3 Intl J of Law in Context 323, at 337.

[31] 492 US 490 (1989).

[32] *Ibid*. See also Poovey, 'The Abortion Question and the Death of Man', in J. Butler and J.W. Scott (eds), *Feminists Theorize the Political* (1992) 239.

[33] F. Ginsburg and R. Rapp (eds), *Conceiving the New World Order: The Global Politics of Reproduction* (1995) 1.

strategy, the focus remains on lowering population to improve economic performance.' The argument, in other words, is that the Cairo document 'leaves intact the assumption that developing countries are poor because they have too many people, and the focus of poverty alleviation remains on population reduction.'[34]

In the next section, I examine a specific aspect of the question 'reproductive choice for whom?' Put crudely, I ask: is reproductive choice gender-specific or gender-neutral, and does it matter?

2. *Reproductive Choice for Women, and for Men?*

Reproduction is widely seen as a matter of more central concern for women than men. For the most part, the same assumption surrounds assisted reproduction, not least because ARTs tend to claim women 'as their primary patients, even when there is male-factor infertility': typically, 'men just aren't around as much because they aren't required for most of the procedures.'[35] These practices reflect and reinforce the view that men's involvement in reproduction is limited. At best, men are 'fathers-to-be', with choices and responsibilities that will kick in around birth (eg the decision on whether to be present during labour and delivery) or thereafter.[36] Law contributes to this picture: in the UK, for instance, there are rules obliging genetic fathers to pay child support, as well as rules that preclude guaranteed anonymity for gamete donors. Rules such as these have been provoking all sorts of controversy; here, I simply want to make the point that they may contribute to a sense of fatherhood as a status that is compelled or forced on men, not chosen by them.

It seems to me that assumptions about reproduction can also reflect and reinforce the view that, prior to the birth of a child, what men have is choice (or choices) rather than *reproductive* choice. In other words, our assumptions tend to reflect and reinforce the view that reproductive choice is gender-specific not gender-neutral. To put it crudely, reproductive choice is widely seen as a 'woman

[34] Buss, 'Racing Populations, Sexing Environments' (2000) 20 LS 463, at 481. See also Greenhalgh, 'Globalization and Population Governance in China', in A. Ong and S.J. Collier (eds), *Global Assemblages: Technology, Politics, and Ethics as Anthropological Problems* (2005) 354, at 355, commenting on the commonplace assumption that government coercion generated and sustained China's 1983 sterilization campaign: 'With its orientalising, Otherizing thrusts, the coercion narrative... keeps us from seeing the place of Western ideas and technologies, and of Chinese dreams of catching up with the West to become a global power, in that complex, continuously evolving assemblage around population... Could it be that the selective borrowing of Western science and technology played a role in the creation of the troubling 1983 campaign.'

[35] Thompson, *supra* note 2, at 165. But see *infra* at 205 for discussion of the myriad ways in which professionals and would-be parents work to counteract this.

[36] See further Callahan, 'Bioethics and Fatherhood' (1992) Utah L Rev 735; Sheldon, 'Reproductive Choice: Men's Freedom and Women's Responsibility?', in J.R. Spencer and A. Du Bois-Pedain (eds), *Freedom and Responsibility in Reproductive Choice* (2006) 175.

thing'—although it is important to note, as John Harris does, that this choice has tended to be 'respected more in the breach than in the observance.'[37]

There is evidence though that the overall picture is changing. For example, in her recent work on men's reproductive bodies, Sally Sheldon emphasizes that men increasingly 'find themselves called upon to make "responsible" reproductive choices, not just in avoiding unwanted pregnancy, but also in taking appropriate measures to safeguard their own reproductive health and to avoid conception of a child likely to suffer from illness or disability.'[38] There is also increasingly widespread use of the technique of intracytoplasmic sperm injection (ICSI) to tackle severe male-factor infertility, thereby rendering men's bodies more visible within assisted reproduction practices.[39] And ethnographic studies have demonstrated that, in ART clinics, it is common for female partners in heterosexual couples, as well as other women patients and staff, to work hard at finding ways to 'include' men—in particular because of the perceived need to restore and secure 'manliness'.[40]

A range of legal rules adds to the sense of a changing picture. As noted earlier, in the UK, there are rules obliging genetic fathers to pay child support, as well as rules designed to capture the identity rights, or 'right to genetic truth' of children born via reproductive technologies.[41] These rules can be said to compel fatherhood but they are also an indication of increased public interest in fathers' roles in families. The question of reproductive choice for men is a live matter in social-movement circles, especially in the US where a number of men's groups have campaigned against both sperm 'banditry' (that is, women who trick men into unprotected sexual intercourse with the intention of trying to become pregnant) and the imposition of financial support obligations on genetic fathers in circumstances where women, who have the opportunity to have a safe, legal abortion, decide to bring the pregnancy to term.[42] There is also important case

[37] See Harris, *supra* note 5, at 5. He continues: 'It is when people, *particularly women*, actually want to exercise choice that the trouble starts' (emphasis added). On the right not to procreate, see eg Sutherland, 'Is There a Right Not to Procreate?', in S.A.M. McLean (ed), *First Do No Harm* (2006) 319.

[38] Sheldon, *supra* note 36, at 195. She also speculates that 'as a rhetoric of more involved fathering gains dominance in discussions of parenting practices, a similar trend may be seen in understandings of reproductive responsibilities.'

[39] Thompson, *supra* note 2, at 124 raises the concern that enthusiasm about ICSI may be leading to overuse; in particular, its use not just for a range of male factors but also as a 'default fertilization method in IVF.'

[40] See eg Inhorn, *supra* note 6; Shaw, 'The Contingency of the "Genetic Link" in Constructions of Kinship and Inheritance: An Anthropological Perspective', in Spencer and Du Bois-Pedain (eds), *supra* note 36, 73; Thompson, *supra* note 2.

[41] For example, anonymity has been removed from gamete donors, and deceased fathers can now be registered on the birth certificates of children conceived posthumously: see, respectively, the Human Fertilisation and Embryology Authority (Disclosure of Donor Information Regulations) 2004, and the Human Fertilisation and Embryology (Deceased Fathers) Act 2003.

[42] See, respectively, Sheldon, '"Sperm Bandits", Birth Control Fraud and the Battle of the Sexes' (2001) 21 LS 460, and Sheldon, 'Unwilling Fathers and Abortion: Terminating Men's Child Support Obligations?' (2003) 66 MLR 175.

law from a range of domestic courts and from the European Court of Human Rights. One such case, *R v Human Fertilisation and Embryology Authority, ex parte Blood*, decided in the UK in 1997, does not augur all that well for men's reproductive choice.[43] The case arose when Mrs Diane Blood sought judicial review of the decision of the Human Fertilisation and Embryology Authority to refuse her permission to use her deceased husband Stephen's sperm for the purposes of trying to conceive via assisted reproduction in the UK or elsewhere in the EU. Mrs Blood's application succeeded in the Court of Appeal and, having been directed to reconsider the issue taking proper account of Mrs Blood's EU free movement rights, the Authority gave her permission to export her deceased husband's sperm to Belgium where, unlike the UK, the legal regime permitted treatment in the absence of Stephen Blood's written consent. Mrs Blood has since had two children via ARTs using Stephen's stored sperm.

The case has been read in different ways. The Court of Appeal was at pains to emphasize that it was a one-off set of facts and that its judgment could not be treated as precedent. The dominant popular account applauded the Court, contrasting its decency and common sense in helping Mrs Blood to give effect to her and her dead husband's desires and intentions, with an alleged lack of humanity on the part of the Human Fertilisation and Embryology Authority. A third account takes a different line, however. It queries whether there was proper respect for Stephen Blood's reproductive choices and, more generally, for him as an incompetent patient. As regards the export decision, it suggests that the Court of Appeal and, in turn, the Authority seem to have given insufficient attention to the grounds on which EU free movement rights can be limited. It also emphasizes that it is possible that the doctors who extracted the sperm from Mr Blood whilst he was in a coma may have acted unlawfully: English law provides that incompetent patients are to be treated in their best interests and, as a range of commentators have pointed out, it is surely difficult to see how having children after one's death can be said to be in a patient's best interests.[44]

What if a woman and a man store frozen embryos but at a later point disagree about their use? Should a court proceed on the basis that there is a right of veto and, if so, should women and men be given equivalent rights of veto? Facts such as these demonstrate very clearly how ARTs have the capacity to challenge long-standing assumptions about reproductive choice.[45] In addition, as Derek Morgan

[43] *Supra* note 21.

[44] Harris, *supra* note 5, at 18: 'this surely over-precious attitude to the destiny of male gametes is not one reflected in custom or practice. Men are notorious for leaving their gametes behind in all sorts of places, some of which may well result in the creation of life. They do so almost always without all these requirements for opportunities for counselling, formal consent, and time for reflection, and usually without missing them and without mishap to them. They cannot control their partner's use of contraception, nor insist on an abortion if they want their genes to survive in this way, or be mixed with those of the woman.'

[45] This impact should come as no surprise: see, in particular, M. Strathern, *Reproducing the Future: Essays on Anthropology, Kinship and the New Reproductive Technologies* (1992) for the argument that reproductive technologies have the capacity to remake allegedly natural facts and assumptions about sex and reproduction, and about kinship.

has noted, they raise the question of whether arguments grounded in reproductive choice require 'a commitment to gender neutrality which is difficult to achieve in all cases and may be undesirable in some.'[46]

The easiest solution might seem to be to avoid the problem in the first place by freezing gametes rather than embryos.[47] But success rates for egg freezing are lower than for sperm (though reports suggest they are improving), and women who are at risk of being unable to produce any more eggs may therefore be advised to freeze embryos rather than their own eggs. There are other problems too: most obviously, if the woman and man disagree, and it is the man who wants to proceed with treatment (he might have become infertile, for example), there has to be another woman who is willing to undergo treatment to be implanted with one or more embryos and carry any ensuing pregnancy to term. The female partner, by contrast, will generally be able to proceed without involving a third party.[48]

The European Court of Human Rights had to consider the issue of veto in the *Evans* case which concerned six embryos stored in the UK in accordance with the provisions of the Human Fertilisation and Embryology Act 1990. The embryos were Natallie Evans' only chance of having genetically related offspring: Ms Evans could carry a pregnancy but, as a result of treatment for ovarian cancer, she could produce no more eggs. The embryos had been created immediately prior to the cancer treatment, using her eggs and the sperm of Howard Johnston, to whom she was engaged at the time. When the relationship between Evans and Johnston broke down, Mr Johnston wrote to the clinic, informing them that he wished to withdraw his consent to the continued storage or use of the embryos. The clinic then refused to provide treatment to Ms Evans, who issued proceedings against both the clinic and Mr Johnston seeking declaratory and injunctive relief.

The core of Evans's argument was that her right to respect for private life under Article 8 of the European Convention on Human Rights (ECHR) (which has been incorporated in domestic law via the Human Rights Act 1998) was violated by the consent provisions of the 1990 Act. The Article 8 argument was grounded in part in the claim that the application of the consent provisions to

[46] Morgan, 'Review of Jackson, Regulating Reproduction' (2003) 11 Med L Rev 262, at 263. Of potential interest here is the statement in the Universal Declaration on Human Cloning (2005) that Member States are 'called upon to take measures to prevent the exploitation of women in the application of the life sciences.'

[47] Another would be to introduce a regime of enforceable written agreements detailing what should happen to stored embryos if a relationship comes to an end.

[48] See further the dissenting opinion of Judges Traja and Mijovic in the decision of the Chamber of the ECtHR in *Evans v UK* (2006) 43 EHRR 21, proposing a 'case-specific' test to balance the rights of gamete providers, including a requirement that the party who wishes to use the embryos 'does not intend to have recourse to a surrogate mother.' But see Annett, 'Balancing Competing Interests Over Frozen Embryos: The Judgment of Solomon?' (2006) 14 Med L Rev 425, at 430, who argues that this is not a 'neutral' test if seen from the perspective of a man who becomes infertile after donating his sperm and whose only prospect of a child who is genetically related to him may be by using a surrogate mother.

male and female gamete providers in identical terms admitted 'no true comparison between the situation of a woman in refusing consent to the implantation of an embryo in her body or refusing to carry it to term and that of a man in withholding his consent to such implantation.'[49] Specifically, overriding a man's objection to implantation would entail no further medical interventions for him. Evans also argued that the aims of the 1990 Act could be just as well served by allowing gamete providers to give an irrevocable consent at fertilization, or by allowing the male gamete provider's withdrawal of consent to be overridden in exceptional cases.

Evans made a second, intertwined argument based on discrimination contrary to Article 14 ECHR.[50] The judgments of the various courts who heard the case contain a good deal of discussion as to the appropriate comparator to be used to assess such a claim. Evans argued that the comparator was a woman with intact ovaries, ie a fertile woman who could either conceive without assistance or produce sufficient eggs for repeated attempts at IVF. The consent provisions of the 1990 Act were therefore discriminatory, she claimed, because they subjected her, an infertile woman, to a 'male veto', whereas a fertile woman who conceived one or more embryos inside her body could make reproductive decisions without reference to the genetic father.

Evans's arguments were considered by four different panels of judges: the High Court and the Court of Appeal in England,[51] and then the Chamber and Grand Chamber of the European Court of Human Rights in Strasbourg. Each of these courts rejected her arguments (though there were dissenting opinions in both the Chamber and Grand Chamber of the European Court of Human Rights).[52] As the Grand Chamber explained, even though the right to respect for the decision to become a parent 'in the genetic sense' fell within the scope of Article 8, there was no violation of that Article here given the lack of European consensus, the fact that the domestic rules were clear and had been brought to the attention of Ms Evans, and also that these rules struck a fair balance between the competing public and private interests involved.[53] As regards the Article 14 argument, the Grand Chamber, echoing both the Chamber and the majority in the English Court of Appeal, noted that the reasons for finding that Article 8 had

[49] *Evans v United Kingdom* (2006) 43 EHRR 21, para 52.

[50] Other arguments, claiming violations of Arts 2 and 12 ECHR, and also promissory estoppel, were disposed of rather swiftly by the courts and are not discussed here.

[51] Leave to appeal to the House of Lords was refused, after which Ms Evans lodged an application against the UK with the ECtHR.

[52] See, respectively, *Natallie Evans v Amicus Healthcare Ltd and Others* [2003] EWHC 2161 (Fam); *Natallie Evans v Amicus Healthcare Ltd and Others* [2004] EWCA (Civ) 727; *Evans v United Kingdom* (2006) 43 EHRR 21; *Evans v United Kingdom* Application No 6339/05, Judgment of 10 April 2007. See also *Nachmani v Nachmani* 50(4) PD 661 (Isr) cited as persuasive authority by Evans.

[53] Cf Sheldon, 'Revealing Cracks in the "Twin Pillars"' (2004) 16 Child and Family LQ 437, who queries whether we would want an anonymous sperm donor to be granted a 'veto' akin to that which Johnston, as a former partner, was accorded by the 1990 Act.

not been violated also formed a basis for holding that there had been no violation of Article 14.[54]

3. The Welfare of the Child

I turn now to what can seem like a constant in the changing picture of ARTs: namely, the requirement to take account of the best interests, or welfare, of the child in decisions about access to treatment. The reality, however, is that the status of the welfare criterion is in flux. On the one hand, its centrality in recently established regulatory regimes in Canada and Ireland suggests that it has ongoing legitimacy and importance; on the other hand, its use in the longer-established regulatory regime in the UK, the Human Fertilisation and Embryology Act 1990, has attracted increasing criticism. In fact, there have been high-profile calls for the abolition of the criterion from the British legislative scheme and reform seems imminent.[55]

This flux makes it difficult to gauge the impact of the welfare criterion on arguments grounded in reproductive choice. Moreover, it is not difficult for advocates of choice to brand the welfare criterion as a ruse; that is, as a way of dressing up potentially controversial limitations on access to ARTs and side-stepping the difficult issues that might elude consensus.[56] The increasing focus on identity rights for ART children also seems to shift the focus away from notions of best interests or welfare. Furthermore, as Cook, Dickens, and Fathalla have emphasized, '[a]pproaches to availability of [ARTs] based on assessment of "the best interests of the child" are confounded by the consideration that prohibiting availability of [ARTs] results in such children not being born.'[57]

It has been suggested that, in assessing the merits of the argument based on reproductive choice, the focus should be on the demands that come with being a family rather than on child welfare *per se*. This is the approach adopted by the philosopher Onora O'Neill, in her argument against the claims for reproductive or procreative autonomy made by writers such as Ronald Dworkin and John Harris. O'Neill acknowledges that appeals to *individual* autonomy are both apposite and convincing when used in debates about contraception and abortion: in these contexts, she says, 'a right to choose can protect responsible reproductive

[54] Cf Lind, '*Evans v United Kingdom*—Judgments of Solomon: Power, Gender and Procreation' (2006) 18 Child and Family LQ 576; Sheldon, *ibid.*

[55] Science and Technology Committee, *Human Reproductive Technologies and the Law*, HC 7-1 (Fifth Report of Session 2004–2005) (2005). At the time of writing, a revised legislative scheme is being debated in Parliament: see *supra* note 26.

[56] See eg Beem and Morgan, 'What's Love Got to Do With It? Regulating Reproductive Technologies and Second Hand Emotions', in McLean (ed), *supra* note 37, 369, at 387 arguing that '[t]he "welfare" criterion is an example of what we call a "second hand emotion"; seldom is it used as a concept with particular force and scope; more often is it adumbrated as a surrogate for other concerns about access to assisted conception and related services.'

[57] *Supra* note 10, at 308.

decisions—for example, to have no children, fewer children or better-spaced children.'[58] But where the issue is reproduction—having a family—O'Neill is not convinced by appeals to autonomy (irrespective of whether such appeals are grounded in *individual* autonomy or in *reproductive* or procreative autonomy). The crux of her argument is that reproduction is 'intrinsically not an individual project': 'nearly always it begins with a relationship' and 'whenever it is successful it demands long-term relationships (whether biological or not) to protect and care for the child who has been brought into existence.'[59] This leads her to conclude that:

> [t]he visions of reproductive autonomy that have attracted so much attention in recent discussions are neither convincing nor attractive because they misconstrue having children as a form of self-expression that demands individual autonomy. Limiting child-bearing may promote individual autonomy, but having children generally curtails it.[60]

She also emphasizes that she cannot see how the harm principle could rescue the autonomy argument. Post-menopausal pregnancies and human reproductive cloning are examples of reproductive choices that some argue should be seen as acceptable 'provided they do not harm': for O'Neill, however, they 'clearly risk inadequate nurture and care of children who may be brought into existence.' She concludes that the question to be asked is 'not just whether reproducing in certain circumstances, or using a particular technology, can sometimes be done without harming':

> It is whether there are reasonable grounds to think that any child brought into existence can expect to have at least an adequate future, cared for by a 'good enough' family (biological or not) who will be present and active for the child across the long run.[61]

4. *An Odd Sort of Choice*

The idea of reproductive choice could also be said to jar with the low success rate of many reproductive technologies.[62] As of 2005, the average success rate in the UK for IVF in women under the age of 35 was 29.6 per cent; moreover, as this was the *average* rate, there would have been variations between providers.[63] Bluntly, 70 to 80 percent of IVF treatments end in failure. PGD is an even more complex procedure than IVF, as evidenced by the following extract from field notes written by the British sociologist Sarah Franklin on her first visit to a PGD clinic in a city in the North of England in 2001. Franklin has extensive experience as an ethnographer of IVF, but this did not prepare her for the way in

[58] O'Neill, *supra* note 24, at 65. [59] *Ibid.* [60] *Ibid*, at 65–6.
[61] *Ibid*, at 66–7.
[62] And, as the UK's Medical Research Council emphasizes in *Assisted Reproduction: A Sound, Safe Future* (2004), there are substantial gaps in knowledge concerning the long-term health implications of IVF.
[63] See the HFEA Guide 2007/8, *supra* note 20.

which staff at the clinic felt obliged to give patients '"the really bleak picture" of PGD, thereby "putting the fear of God into them"':

On my first afternoon, [the PGD nurse coordinator] took me to several meetings between genetic counsellors and prospective PGD patients. I am amazed how much failure is emphasised, how much they go on about how many things can go wrong, how it is all very early days for the technique and for the clinic, and how complicated all of the procedures are.[64]

But low success rates are not the only reason for describing reproductive technologies as an 'odd sort of choice'. The practice of egg-sharing, for example, continues—rightly—to raise deep concerns about consent.[65] In the UK, a woman who needs IVF and who agrees to egg-share—that is, to donate up to half of the eggs retrieved during one cycle—will be offered her own treatment at a reduced cost. But the question that arises is: does the desire for a child—specifically, the chance to try for a child through a procedure that is out of reach financially—place the donor/'would-be' mother in a zero- or no-choice situation about 'sharing'?

'What choice do we have?' seems also to capture the position of parents who want to use PGD and tissue-typing because they have a sick child who might be cured by a saviour sibling. The same holds true with respect to would-be parents who want to use PGD to start a pregnancy in the knowledge that they have done all that they can to avoid passing on a serious genetic disorder, and that they will not have to resort to prenatal diagnosis and abortion. Would-be parents in these circumstances may not readily think of themselves as having reproductive choice.[66] Indeed, as Michael Freeman has noted, amidst the myriad popular questions about babies by design and on demand, 'there is another question which is less commonly raised':

[D]o parents with a sick child who might be cured by a saviour sibling have a moral obligation to create a new child? Could we see this as an element of parental responsibility? . . . Does the chance to save a child trump reproductive autonomy? How far

[64] Kerr and Franklin, 'Genetic Ambivalence: Expertise, Uncertainty and Communication in the Context of New Genetic Technologies', in A. Webster (ed), *New Technologies in Health Care: Challenge, Change and Innovation* (2006) 40, at 50.

[65] This provides further evidence of the need to ask the question 'reproductive choice for whom?' In other words, we need to move beyond stereotypes about young, medical-student sperm donors towards a truer picture of who is 'donating' gametes, and why and how they feel about the range of possible 'lives' of those gametes.

[66] In the English Court of Appeal in *R (on application of Quintavalle) v Human Fertilisation and Embryology Authority* [2003] 3 All ER 257, at 287 Mance LJ emphasized that '[p]references suggests personal indulgence or predilection and the luxury of a real choice'. In his view, the couple at the centre of the case, the Hashmis, who wanted to be allowed to use PGD to select an embryo that could then be tissue-typed (using PTT) to find the best match for one of their existing children, Zain, who had a very serious blood disorder (beta thalassaemia major), were certainly not trying to 'indulge themselves'. The Court said the dual screening could proceed, as did the House of Lords on appeal. On this case, see further the chapter by Roger Brownsword in this volume.

are parents required to go? Would we say that parents . . . , if they refused to attempt to create a saviour sibling, were being unreasonable? Would we censure them? [Have extant attempts] created a new standard which others can be expected to follow?[67]

Infertility-related uses of technology raise questions too, not least because the idea of reproductive choice may not have strong resonance when one is being technologically assisted to try to achieve something that is widely considered to be natural. Put bluntly, needing reproductive technology may feel like an onslaught on one's sense of self.[68] The naturalization of motherhood can make the need for use of ARTs especially hard for women (even if there is also a long-standing association between women's bodies and weakness or vulnerability, and a less long-standing but undeniably powerful understanding of Nature itself as being in need of assistance). Needing ARTs is also hard for men, not least because of the associations between 'fathering a child' and virility or potency, and also because of the linkages between paternity and social standing. Furthermore, for women and men alike, but more so for women because they are the 'primary patients' in ARTs, bodily integrity and more generally the 'fit' between one's body and one's sense of self, can be very battered by the actual techniques of assisted reproduction. Here, for example, is one woman's response to amniocentesis:

I cried for two days after I had the test. I guess I was identifying with universal motherhood; I felt like my image of my womb had been shattered. It still feels like it's in pieces, not like such a safe place as before. I guess technology gives us a certain kind of control, but we have to sacrifice something in return. I've lost my brash confidence that my body just produces healthy babies all by itself, naturally, and that if it doesn't, I can handle whatever comes along as a mother.[69]

One final point: Charis Thompson emphasizes that, for some of the would-be parents she interviewed at US clinics, trying reproductive technology was mandatory in order actually to be infertile. To be infertile you need to have 'tried'; you need to be incapable of both reproduction *and* technologically assisted reproduction. A diagnosis of untreatable infertility may also be seen as the most socially acceptable reason to stop trying. But that diagnosis was rare: the culture at clinics—contributed to by would-be parents and clinic staff—was one of perseverance, in part perhaps because of the difficulties involved in saying 'no' to technology:

Clinic psychologists give couples advice on pursuing alternatives, such as life without children and adoption. Nonetheless, the structuring of treatment is open-ended in the

[67] Freeman, 'Saviour Siblings', in McLean (ed), *supra* note 37, 389, at 405–6.

[68] But see *infra* at 217 for Thompson's argument that the would-be parents she studied in the US built 'agency through objectification'.

[69] Quoted in C. McLeod, *Self-Trust and Reproductive Autonomy* (2002), at 158.

sense that another combination of hormones, another cycle of artificial insemination, another go at IVF can always be embarked on, age and finances permitting.[70]

To summarize: assisted reproduction does not seem like a context where one can 'just say no' or, for that matter, 'yes', or, indeed, 'just do it'.[71] Put shortly, it is an odd sort of choice. But that, as I shall argue later, is precisely part of the reason why the speakability of reproductive choice needs to be secured.

2. Securing the Speakability of Reproductive Choice

To recap: I have suggested that, notwithstanding its array of staunch advocates and its strong credentials in disputes about access to contraception, reproductive choice feels increasingly like a bad argument when used in relation to ARTs. In part, the problem can be traced to overlap between, on the one hand, ARTs and abortion and, on the other, ARTs and genetic technologies. These overlaps tend to cast a shadow over claims for reproductive choice in the field of ARTs. I also emphasized that the range of ways in which reproductive choice does not seem to 'fit the facts' of ARTs could be a further reason for tagging it as a bad argument. In what follows, I want to try to counter the tendency to think of reproductive choice as dangerous, off the point, or otherwise out of place in the field of ARTs. Using two ethnographic studies, I try to show that it would be a serious loss if that tendency limited the speakability of choice in the field of ARTs. The studies I use are Sarah Franklin and Celia Roberts's ethnography of PGD in the UK, and Charis Thompson's ethnography of ARTs in the US.[72]

A. Reproductive Choice 'in Action'

My stepping-off point is the many would-be parents in Franklin and Roberts's study who described opting for PGD as their 'only choice' whilst giving full, thoughtful accounts of their other choices. This messy, mixed-up description jumbles consent and, what I shall call, 'no room for conventional consent'. Moreover, for these would-be parents, the 'only choice' of PGD came with no guaranteed outcome; all that was guaranteed was the need for further, follow-on

[70] *Ibid*, at 95. See, relatedly, S. Franklin, *Embodied Progress: A Cultural Account of Assisted Conception* (1997) describing ARTs as 'hope technologies'.

[71] See also M. Konrad, *Narrating the New Predictive Genetics: Ethics, Ethnography and Science* (2005) Ch 6; K. Throsby, *When IVF Fails: Feminism, Infertility and the Negotiation of Normality* (2004).

[72] These are two amongst an increasing number of ethnographic works on ARTs: see also eg Konrad and Throsby, *ibid*. In the UK, legal scholars have begun to produce empirical work on ARTs: see eg Callus, 'Patient Perception of the Human Fertilisation and Embryology Authority' (2007) 15 Med L Rev 62; Scott *et al*, 'The Appropriate Extent of Pre-Implantation Genetic Diagnosis: Health Professionals' and Scientists' Views on the Requirement for a "Significant Risk of a Serious Genetic Condition"' (2007) 15 Med L Rev 320.

reproductive choices in conditions of hope and anxiety.[73] To my mind, however, these messy, mixed-up descriptions ought to be seen as evidence that choice is thriving and, crucially, that it is what I describe as resonant choice. Where choice is resonant it is deep, rich, and reverberating, and it is strikingly different to the 'for or against' model of choice that has dominated popular, legal, and scholarly discourses. Resonant choice is also far away from bloated consumerism, including what has been described as 'parental eugenics'. Rather, it speaks of a form of decision-making or judgement that is shot through with obligation (to a potential future child or children, and to family, clinic staff, other patients (past, present, and future), community, and society). It also points to choice as a process—to recurrences of the need to judge and decide in this way.

Anne, one of the participants in Franklin and Roberts's study, captures what I mean by resonant choice better than I can. Here, then, is Anne explaining how PGD was her 'only choice', and referring to herself and also to a double 'we' (ie herself and her partner, and latterly herself or them as part of a larger community):

> I can understand it is a very grey area. Because obviously . . . we've used PGD because we didn't want to have another child that was going to die within 12 months. But I mean, . . . at what point do you draw the line? At a child that dies at 2 years, 5 years, 10 years, 20 years, 30 years? Where? . . . What conditions are we going to allow PGD to be used for? . . . I don't know where the line should be drawn.[74]

Anne's multiple frames for her reproductive choice indicate very clearly that she sees decision and judgement in terms of potentially extensive chains of responsibility and obligation. Moreover, she sees these matters as being for her, her and her partner, and all of us, not just for the 'experts'. To describe Anne, or the other would-be parents in Franklin and Roberts's study, as 'choosy'—or as 'would-be parents "for choice"' in the fashion of the lapel pins and placards that are familiar from abortion politics in the US—would be both inaccurate and insulting.

Franklin and Roberts's findings do not stand alone: they chime not just with those of Charis Thompson, which I discuss below, but also with those of Rayna Rapp on would-parents in the US faced with decisions about whether to use the technology of amniocentesis.[75] Rapp's description of the would-be parents she studied as 'moral pioneers' is particularly instructive. Indeed, as Nikolas Rose has emphasized, Rapp seems to have captured 'something crucial':

[73] Indeed, it could be argued that the label 'would-be parents' is misleading in that it suggests that they are a group. Uncertainty and contingency are all that they really share; their presence in the clinic as users of PGD is evidence of a shared present but there are no guarantees of pregnancies or of shared futures more generally. See, relatedly, J. Stacey, *Teratologies: A Cultural Study of Cancer* (1997).

[74] Franklin and Roberts, *supra* note 2, at 1.

[75] R. Rapp, *Testing Women, Testing the Fetus: The Social Impact of Amniocentesis in America* (2000).

These women and men were pioneers because, in their relation with their bodies, with their choices, with experts, with others in analogous situations, and with their destiny, they had to shape new ways of understanding, judging, and acting on themselves, and also to engage in a kind of reimagining of those to whom they owe responsibilities—their progeny, kin, medical helpers, co-citizens, community, society.[76]

In Franklin and Roberts's study, being a moral pioneer comes through particularly clearly in the would-be parents' decision-making around surplus embryos. The would-be parents' willingness to donate embryos to research was high and tended to be framed in terms of trying to 'give something back'. Donation did not represent a lack of concern for the embryos: rather, it was in part a reflection of their extraordinary value or preciousness.

Franklin and Roberts's work also corrects for the near-exclusive preoccupation with questions of access. Access is hugely important but, standing alone, it does not provide a full enough picture of reproductive choice. We need also to understand the different meanings that are being given to individual technologies. These latter points come through very clearly in Rapp's analysis of the use of amniocentesis in the US. Her ethnographic work left her with no doubt about the fact that race and class determine not just one's access to technology, but also its meaning. Marcia Inhorn's work on ARTs in the Middle East makes a similar point:

[T]he assumption on the part of global producer nations that ARTs—as value-free, inherently beneficial medical technologies—are 'immune' to culture and can thus be appropriately transferred and implemented anywhere and everywhere is subject to challenge once local formulations, perceptions, and actual consumption of these technologies are taken into consideration.[77]

In sum, Franklin and Roberts's ethnography of PGD patients is an essential corrective to stories of 'demanding parents' procuring 'designer babies'. It may also help to counteract the fear of 'anything goes' doctors who believe in nothing bar 'patient knows best' and 'rogue' scientists hell-bent on practices that will strip away human dignity:[78] in Franklin and Roberts's study, as in Thompson's on IVF, clinic staff helped to constitute and reinforce practices of resonant choice in ARTs. Consider, for example, the lab technicians in Thompson's work who describe access to ARTs as a moral issue, linking it to their responsibility as possessors of knowledge to surrender that knowledge in a non-discriminatory

[76] N. Rose, *The Politics of Life Itself* (2007) 146.

[77] *Supra* note 6, at 114. On the challenge that difference presents to human rights, see Brownsword, 'Genetic Engineering, Free Trade, and Human Rights: Global Standards and Local Ethics', in Daniel Wüger and Thomas Cottier (eds), *Genetic Engineering and the World's Trading System* (2008) 287.

[78] See also Scott *et al*, *supra* note 72; Svendsen and Koch, 'Unpacking the "Spare Embryo": Facilitating Stem Cell Research in a Moral Landscape' (2008) 38 Social Studies of Science 93.

way.[79] It is also worth recalling Franklin's surprise at the bluntness of clinic staff in discussing the rigours of PGD with potential patients. My hope therefore would be that ethnographies such as these might help to orient us away from potentially unhelpful rhetoric about science 'racing ahead' and towards extant, but arguably under-discussed, issues of concern including egg-sharing, donor anonymity,[80] and the 'IVF-stem cell interface'.[81] Above all, of course, my hope is that they might flesh out the human rights argument for reproductive choice, interrupting the popular dominion of 'for or against', of ingrained positions on autonomy as mere, sheer choice, and of consenting as an information-transfer transaction.[82] My hope, in short, is that ethnographic work might help human rights advocates to fight back against the claim that we are willing to play fast and loose with human life and, more generally, with human dignity.

B. Is PGD Different?

At this point, however, the question arises: is PGD different—different, that is, to ARTs in which IVF is not acting as a platform technology for genetic science? Putting that another way, is it the *genetic* frame of PGD that engenders what I have described as 'resonant' choice? As commentators such as Isabel Karpin have pointed out, genetic discourses call attention to shared identity and participation in the common genetic pool:

> The individual in the age of the gene always contains a trace of the other; not-one but not-two... The discourse of genetics requires us to lose ourselves (or more correctly to find ourselves) in a genetic code that imbricates us with the other. In this case we recognize a self-hood that is based on interconnection and intermingled identity... We share our genes with others by decreasing degrees of exactitude reaching out from the familial and ending in the common genetic pool of the species. Individuality then is the recognition of ourselves in others. 'He has my eyes.'[83]

Is it the case, then, that ARTs which do not involve genetic science are less likely to be sites of resonant choice? Are they sites where would-be parents are more likely to be self, or even selfishly, focused, and where a market-mindset is more likely to dominate for both patients and clinic staff? I do not have answers to

[79] Compare Art 15(1) ICESCR (1966) which recognizes the right of everyone 'to benefit from the protection of the moral and material interests resulting from any scientific, literary or artistic production of which he or she is the author', and Art 15(2) which requires that States Parties undertake steps necessary for the conservation, development, and diffusion of science and culture.

[80] See, further, Turkmendag, Dingwall, and Murphy, 'The Removal of Donor Anonymity in the UK' (2008) 22 Intl J of Law, Policy & the Family 283.

[81] See, further, Franklin, 'Embryonic Economies: The Double Reproductive Value of Stem Cells' (2006) 1 BioSocieties 71.

[82] On the latter, see, in particular, D. Beyleveld and R. Brownsword, *Consent in the Law* (2007); N.C. Manson and O. O'Neill, *Rethinking Informed Consent in Bioethics* (2007).

[83] I. Karpin, 'Genetics and the Legal Conception of Self', in M. Shildrick and R. Mykitiuk (eds), *Ethics of the Body: Postconventional Challenges* (2005) 195, at 196–7.

these questions but I do want to highlight Charis Thompson's study of IVF clinics in the US, which mounts an interesting challenge to conventional assumptions about the likely effects of an ART market.[84] Thompson insists on the perhaps surprising conclusion that, in the US, ARTs have 'become in some ways a model medical speciality.' This, she says, was not how they started out; initially 'there were few clinics, which had astonishingly low success rates, imposed grueling treatment regimens, and excluded most would-be patients because they were unable to pay or were judged to be [not suitable] as parents.' Today, however, ARTs have become 'unusually accountable to various stakeholders' and genuine sites for activism within medicine (including, 'patient activism, scientific activism, legal activism, and professional self-regulation').[85]

Thompson is not arguing that ARTs in the US are a perfect world; rather her point is that, by focusing so extensively on a particular set of 'problems' concerning access to and the use of ARTs, some interesting developments that challenge conventional thinking seem to have passed us by. I found her account of the impact of privacy to be especially interesting. Her principal claim is that, at least in the US, 'the public agora of reproductive technologies is provided . . . by the politically capacious notion of privacy, which collects assisted reproductive technology's publics, contours its contestations, and propels its innovation.' This 'capacious notion' of privacy encompasses, first, reproductive privacy, second, autonomous medical or scientific research, and, third, the private sector. Thompson's evidence suggests that the confluence of these three different types of privacy in ARTs is what 'makes an important if somewhat paradoxical space for social and technical innovations.'[86] She explains this conclusion in the following way:

> The ART market, by being about all three kinds of privacy simultaneously, contains many checks and balances that reflect explicit and latent currents of interest and opinion of ARTs' publics in its own dynamics.[87]

Thompson's idea of 'checking and balancing' relationships between reproductive privacy, scientific interests, and commercial priorities is fascinating, and seems to throw down a challenge to conventional accounts of the need for particular forms of state regulation through law.[88] More generally, her study of US IVF clinics seems to reverberate with Rapp's idea of 'moral pioneers', suggesting that patient

[84] See also Marcia Inhorn's work on ICSI in the Middle East, eg M.C. Inhorn, *Local Babies, Global Science: Gender, Religion, and In Vitro Fertilization in Egypt* (2003).

[85] Thompson, *supra* note 2, at 25–6. [86] *Ibid*, at 211–12. [87] *Ibid*, at 241.

[88] See relatedly Corneilussen, 'The Impact of Regulations on Firms: A Case Study of the Biotech Industry' (2005) 27 Law and Policy 429, challenging the assumption that small start-up biotech firms are necessarily less compliance-oriented; in fact, they may approach 'the regulatory ideal where the routines, procedures, and precautionary measures prescribed by regulations permeate the organizations.'

activism around ARTs is active not passive, and that would-be parents find forms of agency amidst the very real rigours of treatment.[89]

C. The Challenge of 'Speakability'

Some years back I wrote an essay on the 'hoopla over patient voice' in which I noted that 'although the cultural conversation about abortion may appear increasingly cacophonous, it is, at least from certain speaking positions, less and less inclusive'. I gave the example of a woman who has terminated one or more pregnancies and pointed out that:

> Her voice is infrequently heard. It is also prey to vulgar mishearings. [And when heard, it] is suspiciously uniform: it is tentative and, in my view, unnecessarily tortured, and the substance of its speech seems to have been censored. Today, in giving voice to the decision to abort a pregnancy, one *must* declare oneself saddened by the choice.[90]

In retrospect, I think I might have missed something important. I don't think that I was wrong in what I said[91] but, looking back, I think that I should have thought through the idea of being 'saddened by the choice.' Concern about the saturation level of anti-abortion rhetoric may have stopped me from doing this; to be honest, I suspect that I was entirely caught up in the conventional 'for or against' framing of reproductive choice.

Today, I see 'saddened by the choice' in a different light. It remains clunky and awkward; it is also counter-intuitive given that it resonates most strongly with a standard anti-abortion narrative wherein a woman always regrets terminating any pregnancy. At the same time, however, it is deeply appealing: it doesn't flatten out choice into 'for' or 'against', into 'right' choices and 'wrong' ones, and for this reason it manages to hint at the complexity that can attach to reproductive decision-making. In short, 'saddened by the choice' might have the advantage that it leaves space for complexity. This becomes clearer if we contrast it with Justice Kennedy's assumption in *Gonzales v Carhart* (the US Supreme Court's most recent ruling on partial-birth abortion) that *many* women are '*traumatically remorseful*' after *any* abortion,[92] which seems to me to deny the possibility of complex choices (followed by complex reflections) and instead leans strongly towards the notion that women who have abortions face a future full of deep, interminable, and near-inevitable private guilt and shame.

[89] On agency through 'submission to' consent forms, see Jacob, 'Form-Made Persons: Consent Forms as Consent's Blind Spot' (2007) 30 PoLAR: Political and Legal Anthropology Review 249.

[90] Murphy, 'Health Confidentiality in the Age of Talk', in S. Sheldon and M. Thomson (eds), *Feminist Perspectives on Health Care Law* (1999) 155, at 164.

[91] Indeed, given the tone of the remarks on women's responses to abortion in Justice Anthony Kennedy's majority opinion in *Gonzales v Carhart* 127 S Ct 1610 (2007), I think that my comment merits repetition.

[92] Emphasis added.

After I had finished writing the first draft of this essay, I read an article by an Australian health care lawyer, Roger Magnusson, which bolstered my thinking about the need to create more public space for the complexity of choice. Magnusson's focus is the end of life, not ARTs, and the decision-makers on whom he focuses are professionals rather than patients or a mixed group. What resonated with me, however, was his argument that although it is often assumed 'explicitly or implicitly, that the role of medical ethics and law is to guide us towards the "right choice"... some of medicine's most painful decisions are, in fact, the devil's choices'. The devil's choice is, he says, a better model for explaining practices such as the provision of pain relief in end-of-life care 'than the traditional account found in law and medical ethics':

> The benefit of applying the devil's choice to palliative care is that it permits empathy with the dilemmas physicians face, while still acknowledging the extraordinary power that physicians have over the lives of patients at what is perhaps the most vulnerable time of their lives.[93]

Magnusson's topic is end-of-life care; mine was abortion. Neither of us addressed ARTs. So what then is the relevance of this material for the purposes of this essay? For me, it is that what drives both my account and that of Magnusson is a sense that something is missing in the accounts of choice that dominate contemporary discourse. There isn't, of course, a prescription in any of this as to how ARTs ought to be regulated. Some will see that as a gap in my argument. Others will be frustrated by the fact that I have not provided a 'proper' pro-choice argument, and yet others will wonder whether what I have labelled 'resonant choice' would have a stronger chance of public approval if it had a different name. These are valid points. My response to all of them is as follows: as I see it, securing the speakability of reproductive choice—stopping it being rendered unspeakable in the field of ARTs—has to be the priority.

Before closing this part of the essay, I do, however, need to raise some concerns. As noted earlier, ARTs are not a perfect world and in what follows I reaffirm this by reference to two specific concerns or 'imperfections'. To my mind, these concerns point to the very real need to retain a focus on power, and where it resides, amidst any excitement about resonant choice and moral pioneers. The first concern relates to how we will treat those who make *different* choices in a context in which resonant choice and active moral pioneering are viewed as important ways of safeguarding human rights and human dignity. Recall Michael Freeman's query, mentioned in Part 1, as regards parents who refuse to attempt to create a saviour sibling. He asked: 'Would we say that parents..., if they refused to attempt to create a saviour sibling, were being

[93] Magnusson, 'The Devil's Choice: Re-thinking Law, Ethics, and Symptom Relief in Palliative Care' (2006) 34 J of Law, Medicine & Ethics 559, at 559. See also Nelson, 'Reconceiving Pregnancy: Expressive Choice and Legal Reasoning' (2003/4) 49 McGill LJ 593.

unreasonable? Would we censure them? [Have extant attempts] created a new standard which others can be expected to follow?'[94] My second concern focuses specifically on the idea of moral pioneers. Here what I want to emphasize is that moral pioneers in the field of ARTs come in different guises and that it would be a mistake to focus exclusively on individuals, couples, and families who are 'living with' particular illnesses or genetic disorders. In my view, Comment on Reproductive Ethics (CORE),[95] a non-governmental organization which has recently been very active in the UK in high-profile cases (most notably the challenge to the grant of a licence allowing Mr and Mrs Hashmi to have PGD and PTT screening of their embryos) needs to be seen and studied as another strongly emergent type of moral pioneer, with the capacity to have a significant impact on policy-making.[96]

3. Conclusion

My primary aim in this essay has been to argue, first, that the public understanding of the complexity of reproductive choice must not be contracted further than it already has been, secondly, that the complexity of choice in ART decision-making needs to be a great deal more public than it is at present, and finally, that one way to secure this might be for human rights to pay more attention to ethnographic work on reproductive and genetic technology.

To close, I want to be a little more explicit about how my argument relates to human rights law and practice. I have emphasized that I think human rights lawyers ought to pay greater attention to ethnography as we line up to pass judgement on new technologies. I have called for greater attention to be paid to reproductive choice 'in action'—in particular, reproductive choice in action in unfamiliar sites: for the most part, our standard haunts have been courtrooms, legislatures, and the like. Primary sources from these sites and 'expert' texts by fellow lawyers, philosophers, and bioethicists have been our essential reading (though there are traces of statistics and first-person narratives). I have argued that ethnographic studies of ART users—of would-be parents and clinic professionals, as well as, for example, gamete donors and ART children—deserve to be seen as an essential supplement to our essential reading. I believe that

94 *Supra* note 67. See, also, Rose, *supra* note 76, at 147; Stacey, *supra* note 73, describing a 'discourse of responsibility' that envelopes people with cancer.

95 For detail, see <http://www.corethics.org> (accessed 31 March 2008).

96 Indeed, the House of Commons Science and Technology Committee has described campaigners such as Josephine Quintavalle, one of the founders of CORE, as having made a 'valuable contribution to various debates' and noted that 'it could be argued that her application for judicial review in relation to preimplantation tissue typing is a valuable public service.' See Science and Technology Committee, *Human Reproductive Technologies and the Law*, HC 7-1 (Fifth Report of Session 2004–2005) (2005), at para 206.

ethnographies of decision-making in new technologies could help to change what we notice and, in turn, how we frame issues and conduct arguments. And, in line with others, I see ethnographic work not as a form of fundamental or ultimate truth, or as a route to the 'real world'. Ultimately what attracts me to it is that I believe very strongly that it can be of assistance or an 'aid to a sluggish imagination.'[97]

[97] Garfinkel, *supra* note 3.

7

The International Law of Genetic Discrimination:

The Power of 'Never Again'

Iulia Voina Motoc

The goal of this essay is to offer an approach to genetic discrimination based on international law.[1] Many studies have been written on the topic of genetic discrimination, either from a comparative law perspective or from a utilitarian perspective of medical research development. Others have tackled the problem from the perspective of social justice.[2] Although all of these contain elements of international law, they do not provide a comprehensive overview: hence the need for this study.[3]

The relationship between new technologies and the protection of human rights is complicated. In part, this is because history shapes that relationship: the legal regime of genetic discrimination is influenced by the history of genetic discrimination, specifically its use during the genocide and crimes against humanity during World War II. More controversially, it might also be argued that this history is also the reason why the legal regime relating to genetic discrimination should be analysed more thoroughly than other types of discrimination that may arise from other uses of technology.

Perhaps this can be put another way: imagine if the science underpinning eugenic thought had been right—what would our critique be then? If we could

[1] Some of these ideas were developed in the UN Sub-Commission on the Promotion and Protection of Human Rights, 'Human Rights and the Human Genome', Preliminary report submitted by Special Rapporteur Iulia-Antoanella Motoc, E/CN.4/Sub.2/2003/36 of 10 July 2003; E/CN.4/Sub.2/2004/38 of 23 July 2004; E/CN.4/Sub.2/2005/38 of 14 July 2005. I thank Emmanuel Decaux, Francesco Francioni, and Francoise Hampson for comments on early drafts, Moma Momescu for her constant help with English translation, and Hadas Leffel for research assistance and editing the text.

[2] See eg the account of Gulati's social justice approach, *infra* Part 1.

[3] In addition, although cases of genetic discrimination are still isolated, fear of this type of discrimination is one of the most important factors that impedes the progress of research in genetics: see Hall, 'Patients' Fear of Genetic Discrimination by Health Insurers: The Impact of Legal Protections' (2000) 2(4) Genetics in Medicine 214.

not summarily dismiss eugenics as hegemony based on ignorance and as instrumentalist science, what arguments would we marshal to defeat its programme? What principled distinctions can we make between their institutionalization efforts and ours? What difference is there between their pedigree counselling and our genetic counselling? What distinction can be drawn between their race betterment agendas and our hope for alleviation of suffering through genetic therapy? Such questions are not posed to suggest that there is no difference. Rather, they are posed in order to remind us that we must consciously and deliberately respond to eugenics not only so that we avoid repeating the mistakes of the past, but also so that unwarranted fear and guilt do not render us incapable of properly appropriating and incorporating genetic information into our social structure, our policies, and our jurisprudence.[4]

The main questions that will be asked in this chapter are: What is the meaning of genetic discrimination? How can efficient legislation on privacy protection prevent genetic discrimination? What is the relationship between international law and domestic law in fields such as insurance, employment, and criminal investigation, and what is the best way to protect individuals? How are various groups, such as women, indigenous people, and persons with disabilities, affected by or vulnerable to genetic discrimination?

In Part 1 of this chapter, the main texts in international law will be outlined. Particular emphasis will be placed on the 1997 Universal Declaration on the Human Genome and Human Rights (UDHGHR), the first international instrument in this field. This will lead into an examination of the controversies surrounding the definition of genetic discrimination. By contrast with those who argue that it is not necessary to single out this type of discrimination, I will suggest that it should be legislated for separately within the general human rights framework. Arguments referring to the recent eugenic past, as well as the fundamental principle of human dignity[5] and the rights 'of everyone to enjoy the benefits of scientific progress and its applications',[6] provide further support for adopting an individual approach to this type of discrimination.

[4] Geetter, 'Coding for Change: The Power of the Human Genome to Transform the American Health Insurance System' (2002) 28 American J of Law & Medicine 1.

[5] The Preamble of the Universal Declaration of Human Rights 1948 (UDHR) refers to human dignity as 'the foundation of freedom, justice and peace in the world' and Art 1 states that 'human beings are born free and equal in dignity and rights.' There are similar references to the foundational position of human dignity in both the International Covenant on Civil and Political Rights 1966 (ICCPR) and the International Covenant on Economic, Social and Cultural Rights 1966 (ICESCR). See in addition Arts 2, 10, 11 of the UDHGHR 1997: Art 2, eg, states that '[e]veryone has the right to respect for their dignity and for their rights regardless of their genetic characteristics' and '[t]hat dignity makes it imperative not to reduce individuals to their genetic characteristics and to respect their uniqueness and diversity.' On human dignity and biotechnology, see eg D. Beyleveld and R. Brownsword, *Human Dignity in Bioethics and Biolaw* (2001).

[6] See eg Art 15 ICESCR. See also Arts 17–19 UDHGHR on 'solidarity and international co-operation', including between industrialized and developing countries.

The second part of the chapter analyses the protection of genetic privacy. The 2003 International Declaration on Human Genetic Data (IDHGD) of UNESCO, which has been described as 'a sequel'[7] to the UDHGHR, provides the main international legal framework in this regard. The most sensitive issues arise when consent to genetic data use is involved. Express consent is stipulated by the IDHGD but, as Hélène Boussard explains in her chapter in this volume, there are still many issues which remain unclear, especially as regards the relationship between individual consent and the interests of third parties, and also collective consent.

Part 3 deals with how international law against genetic discrimination becomes operational in domestic law in the fields of employment, insurance, and criminal investigation. In accordance with Article 23 of the IDHGD, '[s]tates should take all appropriate measures, whether of a legislative, administrative or other character, to give effect to the principles set out in [the] Declaration, in accordance with the international law of human rights.' To date, the European tendency has been to prohibit the use of genetic data in employment and insurance; by contrast, in the US, state legislation provides a regulatory approach with variable definitions and degrees of prohibition of genetic discrimination. Criminal investigation is another field which raises concerns about possible discrimination.[8] DNA databases are a valuable instrument for criminal investigations and bring major benefits. Yet, as emphasized by the IDHGD, certain practices bring up rights concerns, including: retaining full DNA samples, rather than just the DNA profiles used for identification; retaining people's records on the database without consideration of the nature of their offence and also including those who have been arrested but not charged, or those who have been acquitted; and using the database for genetic research without consent.

The last part of this chatper deals with the issue of vulnerable groups, specifically, women, children, indigenous peoples, and disabled persons.

1. The Definition of Genetic Discrimination

Discoveries in genetics have opened the door to new notions of discrimination. As a result, people who are afraid of potential genetic discrimination may be discouraged from obtaining genetic information that could bring health benefits to them and their families.

[7] A.A. Yusuf, 'UNESCO Standard-setting Activities on Bioethics: Speak Softly and Carry a Big Stick', in F. Francioni (ed), *Biotechnologies and International Human Rights* (2007) 85 at 91.

[8] See, also, the chapter by Roger Brownsword in this volume discussing what he describes as the 'revolutionary sting': '[t]he revolutionary sting... is precisely that the use of modern technologies, such as biotechnology, as a regulatory instrument might transform our culture and practice of criminal justice—ultimately with profound consequences for our valuation of both legality and human dignity.'

The present environment of suspicion regarding genetics can be traced in part to the early twentieth-century 'eugenics' movement, which utilized the classification of genetic inferiority as a rallying point. The objective of the movement was to preserve and advance genetic superiority, and it sought to achieve this by sterilizing the 'genetically inferior' to prevent further dilution of the gene pool, prohibiting interracial marriage, and limiting immigration. While the most horrible form of eugenics was practiced by Adolf Hitler and the Nazi Party, a variety of eugenics-related movements existed elsewhere. In 1910, a Eugenic Records Office was established in Cold Spring Harbor, New York, which trained field-workers to collect family histories from people around the country. By 1924, data on people had been entered on around three quarters of a million cards and people made inquiries to the Office about whether particular proposed marriages would be eugenically appropriate.[9]

We are once more at a point in history where there is a strong individual and public interest in genetics. To a great extent, it will be the law that will be the custodian of our values as we decide on the right uses of genetic technologies and knowledge. And, in deciding how to shape and deploy the law, it is helpful to analyse the parallels between the previous uses of genetics and those of today. An understanding of previous abuses can act as a check against future abuse.[10] It can also help to facilitate anticipatory or proactive engagement with emergent or potential problems.[11]

After the eugenics horrors during World War II, international law set out to forbid genetics-based discrimination which can lead to a gradual development of eugenics. Let us try to examine the evolution of the law and, from this, the most appropriate definition of genetic discrimination.

A. International Legal Instruments

The Universal Declaration of Human Rights (UDHR), adopted by the United Nations General Assembly in 1948, stipulates in Article 5 that 'No one shall be

[9] Genetic theories were used as a basis for proposals for social and legal reform: the prime thrust of the reforms was to prevent people with allegedly undesirable genes from reproducing. The Chairman of the Department of Psychology at Harvard University advocated 'the replacement of democracy by a caste system based upon biological capacity with legal restrictions upon breeding by the lower castes and upon intermarriage between the castes'. See Geetter, *supra* note 4, at 38. The first eugenics law, enacted in Indiana in 1907, provided for the involuntary sterilization of institutionalized, non-improvable individuals who were idiots, imbeciles, rapists, or habitual criminals. Twenty-nine states passed sterilization laws between 1907 and 1931. Even the United States Supreme Court at one time did not totally reject the idea of genetics-based classes of people. In 1927, the Court held in *Buck v Bell* that a Virginia law authorizing the sterilization of certain inmates with hereditary forms of insanity or imbecility was constitutional. Justice Oliver Wendell Holmes, writing for the Court, stated that 'three generations of imbeciles are enough' and legitimized the sterilization of Carrie Buck, an 18-year-old female. It is this potential for defining 'normalcy'—and by comparison, inferiority—through the detection and selection of genetic propensities that illuminates the great possibility of harm resulting from genetic discoveries.

[10] See Geetter, *ibid.* [11] See Motoc, *supra* note 1.

subjected to torture or to cruel, inhuman or degrading treatment or punishment.' Article 7 of the International Covenant on Civil and Political Rights (ICCPR), adopted in 1966, covers similar ground but makes express mention of medical and scientific experimentation. It provides that 'No one shall be subjected to torture or to cruel, inhuman or degrading treatment or punishment. In particular, no one shall be subjected without his free consent to medical or scientific experimentation.' These provisions are complemented by a range of ethical guidelines concerning biomedical research, in particular the Declaration of Helsinki, which was promulgated in 1964 by the World Medical Association and has subsequently been revised several times. The Declaration has had considerable influence on the formulation of international, regional, and national legislation and codes of conduct. It addresses the ethics of research involving human subjects, and it specifies ethical guidelines for physicians engaged in both clinical and non-clinical biomedical research, including rules for informed consent of subjects and ethical review of research protocols. In similar vein, the World Health Organization (WHO) International Ethical Guidelines for Biomedical Research Involving Human Subjects stipulate that all research involving human subjects should be conducted in accordance with three basic ethical principles, namely respect for persons, beneficence, and justice.[12]

In recent years, the human genome has emerged as a specific concern within international law and policy. For instance, in 1997, the General Conference of the United Nations Educational, Scientific and Cultural Organization (UNESCO) adopted the Universal Declaration on the Human Genome and Human Rights (UDHGHR), which was endorsed a year later by the General Assembly of the United Nations.[13] Article 1 of the Declaration states that '[t]he human genome underlies the fundamental unity of all members of the human family, as well as the recognition of their inherent dignity and diversity. In a symbolic sense, it is the heritage of humanity.'[14] A similar sentiment is found in the Preamble to the Declaration, which emphasizes *inter alia* that:

> the recognition of the genetic diversity of humanity must not give rise to any interpretation of a social or political nature which could call into question the 'inherent dignity and . . . the equal and inalienable rights of all members of the human family', in accordance with the preamble to the Universal Declaration of Human Rights.[15]

[12] See <http://www.cioms.ch/frame_guidelines_nov_2002.htm>. On genetic research, see further the chapter by Hélène Boussard in this volume.

[13] In 1999, in order to facilitate the interpretation and application of the UDHGHR in domestic law, UNESCO adopted a resolution (30 C/Resolution 23) endorsing guidelines for its implementation.

[14] The phrase '[i]n a symbolic sense' has provoked diverse interpretations: see eg R. Pavoni, 'Biodiversity and Biotechnology: Consolidation and Strains in the Emerging International Legal Regimes', in F. Francioni and T. Scovazzi (eds), *Biotechnology and International Law* (2006) and Yusuf, *supra* note 7, at 88–9.

[15] See Preamble, para 4. See also Art 10 UDHGHR which provides that genomic research must respect the human rights, fundamental freedoms, and human dignity of individuals and groups of people.

The Preamble also emphasizes that although research on the human genome has vast potential for improving the health of individuals and of humankind as a whole, such research should fully respect 'the prohibition of all forms of discrimination based on genetic characteristics.' This sentiment is bolstered by several provisions aimed at preventing genetic discrimination. Article 2(a) provides for the right of everyone to have respect for his/her dignity and rights regardless of genetic characteristics. It then states '[t]hat dignity makes it imperative not to reduce individuals to their genetic characteristics and to respect their uniqueness and diversity' (Article 2(b)). Article 6 prohibits discrimination based on genetic characteristics that is intended to infringe or has the effect of infringing human rights, fundamental freedoms, and human dignity. The UDHGHR also proposes strict rules for research on the human genome (see, in particular, Articles 10, 11, and 12), and endorses the principle of prior, free, and informed consent (Article 5) and respect for confidentiality (Article 7). Important principles regarding solidarity and international cooperation are set down in the Declaration (Articles 17, 18, and 19). For example, Article 17 provides that states should respect and promote 'solidarity towards individuals, families and population groups who are particularly vulnerable to or affected by disease or disability of a genetic character. They should foster, *inter alia*, research on the identification, prevention and treatment of genetically based and genetically influenced diseases, in particular rare as well as endemic diseases which affect large numbers of the world's population.' In addition, in Article 18, states are directed to 'continue fostering the international dissemination of scientific knowledge concerning the human genome' and to foster cooperation, 'particularly between industrialized and developing countries.'[16]

Attempts to protect against genetic discrimination have also been made at the regional level, such as through the 1997 European Convention for the Protection of Human Rights and Dignity of the Human Being with Regard to the Application of Biology and Medicine (the Oviedo Convention). This Convention bans all forms of discrimination based on a person's genetic make-up (Article 11) and allows predictive genetic tests only for health or scientific research purposes (Article 12). It also sets out rules for medical research and recognizes a patient's right to know (including a right *not* to know if the patient so wishes).[17] Prior to

[16] See also Art 19. Francioni, 'Genetic Resources, Biotechnology and Human Rights: The International Legal Framework', in Francioni (ed), *supra* note 7, at 3 notes that 'the more genetic tests and therapies are made available, the greater the gap will grow between the fortunate who have access to such tests and therapies and those who do not. This new "discrimination" would run along the fault line that separates the rich world from the less developed world' (p 21).

[17] The Explanatory Report also recognizes that: '[w]hereas the term "discrimination" usually has a negative connotation in French, this is not necessarily the case in English (where one must use the expression "unfair discrimination"); it has, however, been decided to keep the same term in both languages, as it is in the European Convention of Human Rights and in the case law of the Court. Discrimination here must, therefore, in French as in English, be understood as unfair discrimination. In particular, it cannot prohibit positive measures which may be implemented with the aim of re-establishing a certain balance in favour of those at a disadvantage because of their genetic inheritance.'

ratification, each state has to bring its laws into line with the Convention. This may require a change in the law or a new law, and any such legislation must include legal sanctions and require compensation for individuals who have suffered undue harm following medical treatment or research.

It is now clear that the Oviedo Convention and the UDHGHR were the first instruments in an ongoing engagement by international law with the human genome and questions of genetic discrimination. For instance, in Resolution 2001/39 on Genetic Privacy and Non-Discrimination, the Economic and Social Commission (ECOSOC) urged states to ensure that no one shall be subjected to discrimination based on genetic characteristics and to take measures to prevent the use of genetic information and testing leading to discrimination or exclusion against individuals, particularly in social, medical, or employment-related areas, whether in the public or the private sector. A later resolution, of 21 July 2004, on the same topic, contains similar injunctions.[18] In addition, the World Health Organization (WHO) International Ethical Guidelines for Biomedical Research Involving Human Subjects, which were revised in 2002, state in Guideline 18, on issues of confidentiality in genetic research, that an investigator who proposes to perform genetic tests of known clinical or predictive value on biological samples that can be linked to an identifiable individual must obtain the informed consent of the individual or, when indicated, the permission of a legally authorized representative.[19]

In October 2003, UNESCO adopted the International Declaration on Human Genetic Data (IDHGD). This normative instrument establishes the principles that should govern the collection, processing, use, and storage of human genetic data.[20] Article 1(c) of the Declaration establishes an important exception: it provides that the IDHGD applies to the collection, processing, use, and storage of human genetic data, human proteomic data, and biological samples, *except* in the case of investigation, detection, and prosecution of criminal offences and in parentage testing that are subject to domestic law that is consistent with the international law of human rights.[21]

The aims of the IDHGD are threefold: first, to guarantee the respect of human dignity and protection of human rights and fundamental freedoms in the collection, processing, use, and storage of human genetic data, human proteomic data, and biological samples, in keeping with the requirements of equality, justice, and solidarity, while giving due consideration to freedom of thought and expression, including freedom of research; second, to set out the principles which should guide states in the formulation of their legislation and their policies on

[18] See ECOSOC Res. 2004/9 on Genetic Privacy and Non-Discrimination.

[19] *Supra* note 12.

[20] Article 2(i) defines 'human genetic data' as '[i]nformation about heritable characteristics of individuals obtained by analysis of nucleic acids or by other scientific analysis.'

[21] See also Art 5.

these issues; and finally, to form the basis for guidelines of good practice in these areas for institutions and individuals handling genetic data.

Article 3 is one of the most important articles in the IDHGD. It states that '[e]ach individual has a characteristic genetic make-up. Nevertheless, a person's identity should not be reduced to genetic characteristics, since it involves complex educational, environmental and personal factors and emotional, social, spiritual and cultural bonds with others and implies a dimension of freedom.' Article 7 also deserves mention. It addresses non-discrimination and non-stigmatization, and provides that:

(a) Every effort should be made to ensure that human genetic data and human proteomic data are not used for purposes that discriminate in a way that is intended to infringe, or has the effect of infringing human rights, fundamental freedoms or human dignity of an individual or for purposes that lead to the stigmatization of an individual, a family, a group or communities.
(b) In this regard, appropriate attention should be paid to the findings of population-based genetic studies and behavioural genetic studies and their interpretations.

An interdiction of discrimination and stigmatization close to that in the IDHGD can be found in Article 11 of UNESCO's Universal Declaration on Bioethics and Human Rights (UDBHR) adopted in 2005.[22] Of course, as is the case when the term 'grounds of discrimination' is used in human rights documents, further interpretation is needed. The aim of the anti-discrimination provisions outlined above is to prohibit discrimination that impacts on human rights, fundamental freedoms, and human dignity. Academic literature, however, is divided on the issue of how to define genetic discrimination. One early proposal defined genetic discrimination as 'discrimination against an individual or against members of that individual's family solely because of real or perceived differences from the "normal" genome of that individual',[23] distinguishing it from 'discrimination based on disabilities caused by altered genes.' On this view, the criterion for genetic discrimination is whether the disease has actually occurred or not. If the disease has not yet occurred, the discrimination is termed 'genetic'; but, if the disease is expressed, it becomes another type of discrimination, most likely discrimination based on someone's health condition.

The question of whether genetic discrimination is significantly dissimilar from discrimination on the basis of general health status has been prevalent in the

[22] On the UDBHR, see eg Boussard, 'The "Normative Spectrum" of an Ethically-inspired Legal Instrument: The 2005 Universal Declaration on Bioethics and Human Rights', in Francioni (ed), *supra* note 7, at 97.

[23] Billings *et al*, 'Discrimination as a Consequence of Genetic Testing' (1992) 50 American J of Human Genetics 476. Rothstein and Anderlik, 'What Is Genetic Discrimination, and When and How Can It Be Prevented?' (2001) 3 Genetics in Medicine 354 use the term to describe 'differential treatment based on genetic status'. In contrast, in one of the first studies analysing occurrences of genetic discrimination, Yesley mentions that 'the quintessential feature of genetic discrimination is the use of genetic information about an asymptomatic person.'

literature. Critics of laws against genetic discrimination argue that genetic discrimination is no different from discrimination on the basis of health status and that such discrimination is essentially at the centre of the suitable management of insurance.[24] Some authors consider that there is no real cause to distinguish between those with a genetic predisposition to disease and those who become ill with disease, though, to a certain extent, rather than concluding that no anti-discrimination laws are essential, they see existing laws as merely a first step in the right direction: the second step, then, would be a generic law that protects people from discrimination on the basis of health status.[25]

Another proposal is that legislation should sketch a difference between discrimination on the basis of genotype and discrimination on the basis of phenotype. It has been suggested that such a law would forbid health insurers from charging higher rates on the basis of information about unexpressed genetic traits, regardless of the source of the information. In other words 'genetic information should thus be defined broadly to encompass any . . . information that provides probabilistic information about a person's genotype . . . from genetic tests, other medical tests, family history, diagnoses of traits or conditions, or the taking of (or even making inquiries about) a genetic test.'[26] In order to avoid also forbidding discrimination on the basis of health status generally, Greely proposes allowing discrimination when the trait has already manifested itself in the form of illness or disability. He considers that once the genetic predisposition is evident as illness, or in the form of a medically relevant symptom, it should not be forbidden for it to be taken into account.[27]

A different perspective, focusing on the US context, is offered by Gulati, who uses social justice considerations to make a case against anti-discrimination law focusing on genetics. He argues that genetic differentiation should not be tolerated, not because there is anything unique about genes, but because no one should be a citizen of the United States and be forced to sacrifice their economic security for the sake of obtaining necessary medical services.[28]

[24] See eg Rothstein and Anderlik, *ibid*; Suter, 'The Allure and Peril of Genetic Exceptionalism: Do We Need Special Genetics Legislation?' (2001) 79(3) Washington ULQ 669; Wolf, 'Beyond "Genetic Discrimination": Toward the Broader Harm of Geneticism' (1995) 23(4) J of Law, Medicine & Ethics 345.

[25] See Hellman, 'What Makes Genetic Discrimination Exceptional?' (2003) 29(1) American J of Law and Medicine 77.

[26] See Greely, 'Genotype Discrimination: The Complex Case for Some Legislative Protection' (2001) 149 U Pennsylvania LR 1483, at 1495.

[27] Greely's resolution of the definitional dilemma adopts the meaning originally proposed by Billings *et al*, *supra* note 23, in an early piece on genetic discrimination. In Greely's view, the definition is 'conceptually straightforward though perhaps complicated to implement.' See also Hellman, *supra* note 25.

[28] As long as the oppression of having to live without health insurance is limited to the working poor and others at the margin, the goal of universal health insurance will remain unrealized. However, once interests converge and the oppression threatens to afflict those in the higher socio-economic classes, social movements for political change become possible.

Gulati advances three main points in support of his argument: first, that genetic anti-discrimination legislation, standing within a commercial health insurance system, is normatively inferior; second, that ultimately genetic anti-discrimination laws will fail and spur a social movement for a radical reorientation of the health insurance system in the United States; and finally, that genetic anti-discrimination legislation delays the formation of the social movement and in so doing helps to maintain the current commercial health system.[29]

In my view, singling out genetic discrimination is necessary for at least two reasons. On the one hand, it is necessary from the perspective of human rights that acknowledges the increased protection that could be provided by a specific body of law. On the other hand, it appears necessary from a utilitarian perspective, as it could help to reduce fear and thereby facilitate scientific research that contributes not just to the progress of knowledge, but also helps to relieve human suffering and improve the health of individuals and humanity as a whole. Also, although I acknowledge what Gulati describes as the 'normative inferiority' of legislation that forbids genetic discrimination, I do not believe that genetic anti-discrimination laws will slow or stop reform of the US health system. In addition to the fact that such a development seems unlikely in practice, it would be wrong to accept the acknowledgement of a human right at the expense of another right. In sum, then, although in general in the international law of genetics, human dignity and utilitarian perspectives stand in direct contrast,[30] in this particular circumstance they converge.

2. The Challenge of Privacy in the Genetic Context

In the past, in the arena of genetics, most commentators, advocates, and policy-makers have tended to separate privacy from non-discrimination, using different approaches to tackle these issues.[31] In my view, however, the best way to avoid discrimination of all kinds is to use a two-sided approach. First, interrupt the right to use information concerning the trait in question, whether national origin, religion, disability, or genetic predisposition, where suitable and possible. This demonstrates a strict 'privacy' method. Then, forbid the use of any information obtained despite the attempt to shut down the flow of information.

In an era in which technology has enhanced the aptitude of physicians and others to gather and distribute patient data, it is of enormous significance that we protect individual privacy and the confidentiality of genetic information in the

[29] Gulati, 'Genetic Antidiscrimination Laws in Health Insurance: A Misguided Solution' (2001) 4 Quinnipiac Health LJ 149, at 168.

[30] For accounts of this contrast, see the chapters by Roger Brownsword and Han Somsen in this volume.

[31] See eg the US Health Insurance Portability and Accountability Act 1996 (HIPAA) which contains entirely separate titles and tracks for protecting privacy and protecting against inappropriate uses of medical information, including genetic information, by group health plans and health insurers.

health care setting.[32] As one scholar has stated, 'restrictions on disclosure will become more desirable from the viewpoint of patients as medicine is able to determine distant future risks of developing diseases' by means of genetic testing.[33] There are, however, legitimate reasons for disclosure of patient information in the absence of specific consent. Consequently the law permits and occasionally requires such disclosure, for instance when third parties may be at risk of injury due to their lack of knowledge of such information. However, although in some cases the danger may stand out as unambiguous, in many cases there will be substantial doubt and the decision about disclosure will be extremely difficult. For example, if a person tests positive for a foremost genetic trait[34] which carries a serious forecast, and which also carries a certain risk for first-degree relatives, would this be a basis for disclosure to such relatives even in the face of the tested person's objection?[35]

In the case of genetic screening at work, if the employee does consent, then such consent should be construed narrowly.[36] In addition, provisions should be enacted to guarantee that genetic information is not distorted. Such a stipulation might forbid employers or unions from making employment decisions based on genetic information if the genotype has no clear relation to the capability of performing the job. Further protections from misuse could prohibit employers from using the information for health benefit decisions.[37]

The IDHGD is the most important international instrument on genetic privacy. As noted earlier, it applies to the collection, processing, use, and storage of human genetic data, human proteomic data, and biological samples, except in the case of the investigation, detection, and prosecution of criminal offences and in parentage testing, which are provided for by domestic law that is consistent with the international law of human rights.[38] Not surprisingly, the IDHGD places particular emphasis on the need for consent.[39] Article 8 indicates that

[32] See Balint, 'Issues of Privacy and Confidentiality in New Genetics' (1998) 9 ALJST 27, at 27.

[33] Medical records are now routinely examined in the US by auditors from insurance companies to see if they fully document the services rendered and billed for by the physician. This occurs exclusive of the patient's express permission, and thus again is a breach of privacy and confidentiality. In addition, in the hospital setting, increasing numbers of individuals regularly look at the patient's record. See Kotval, 'Market-Driven Managed Care and the Confidentiality of Genetic Tests: The Institution as Double Agent' (1998) 9 ALJST 1.

[34] Such situations might arise with genetic testing for Huntington's disease or hereditary breast cancer.

[35] See, further, the chapter by Hélène Boussard in this volume.

[36] See, generally, King *et al*, 'Workplace Privacy and Discrimination Issues Related to Genetic Data: A Comparative Law Study of the European Union and the United States' (2006) 43 American Business LJ 79. For discussion of European data protection measures, see the chapter by Francesca Bignami in this volume.

[37] Andrews and Jaeger, 'Confidentiality of Genetic Information in the Workplace' (1991) 17 (1–2) AJLM 75.

[38] Article 1(c).

[39] See similarly the non-binding Charter of Fundamental Rights of the European Union, Art 3 which provides that in order to ensure the integrity of the person, it is essential that in the fields of medicine and biology: 'the following must be respected in particular: the free and informed consent of the person concerned, according to the procedures laid down by law; the prohibition of eugenic practices, in particular those aiming at the selection of persons.'

'[p]rior, free, informed and express consent, without inducement by financial or other personal gain, should be obtained for the collection of human genetic data, human proteomic data or biological samples, whether through invasive or non-invasive procedures, and for their subsequent use and storage, whether carried out by public or private institutions.' It also makes it clear that any limitations on the principle of consent 'should only be prescribed for compelling reasons by domestic law consistent with the international law of human rights.'[40]

Privacy and confidentiality are addressed in Article 14 of the IDHGD.[41] Article 14(a) indicates that states should work to protect the privacy of individuals and the confidentiality of human genetic data linked to an identifiable person, family, or group in accordance with domestic law consistent with the international law of human rights. Article 14(b) provides that:

> Human genetic data, human proteomic data and biological samples linked to an identifiable person should not be disclosed or made accessible to third parties, in particular, employers, insurance companies, educational institutions and the family, except for an important public interest reason in cases restrictively provided for by domestic law consistent with the international law of human rights or where the prior, free, informed and express consent of the person concerned has been obtained provided that such consent is in accordance with domestic law and the international law of human rights. The privacy of an individual participating in a study using human genetic data, human proteomic data or biological samples should be protected and the data should be treated as confidential.

Thereafter, Article 14(c) and (d) address human genetic material collected for medical and scientific purposes and Article 14(e) closes out the provision by emphasizing that human genetic data and human proteomic data should not be kept in a form which allow the data subject to be identified 'for any longer than is necessary for achieving the purposes for which they were collected or subsequently processed.'

3. From International Law to Domestic Implementation: Insurance, Employment, and Criminal Investigation

A. Insurance and Employment

As Francesco Francioni has pointed out, insurance and employment are areas where 'the risk of discrimination on a genetic basis is the highest and most disturbing'. And, as he goes on to explain, there is a range of questions that need to be addressed in order to prevent discriminatory practices, including '(1) whether

[40] For detailed analysis of the problems concerning consent in the field of genetic research, see the chapter by Hélène Boussard in this volume. See, more generally, D. Beyleveld and R. Brownsword, *Consent in the Law* (2007).

[41] See also Art 15 which addresses the security of genetic data.

insurers and employers may be allowed to require genetic tests as a condition of insurance or employment; (2) whether insurers or employers may require disclosure of prior genetic tests by the applicant; and (3) whether insurers or employers may give weight for business purposes to genetic information voluntarily provided by applicants.'[42]

Legislation that regulates the use of genetic data in insurance and employment needs to be understood within the broader contexts of the right to work, the right to health, and citizenship rights. In most European states, for example, health insurance is linked to citizenship rights; hence, genetic risks neither contribute to an increase in insurance rates, nor do they determine exclusion from the insurance system. It is only in the case of life, injuries, or disability insurance that they may incline the balance in a certain way. In the US, however, where many people do not benefit from health insurance, the situation is different and it has been argued that focused protection against genetic discrimination is unfair when so many are excluded from health insurance on other grounds. Moreover, although the consequences are less severe when it comes to life insurance, it has also been argued that prohibiting the use of genetics in insurance underwriting, while allowing similar non-genetic risk factors to be used, is unfair.[43]

Explicit legislation prohibiting the use of genetic testing for insurance purposes exists in Austria, Belgium, Denmark, and Norway.[44] In these countries, insurers can neither request genetic testing, nor use the test results of applicants available in medical records. Norwegian law further prohibits insurers from even asking whether a genetic test has been carried out.[45] In the Netherlands, a 1997 statute on medical examinations may pose a barrier to the use of genetic testing for insurance purposes. Article 3 provides that a medical test must not cause an excessive invasion into the privacy of the individual being tested. It further provides that a medical examination cannot be undertaken if the risks of the test, as well as the risk of receiving information on an untreatable illness, prevail over

[42] Francioni, *supra* note 16, at 22. He concludes that '[p]rima facie, the answer to these questions appears to be negative in the light of the norms contained in universal and regional instruments on bioethics.'

[43] Lemmens, 'Selective Justice, Genetic Discrimination, and Insurance: Should We Single Out Genes in Our Laws' (2000) 45 McGill LJ 347, at 383 argues that statutes singling out genetic susceptibility as a category, and offering it much wider protection than other similar health conditions, although intended to promote equity in access to social goods, may themselves be ineffective and to some extent even inequitable. Lemmens suggests that many policy analysts in the US have embraced anti-discrimination and privacy legislation focusing on genetics as an imperfect way of protecting an already insufficient degree of access to health care.

[44] The European Trade Union Confederation (ETUC) has called for such a prohibition to be included explicitly in a European Commission directive on the protection of workers' personal data. ETUC claims that gene testing is not yet a problem in EU workplaces and EC legislators should make sure this remains the case. It argues that tests would reduce efforts to prevent exposure to occupational hazards, in particular in the chemical field, and introduce discrimination among workers according to certain genetic characteristics.

[45] Lemmens, *supra* note 43, at 383.

the expected advantage of the test, thereby placing a disproportionate burden on the individual. The law thus embraces a proportionality test for the determination of the acceptability of genetic tests.[46] In France, the National Ethics Committee (*Comité National d'Ethique*) recommended in October 1995 that insurers should be forbidden from using any genetic information, even if applicants submit it freely. A year earlier, a law 'On respect for the human body' introduced new provisions on genetic testing and DNA identification into the French Civil Code. According to Article 16-10, a genetic report of the characteristics of a person may be undertaken only for medical purposes or for scientific research. This seems to forbid insurers from using genetic tests for underwriting purposes but it does not prevent them from obtaining genetic test information from medical files.[47]

In Austria, Denmark, Norway, the Netherlands, and Belgium, genetic testing for employment purposes is illegal.

In the United States, most of the legislative activity involving issues of genetic privacy and genetic discrimination has occurred at the state,[48] as opposed to the federal, level.[49] About forty states have laws that address some aspect of genetic privacy and discrimination. Sixteen states require informed consent for a third party to perform or require a genetic test or obtain genetic information, while twenty-three require the informed consent of the subject before disclosing genetic information to a third party. Thirty-two states have passed laws concerning

[46] *Ibid.*

[47] France, Law no 2002-303 of 4 March 2002 relating to the rights of the patients and the quality of the system of health (1) Art 98: '*Les entreprises et organismes qui proposent une garantie des risques d'invalidité ou de décès ne doivent pas tenir compte des résultats de l'examen des caractéristiques génétiques d'une personne demandant à bénéficier de cette garantie, même si ceux-ci leur sont transmis par la personne concernée ou avec son accord. En outre, ils ne peuvent poser aucune question relative aux tests génétiques et à leurs résultats, ni demander à une personne de se soumettre à des tests génétiques avant que ne soit conclu le contrat et pendant toute la durée de celui-ci.*' See also Lemmens, *supra* note 43.

[48] Because of federal law pre-emptions, state laws do not protect the nearly one-in-three Americans who get their health insurance through their employer.

[49] On 24 April 2008, the US Senate passed the Genetic Information Nondiscrimination Act of 2008. But similar legislation has been introduced in every Congress since 1995, and President George W. Bush announced his backing for a law against genetic discrimination as early as June 2001. The House of Representatives is the sticking point: knowledgeable observers consider the prospects for passage of the House equivalent of this bill to be only slightly better than they were in 2003, when the Genetic Information Nondiscrimination Act of 2003 passed the Senate by a vote of 95 to 0 but never reached the House floor. See Greely, 'Banning Genetic Discrimination' (2005) 353(9) New England J of Medicine 86. The bill would prohibit health insurance plans from denying enrolment or charging premiums on the basis of an individual's or family member's genetic information. It also prohibits health insurers from basing premiums of a group health plan on the basis of genetic information of plan members or their families. The bill prohibits disclosures or collection (requesting, requiring, or purchasing) of genetic information for underwriting purposes. In addition, it prohibits the use of genetic information in employment decisions and applies the same procedures and remedies as apply to other forms of employment discrimination. Following the model of the HIPAA Privacy Rule, it provides basic protections for genetic information while permitting greater protection under other federal and state measures. Congress has certainly taken notice of the issue.

genetic discrimination in employment. At least thirty-four states prohibit the use of genetic information for certain health insurance purposes. Additionally, three states require actuarial justification for the use of genetic information. Thirty states restrict employers from discriminating based on genetic information, but the laws are varied and contain different definitions and loopholes.

Looking at the employment context at the federal level, there is some protection against discriminatory use of genetic information in the workplace. The existing protection, however, falls short of the comprehensive law needed to protect the rights of workers. In 1995 the Equal Employment Opportunity Commission (EEOC) interpreted the Americans with Disabilities Act (ADA)[50] as protecting a person's genetic predisposition to disease, but the interpretation did not address whether someone can be denied a job because he or she is a carrier of a recessive disorder (and the potential employer does not want to pay health costs associated with potential or actual affected children), nor did it prevent employers from requiring that potential employees undergo genetic testing. In 2002, the EEOC filed a petition for a preliminary injunction against Burlington Northern Santa Fe Railroad, looking for an end to the practice of genetic testing of all employees who filed claims for work-related injuries based on carpal tunnel syndrome. This was the first suit by the EEOC challenging genetic testing. It alleged that the employees were not told of the genetic test or asked to consent to it. Furthermore, in at least one instance, an employee who refused to give a blood sample because he suspected that it would be used for genetic testing was purportedly threatened with expulsion if he failed to provide the sample. Burlington Northern asked employees to agree to blood tests to see if they carried a genetic trait called Chromosome 17 deletion; some studies imply that a person with that attribute is more likely to experience some forms of carpal tunnel syndrome. Within days of the EEOC action, Burlington Northern agreed to stop requiring genetic testing of employees who filed claims for carpal tunnel syndrome.[51]

As noted earlier, at the state level there are laws that prohibit employers from discriminating against people on the basis of genetic tests but the specifics of what constitutes unlawful discrimination vary widely. Some definitions of genetic information only include the employed individual, while others include family members as well; some only protect information from genetic tests and others include family histories and other sources of genetic information. The laws vary

[50] Americans with Disabilities Act, 42 USC section 12112 (1992).

[51] *EEOC v Burlington Northern Santa Fe Railway Co*, No 02-C-0456 (ED Wis 2002). A similar case occurred in Hong Kong where a court ruled in favour of compensation for three persons who had been refused employment on the ground that their parents suffered from schizophrenia: see ILO, 'Equality at Work: Tackling the Challenges. Global Report under the Follow-up to the ILO Declaration on Fundamental Principles and Rights at Work.' International Labour Conference, 96th Session 2007, Report I(B), at 49.

as to whether employers may request or require genetic testing and how, if at all, that information may be disclosed or used. Most of the state laws do not protect people from discrimination based on genetic information about their family members, thereby allowing insurers to easily circumvent the reach of the laws by basing their decisions on family histories. Few of the laws would prohibit an insurer from circumventing the law by discriminating against people based not on their test result, but on the fact that they previously requested genetic services. Some laws focus on tests of DNA and RNA, others only on tests of genes or chromosomes, or on gene products such as proteins or enzymes. Some protect people from discrimination based only on genetic test results, but would not prohibit discrimination based on genetic information gathered by other means. Others define a 'genetic test' narrowly so that protections apply only to testing for the presence or absence of genes. At the federal level, the Health Insurance Portability and Accountability Act of 1996 (HIPAA) provides some protection against genetic discrimination, such as barring the use of genetic information in denying or limiting health insurance coverage for members of a group plan.

B. Criminal Investigation

DNA forensics can be seen as merely the latest in a long line of biologically based identifying law enforcement technologies that incorporate fingerprints and serotyping. Yet, DNA has qualities that make it meaningfully dissimilar from its predecessors with respect to the human rights apprehension it raises. DNA is predictive for sensitive information such as an individual's hereditary diseases, phenotypic propensities, and familial relations. Secondly, it can be amplified from minute quantities, and because it is shed in detectable quantities in the form of sloughed skin, hand smudges, hair follicles, and saliva residues, it is more abundant as a source of evidence than fingerprints and serotypes. Thirdly, DNA profiles are partially shared with biological relatives, which means an offender's profile cannot be collected without consequentially obtaining incomplete information about non-offending relatives' profiles. Moreover, certain alleles and allele frequencies are predictive for biologically defined racial and ethnic categories. Finally, DNA is a relatively durable material, and therefore retains its 'informational content' much longer than would other sources of biological evidence. Together, these properties qualitatively distinguish DNA samples and profiles from fingerprints and militate against convenient analogies to fingerprints.[52]

DNA evidence raises several questions related to genetic discrimination and the protection of genetic privacy. Human rights organizations, such as Human

[52] See Kimmelman, 'Risking Ethical Insolvency: A Survey of Trends in Criminal DNA Databanking' (2000) 28 J of Law, Medicine & Ethics 28.

Rights Genes, have recognized the important role played by DNA evidence in criminal investigations, but remain opposed to the permanent retention of all DNA samples. They argue, first, that retention brings an increasing threat to genetic privacy if information is revealed about health or family relationships, not just identity; secondly, that it creates a permanent 'list of suspects', including anyone arrested for a wide range of offences; and thirdly, that it increases the potential for discrimination (particularly against ethnic minorities) by permanently retaining samples taken before an individual is charged, even when DNA evidence is not relevant to the investigation.

Using DNA as an investigative tool also raises the issue of how to obtain a suspect's DNA. Currently there are three ways to get a DNA sample from a known suspect: voluntary submission; DNA abandonment; or a court order. Getting a DNA sample by voluntary submission is the least problematic of the three. Although obtaining a DNA sample from a person's body is considered a search, an individual may consent to a search. DNA abandonment occurs when a person discards a personal item containing DNA. Abandonment is also usually not problematic as long as the evidence was truly abandoned, and not stolen or retrieved by an illegal search and seizure. Obtaining a court order is more difficult and usually implicates a seizure clause.

In England and Wales, where DNA forensics originated, the police are authorized to collect 'non-intimate' samples (such as hair follicles) from anyone questioned or arrested for any recordable offence, which is generally defined as an offence that could lead to imprisonment. The practice of 'mass intelligence screening', colloquially known as 'sweeps' or 'bloodings', is also permitted.[53] The samples are linked to the person's record on the National DNA Database, which includes their DNA profile (a string of numbers obtained from part of the sequence of the DNA). The samples are stored permanently by the companies that analyse them, for an annual fee.

In the US, federal criminal DNA databanking legislation was first passed in the 1994 Violent Crime Control and Law Enforcement Act, which authorized the

[53] Persons living near a crime scene are asked to voluntarily submit DNA samples that can be checked against a crime scene sample, allowing police quickly (if expensively) to rule out suspects. Other countries have conducted genetic sweeps as well, including Australia, Canada, and Germany; the latter collected DNA from 12,000 volunteers for a single crime in 1998. Genetic sweeps raise important questions about privacy and personal integrity. According to British authorities, samples collected from volunteers are not entered into the general database but are instead compared against the profiles of a specific crime scene sample. Profiles and tissue samples are destroyed after completion of an investigation, though they may be retained until conviction. However, the notion that sweeps are voluntary is unrealistic given the coercive pressures of heavy media attention, widespread community participation, and the fact that failure to submit a tissue sample may in itself invoke police submission. The Police Federation of Australia opposed proposed legislation, requiring police to provide genetic samples for the purpose of eliminating them as a source of crime scene contamination. Concerns were expressed that such a requirement is quite different in scope from the requirement to provide fingerprints on file, in that it is capable of being used for potentially discriminatory purposes by the employer.

FBI to establish a software system for sharing information contained in state DNA databases (the Combined DNA Identification System, or CODIS).[54] The Act provided a substantial boost to state DNA databanks and, since 1994, thirty-five new databanking statutes have been passed. The argument can be made that removal of individual identifiers prevents any harms to the offender providing the tissue sample, since third parties like employers would be unable to link a sample to an individual, and stigmatizing or medical information would not be traceable to an individual. On the other hand, release of tissue samples, anonymized or otherwise, involuntarily involves offenders in research they might not have consented to; and requiring authorization before samples are anonymized and released for research recognizes and respects the diversity of attitudes individuals may have about supporting certain types of research. Clearly, there are compelling reasons for collecting tissue samples from offenders without their consent, but there are no clear reasons why offenders should be denied the right to exercise their choice to opt out of biomedical or forensic research—in particular, research that could be objectionable on racial or social grounds. Anonymizing DNA samples for research purposes without first obtaining consent from their source does not violate any US federal regulations. It is, nevertheless, inconsistent with the recommendations of several commentators as well as the American College of Medical Genetics. One can furthermore question whether it is truly possible to anonymize tissue samples that have been collected for the contradictory purpose of maintaining identifying information.[55]

Defendants have challenged the constitutionality of DNA collection as an unreasonable search and seizure under the Fourth Amendment of the US Constitution.[56] The US Supreme Court has deemed the involuntary taking of a biological sample a 'search' under the Fourth Amendment; therefore, the forced collection of DNA must fall within an exception that allows for the collection of DNA without the existence of probable cause or reasonable suspicion. One such

[54] CODIS includes a Convicted Offender Index that stores the DNA profiles of people convicted of felony sex offences as well as other violent crimes, giving law enforcement agencies a powerful tool to identify repeat offenders and link together crime scenes. CODIS also contains a Forensic Index, which compiles DNA profiles developed from evidence left at crime scenes. Matches made in the Forensic Index can link crime scenes together and identify the possibility of a serial offender. Another feature of CODIS is its Population File, an anonymous database of DNA profiles of the general population that determines the statistical significance of a match. CODIS also includes a National DNA Index System (NDIS), which enables participating agencies to exchange and compare DNA profiles on a national level. Currently, all fifty states, as well as Puerto Rico and the United States Army, participate in NDIS.

[55] See Lewis, 'The Role Genetic Information Plays in the Criminal Justice System' (2005) 47 Arizona L Rev 519.

[56] 'The right of the people to be secure in their persons, houses, papers, and effects, against unreasonable searches and seizures, shall not be violated, and no Warrants shall issue, but upon probable cause, supported by Oath or affirmation, and particularly describing the place to be searched, and the persons or things to be seized.'

exception is the 'special needs exception'.[57] Most of the courts that have addressed this issue, including the US Supreme Court, have characterized the collection of a DNA sample from a prisoner or probationer as a search and seizure, but accepted that the collection is reasonable in relation to the compelling state interests of reducing recidivism and ensuring accuracy, in light of the reduced privacy expectation of prisoners and probationers, and the minimal bodily intrusion involved in the procedure.

In the US, defendants could argue that a discriminatory purpose can be inferred based on the arrest and conviction rates that vary greatly between the sexes and among different races. In some, predominantly black, neighbourhoods, up to ninety per cent of adult males have been arrested. Nationally, a black person is five times more likely to be arrested than a white person. Therefore, if DNA is collected from arrestees, or even just felons, convicted of certain crimes, then a black person is more likely than a white person to be identified from DNA left at future crime scenes. But while this disparity shows a clear discriminatory impact on black males, as they will more often be subjected to DNA profiling, it does not show discriminatory purpose and statistical evidence of disparate impact, without more, is not enough. It is unlikely, however, that defendants will be able to prove discriminatory intent behind the legislation unless evidence of a discriminatory purpose is expressed either on the face of the statute or in its legislative history. Canada's law shows greater solicitude on the issue of privacy. Canada's databanking system divides crimes into two categories: primary offences, which include murder and rape, and secondary offences, which include robbery, assaults, and arson. Canadian law mandates databanking all individuals convicted of primary offences and provides provincial judges with the option of collecting tissue samples from individuals convicted of secondary offences. Together, primary and secondary offences cover a range of criminal offences similar to that of most US states: serious property crimes such as arson, sexual offences, and violent crimes against persons. Also, as is the case in the US and the UK, samples are retained indefinitely, on the premise that forensic technologies are rapidly developing. However, in contrast to the US, Canada automatically expunges records and samples from any acquitted individuals, and provides a schedule for elimination of juvenile samples and records. Finally, the Canadian databanking law does not authorize using samples collected from offenders for research of any kind, including construction of a population database from anonymized samples of those who have been acquitted.

[57] This exception allows for a search when designed to serve 'special needs, beyond the normal need for law enforcement.' Courts have utilized this exception in cases involving random drug screening in high schools and sobriety checkpoints. Unsurprisingly, defendants have argued that a DNA sample collection for a government database does not come within the special needs exception. In making this determination, courts must examine whether the purpose of the challenged search, in this case the collection of DNA, goes 'beyond the normal need for law enforcement.'

The issue of *collection* of human genetic and proteomic data is addressed in Articles 8–12 of the IDHGD, *processing* in Articles 13–15, *use* in Articles 16–19, and *storage* in Articles 20–22. As regards collection of samples for forensic medicine or in civil, criminal, and other legal proceedings, Article 12 provides that '[w]hen human genetic data or human proteomic data are collected for the purposes of forensic medicine or in civil, criminal and other legal proceedings, including parentage testing, the collection of biological samples, *in vivo* or *post-mortem*, should be made only in accordance with domestic law consistent with the international law of human rights.' As regards destruction of samples, Article 21(b) and (c) emphasize that:

(b) Human genetic data, human proteomic data and the biological samples collected from a suspect in the course of a criminal investigation should be destroyed when they are no longer necessary, unless otherwise provided for by domestic law consistent with the international law of human rights.
(c) Human genetic data, human proteomic data and biological samples should be available for forensic purposes and civil proceedings only for as long as they are necessary for those proceedings, unless otherwise provided for by domestic law consistent with the international law of human rights.

As noted earlier, the IDHGD does stipulate an exception for criminal investigations (Article 1(c)). However, it also specifies that any such exception has to be 'subject to domestic law that is consistent with the international law of human rights'. From this optic, therefore, I would argue that time limits on the retention of records on a database could provide an important safeguard, limiting the potential for misuse by future governments. For example, records for convicted murderers and rapists could be kept permanently, but other records removed after fixed time-periods, depending on the nature of the offence.

4. Vulnerable Groups

The final issue concerning genetic discrimination to be addressed in this essay is vulnerable groups. Women, children, the disabled, and indigenous peoples are amongst the groups that have been singled out as especially vulnerable to genetic discrimination. As regards the first of these, women, it has been argued that the patenting of human gene sequences represents a new enclosure of a long-accepted argument made by many pro-choice feminists and women's health advocates, who oppose new genetic and reproductive technologies that put corporate profits over women's health and well-being. Feminists have also raised concerns about the commodification of reproduction and human relationships that may follow from increased recourse to genetic reproductive technologies. In particular it has been argued that genetic testing and screening, both prenatally and in combination with IVF procedures, threaten new forms of discrimination and stigmatization. Sex selection for 'non-medical' reasons is prohibited by Article 21 of the WHO Draft Guidelines for Bioethics, presented at the World Health Assembly in 1999,[58] Article 14 of the Oviedo Convention,[59] and Article 68 of the Report of the International Bioethics Committee of UNESCO on Pre-Implantation Genetic Diagnosis and Germ-Line Intervention.[60]

One of the most sensitive issues in genetic discrimination is that of 'experiments', or non-therapeutic research, on children, which leads us back to the Nazi eugenics age. In order to prevent this, the UNESCO Universal Declaration on Bioethics and Human Rights (UDBHR), from 2005, dedicated a special article, Article 7, to persons who do not have the capacity to consent. The Article not only repeats the stipulations of other international documents that say that authorization for such research should be based on the best interests of the person, it also stipulates that, as a general rule, research should only be carried out where it is for the direct benefit of the health of a person who does not have the capacity to consent.[61] This Article therefore lends some support to the point of view according to which non-therapeutic research on children should be banned altogether.

Concern about genetic discrimination and vulnerable groups has also been raised by others. Civil rights and human rights leaders, for instance, are wary of a new free-market eugenics that could stoke the fires of racial and ethnic hatred. Many environmentalists see human genetic modification as another potentially disruptive technology being approved before long-range consequences are considered. And disability rights leaders charge that a society obsessed with genetic perfection could come to regard disabled people as mistakes that should have been prevented. At the international level, the Declaration on the Rights of Disabled Persons, adopted in 1975, stipulated that 'whatever the origin, nature and seriousness of their handicaps and disabilities, [disabled persons] have the

[58] 'Sex is not a disease. Except for severe sex-linked genetic disorders, the use of genetic services for the purpose of sex-selection is not acceptable.'

[59] 'The use of techniques of medically assisted procreation shall not be allowed for the purpose of choosing a future child's sex, except when serious hereditary sex-related disease is to be avoided.'

[60] 'Destruction of embryos for non-medical reasons or termination of pregnancies because of the specific gender are not "counterbalanced" by avoiding latter suffering by severe disease.'

[61] Article 7 UDBHR: 'In accordance with domestic law, special protection is to be given to persons who do not have the capacity to consent: (a) authorization for research and medical practice should be obtained in accordance with the best interest of the person concerned and in accordance with domestic law. However, the person concerned should be involved to the greatest extent possible in the decision-making process of consent, as well as that of withdrawing consent; (b) research should only be carried out for his or her direct health benefit, subject to the authorization and the protective conditions prescribed by law, and if there is no research alternative of comparable effectiveness with research participants able to consent. Research which does not have potential direct health benefit should only be undertaken by way of exception, with the utmost restraint, exposing the person only to a minimal risk and minimal burden and, if the research is expected to contribute to the health benefit of other persons in the same category, subject to the conditions prescribed by law and compatible with the protection of the individual's human rights. Refusal of such persons to take part in research should be respected.'

same fundamental rights as their fellow-citizens of the same age, which implies first and foremost the right to enjoy a decent life, as normal and full as possible.'[62] And the recent Convention on the Rights of Persons with Disabilities, adopted by the UN General Assembly in late 2006 and opened for signature in March 2007, provides in Article 17 that '[e]very person with disabilities has a right to respect for his or her physical, and mental integrity on an equal basis with others.' Yet, the social model of disability, which shifts responsibility away from the disabled person's biological, psychic, or cognitive equipment and towards the social, institutional, and physical world designed with the characteristics and needs of the non-disabled in mind, has yet to gain wide currency in many societies. A public understanding of disabilities as 'illnesses' or 'deformities' to be eradicated could become especially problematic when the eradication of difference through genetic manipulation is within reach.[63] It could lead to an attempt to manipulate science to carry out a form of 'disability cleansing'. To reduce this risk, it will be important to listen to what patients' associations have to say on the subject of genetics. Patients and their families have concrete expertise of their own which must be taken into account. It is also the case that some disabled people are now seeking the right to choose to have disabled children, on the basis of new genetic screening tests that are becoming more widely available. Their argument is that they should be allowed to choose children more like themselves. The issue has divided opinions amongst obstetricians and gynaecologists, some of whom regard the notion of deliberately choosing an embryo manifesting genes for deafness or dwarfism as pandering to the desires of parents rather than reflecting the best interests of a future child.

'Scientists say it's just DNA. For an Indian, it is not just DNA, it's part of a person, it is, with deep religious significance. It is part of the essence of a person. To us, any part of ourselves is sacred.'[64] It is widely known that the genes of some indigenous peoples are of special interest to researchers because their relative homogeneity facilitates the search for correlations between specific genes and phenotypic traits. During the 1990s, the Human Genome Diversity Project undertook to collect DNA samples from hundreds of indigenous groups for this purpose. The project has since been disbanded, but indigenous peoples are still the subjects of genetic research[65] and they have raised serious concerns about a

[62] GA Res 3447 (XXX), 30 UN GAOR Supp (No 34) at 88, UN Doc A/10034 (1975).

[63] Koh, 'Different but Equal: The Human Rights of Persons with Intellectual Disabilities' (2004) 63 Maryland L Rev 1.

[64] D. Harry and F. Dukepoo, *Genes and Genetics: What Indians Should Know About the New Biotechnology* (1998).

[65] Despite the increased measures of protection, the HapMap is still vulnerable to misuse. The public nature of the project has removed questions of individual ownership and commercialization, yet unrestricted access to the database creates an opportunity for abuse. HapMap cannot guarantee that the public will not mischaracterize results of the study and associate negative results with certain groups. Unless there are regulatory measures to enforce sanctions, groups and individuals within those groups will be unprotected against discrimination and social marginalization.

range of issues, including: the patenting and commercialization of the information derived from their samples; the lack of fully informed consent by many of those from whom samples were taken; the potential for genetic discrimination based on the identification of group differences; and the disproportionate allocation of public funds to genetic research rather than to direct health care and prevention programmes. There is also a long-standing problem concerning individual consent and the usurpation of group identity in favour of an individualistic one that may not be reflected in the local culture. By conferring only with individual participants, researchers may diminish the authority of the group to make decisions concerning its members. This violates the draft United Nations Declaration on the Rights of Indigenous Peoples by compromising the right to self-determination and cultural independence. The challenge is to incorporate all levels of consent in a way that conforms to the cultural norms of a particular group.[66]

It is also the case that very little genetic diversity research has focused on the health of indigenous peoples. Yet, across populations, different genes may be implicated in what appears to be the same syndrome, for example as non-insulin-dependent diabetes mellitus. This means that diversity research could play a very important role in ensuring that therapeutic methods are effective in diverse populations. Another real danger is that research will neglect the genetic bases of disease among relatively isolated, traditional communities. A brief glance at the future plans of major pharmacogenomic companies shows that some of their work is relevant to the specific health concerns of indigenous peoples. To date, however, most pharmacogenomic research has been directed at cancer, asthma and allergies, cardiovascular disease, and neurodegenerative conditions such as Parkinson's and Alzheimer's diseases. Hence, one of the main concerns of indigenous peoples is assuring the responsiveness of the pharmacogenomic research to the needs of genetically distinct but marginalized societies, so that they fully benefit from new technologies rather than remaining—insofar as genetic research is concerned—an academic sideshow.[67]

5. Conclusion

General laws governing personal health information can frequently protect genetic information, even if these laws are sometimes inadequate. Nonetheless the existence of genetic information imparts new intensity to the need to protect personal health information. As a result, legislation to address specific issues relating to genetic testing may be required to complement existing legislation. In

[66] Khan, 'Colonialism Revisited: Insights into the Human Genome Diversity Project' (1999) 3 J of Law & Social Challenges at 345.

[67] Barsh, 'Pharmacogenomics and Indigenous Peoples: Real Issues and Actors' (2003) 11 Cardozo J of Intl and Comparative L 543.

this respect, guidelines regarding genetic discrimination should be adopted at the international level. In addition, legislation related to the taking of DNA from criminal suspects and the establishment of DNA databanks relating to convicted offenders must be carefully monitored to prevent an unwarranted enlargement of its scope.

More generally, public education is necessary in order to protect genetic privacy and prevent discrimination. Governments, in particular, have a duty to explain the uses of genetic information and their possible impact on society, and to educate the public and researchers about what the results of genetic studies mean and do not mean.[68]

Research should focus on individual variations within populations, not among populations. And researchers need to shape their studies and present their results cautiously. They should define the population being studied precisely. They should also explain to what extent the threat of disease can be ascribed to genetic variants and how such variants interrelate with environmental factors. Where these matters are not well understood, doubts should be acknowledged.

To conclude, mention needs to be made of one of the most important forms of discrimination that has not been discussed here: namely, discrimination that limits or forbids access to research results for the states of the South. Put crisply, although international law stipulates the obligation to solidarity insofar as the availability of genetic research results, it still has no correspondence in reality.[69]

[68] See eg Art 6 IDHGD which provides that: '[s]tates should endeavour to involve society at large in the decision-making process concerning broad policies for the collection, processing, use and storage of human genetic data and human proteomic data and the evaluation of their management, in particular in the case of population-based genetic studies. This decision-making process, which may benefit from international experience, should ensure the free expression of various viewpoints.'

[69] See Francioni, *supra* note 16.

8

Individual Human Rights in Genetic Research:

Blurring the Line between Collective and Individual Interests

Hélène Boussard

Biomedicine[1] and human rights pursue a common end—namely, the betterment of human life conditions—from different premises. While human rights tend to protect the intrinsic value of the individual, biomedicine increases the medical (or, instrumental) value of human body and body parts. As a result of this tension, breakthroughs in biomedicine have forced a shift from a deontological to a teleological model of protection of the individual, where intrinsic and instrumental values of the human body are balanced.

This chapter presents human genetics as the new 'engine' of the modern construction of human rights. It scrutinizes some of the key legal[2] and ethical[3] instruments, which taken together refine international human rights law to cope with medical genetics.[4] In this fragmented and inconsistent regulatory framework, two instruments have a particular authority as to what the human rights in genetics are: the Universal Declaration on the Human Genome and Human Rights (UDHGHR) and the Oviedo Convention on Human Rights and

[1] 'Biomedicine' is defined as life sciences applied to human beings for medical purposes. See Feuiller-Le Mintier, 'La biomédecine, une nouvelle branche du droit', in B. Feuillet-Le-Mintier (ed), *Normativité et biomédicine* (2003), at 2.

[2] In particular, the 25 Recommendations on the Ethical, Legal and Social Implications of Genetic Testing, European Commission, 2004; Recommendation (97)5 of the Committee of Ministers of the Council of Europe on the Protection of Medical Data; and Recommendation 92(3) of the Committee of Ministers of the Council of Europe on Genetic Testing.

[3] In particular, the International Ethical Guidelines for Biomedical Research Involving Human Subjects of the Council for International Organizations of Medical Science (CIOMS), Geneva, 2002, and the World Medical Assembly (WMA) Statement on Genetics and Medicine, Geneva, 2005.

[4] Medical genetics is the application of genetics to medicine. Medical genetics research seeks to find and study the genetic causes of human diseases.

Biomedicine (Oviedo Convention).[5] The UDHGHR was adopted by the United Nations Educational, Scientific and Cultural Organization (UNESCO) in 1997. At the time of its drafting, it was presented explicitly as a 'companion instrument to the 1948 Universal Declaration of Human Rights to establish ethical principles in a field which had not yet been explored'.[6] It has served as the building block for the UNESCO 2003 International Declaration on Human Genetic Data (IDHGD) and the 2005 Universal Declaration on Bioethics and Human Rights (UDBHR). The second major instrument, the Oviedo Convention, is the only *legally binding* instrument (for state parties) and, as such, it has a special authority to create new human rights or interpret positive rights (ie rights enshrined in human rights treaty or customary international law) in genetics.

The chapter that follows aims to identify the human rights accruing to the individual volunteering in genetic research, participating in gene therapy[7] and stem cell research,[8] or undertaking genetic testing in medical settings,[9] in a way consistent with the block of positive human rights. At the same time, the chapter highlights the limited capacity of international human rights law to adapt to new subject areas. In fact, the quite rigid structure of the human-rights based approach does not readily permit the protection of either group or future interests. International human rights law constitutes a pattern of relationship between the state and the individual: a human right is a fundamental right held *vis-à-vis* the state to protect the intrinsic value of the person, that is human dignity.[10] However, because genetic information is shared to a lesser or greater extent by all human beings, group interests and the interests of future generations are at stake in genetics. The family, the community, and society may have a comparable interest in obtaining genetic information. And the perceived interest for these groups is all the greater given that genetic information is widely seen as

[5] Convention for the Protection of Human Rights and Dignity of the Human Being with Regard to the Application of Biology and Medicine, ETS No 164. Three additional protocols to the Convention were adopted against cloning (1998), on transplantation (2002), and concerning biomedical research (2005).

[6] UNESCO, Res 29C/21, 24 September 1997, 'Drawing Up of a Declaration on the Human Genome, Report by the Director-General', 5.

[7] Gene therapy is a rapidly growing field of medicine in which genes are introduced into the body to treat diseases. Genes control heredity and provide the basic biological code for determining a cell's specific functions. Gene therapy seeks to provide genes that correct or supplant the disease-controlling functions of cells that are not, in essence, doing their job. Somatic gene therapy introduces therapeutic genes at the tissue or cellular level to treat a specific individual. Germ-line gene therapy inserts genes into reproductive cells or possibly into embryos to correct genetic defects that could be passed on to future generations.

[8] Stem cells are primal cells that are undifferentiated and whose daughter cells may differentiate into other cell types. Stem cell research is experimental research that typically uses aborted foetuses and embryos to help cure certain illnesses and diseases.

[9] Genetic testing denotes tests done for clinical genetic purposes. Genetic tests may be done for diverse purposes pertaining to clinical genetics, including the diagnosis of genetic disease in children and adults; the identification of future disease risks; the prediction of drug responses; and the detection of risks of disease to future children.

[10] R. Higgins, *Problems and Process: International Law and How We Use It* (1994), at 98.

having a highly predictive power as to future identity and health status. A coherent legal construction is therefore needed that protects not only individual *but also* group interests in genetics.

The structure of the chapter is as follows. Part 1 sketches an initial conceptual distinction among human rights allocated over an individual's genetic resources: specifically, it distinguishes between those rights that protect the individual *per se* and those that protect the individual as depositary of the genetic heritage of the human species. Thereafter, in Part 2, the legal constructions at hand to protect the interests of genetic groups will be explored.

1. Candidate Human Rights in Genetics

The refinement of human rights in scientific research must respond to a certain perception of the role of the individual therein. Until recently, the individual was perceived as the vulnerable subject in medical and research relationships. Accordingly, human rights in biomedical research and medical treatment were *freedom-rights*, protecting the individual's privacy and physical integrity and right to non-discrimination in medical health care. Informed consent of research participants was to ensure freedom to participate in research.[11] A shift was initiated in the 1970s. In fact, at the time genetic research started expanding, the individual began to be perceived as a partner in both the scientific project and in medical practice. As a result, in addition to freedom-rights, the *claim-right* to enjoy the benefits of science[12] was invoked increasingly in order to access treatments and medicines. Furthermore, informed consent, together with benefit-sharing, came to be viewed as the flip side of the contribution that research participants are making to the advancement of knowledge. In fact, informed consent may be required, *inter alia*, for tissue-taking and stocking, DNA analysis, the disclosure of health information, and the commercial use of genetic data. It seems therefore correct to say that informed consent does not only ensure freedom to participate in research, but also the participant's control over the use of his/her genetic material and genetic information.[13]

[11] Article 7 of the International Covenant on Economic, Social and Cultural Rights (ICESCR).

[12] Article 27 of the Universal Declaration of Human Rights (UDHR) and Art 15 ICESCR recognize the right to participate in the progress of science and to enjoy the benefit of science. See also, the Recommendation on the Status of Scientific Researchers, UNESCO, 20 November 1974, Art 8. Declaration on the Use of Scientific and Technological Progress in the Interest of Peace and for the Benefit of Mankind, UN GA Res 3384 (XXX), 10 November 1974.

[13] On the issue of informed consent in genetic research, see eg Strouse, 'Informed Consent to Genetic Research on Banked Human Tissue' (2005) 45 Jurimetrics 135, at 135–52; Wendler, 'Informed Consent, Exploitation and Whether it is Possible to Conduct Human Subjects Research Without Either One' (2000) 14 Bioethics 310; Miller and Weijer, 'Trust and Exploitation in Clinical Research', SSRN Paper (2005) (accessed 1 October 2008).

Conceptually, informed consent is a bioethical requirement, not a legal human right. That is to say, there is no 'right' to informed consent, just as there is no right to control genetic information as such. Instead the individual may invoke a set of rights[14] in which informed consent is embedded as the possibility to waive his/her right. As such, Article 7 of the International Covenant on Civil and Political Rights, which was until recently the only human rights provision requiring the free consent of the person subjected to 'medical and scientific experimentation', refines the right to physical integrity in scientific research.[15] The rational underpinning is that medical acts constitute an infringement of the physical integrity of the patient or research participant and can be justified only by a waiver of his/her right.

This conceptual articulation between human rights and informed consent remains valid today. From its introduction in the Oviedo Convention and in Article 5a of the UDHGHR, some authors seemingly consider informed consent as an autonomous right[16] rather than a right component. However, it is still far from clear that the nature of informed consent has radically changed in any way: indeed, the reference to different regimes of informed consent depending on the context and the person invites us to still consider it as a right 'modality' rather than a right *per se*. The conversion into a right is tempting only because it allows the delicate issue of the right secured by informed consent to be overridden. This is precisely why the following paragraphs, while contending that there is no *legal human right* to informed consent, identify the autonomous human rights which are currently crystallizing in the field of genetic research. The analysis takes the protection of genetic identity as the new rationale informing the protection of the individual *per se,* on the one hand, and of the human species, on the other.

A. The Protection of Genetic Identity as New Rationale for Human Rights

The refinement of human rights in genetics reveals a concern to protect the genetic identity of the individual, as well as the genetic identity of the human species. Put differently, the protection of genetic identity appears as the new rationale of the right to identity (the genetic identity of the individual) and the right to uniqueness of the individual (the genetic identity of the human species).

14 R.P. Claude, *Science in the Service of Human Rights* (2002), at 26–39 noting that among the 'many ways whereby the moral fiber of the UDHR stiffens and strengthens contemporary issues on science and human rights', the principle of non-discrimination, the right to life, and the right to physical integrity protect research participants.

15 Clayton, 'Informed Consent and Biobanks' (2005) JLME 15, at 18.

16 See eg Roucounas, 'Le droit au consentement et ses restrictions dans la convention sur les droits de l'homme et la biomédecine 1997' in R. J. Dupay (ed), *Mélanges en l'honneur de Nicolas Valticos* (1999) 479, at 482; Feuillet-Le Mintier, 'Le devenir des Comités régionaux d'ethique' (1999) 10 JIB 75.

1. Genetic Identity of the Individual and the Human Right to Identity

The protection of genetic identity must be understood as the right to identity encompassing genetic characteristics: genetic information is a component of, and therefore does not equate to, identity.[17] There is a risk, however, that while singling out genetic characteristics, the whole-scale geneticization of identity that began in the 1990s resurfaces. This drift is also referred to as 'genetic essentialism' or 'genetic reductionism' and leads to the reduction of the human subject to a mere expression of their genetic components.[18] Laws that treat genetic information as identity transform the social meaning of all genetic information by looking exclusively at the genetic attributes of the information, rather than by examining the substantive content and context of genetic information.[19] This is the reason why in Article 3 of the IDHGD the 'person's identity' refers not only to the genetic make-up of each individual but also to the 'complex educational, environmental and personal factors and emotional, social, spiritual and cultural bonds with others and implies a dimension of freedom.'

The right to genetic identity—that is, the right to identity encompassing genetic characteristics—may be protected under two different perspectives. First, from a civil law perspective, the right to identity serves to identify the individual from others. The legal 'right' to identity thus encompasses merely the sum of the elements that the society needs to individualize its members to grant citizenship and impose taxes: name, parentage, nationality, domicile, birthdate, and sex.[20] The emphasis thus placed is on the way in which society, rather than the individual, is made up: even if the two interact, self-perception and self-development are seen as philosophical matters that are not of legal significance.[21] From a civil law perspective, genetic characteristics are constitutive components of so-called

[17] On genetic ancestry and the risk of considering genetic testing as the final arbiter of kinship, see Elliott and Brodwin, 'Identity and Genetic Ancestry Tracing' (2002) 325 BMJ 469, at 469–1471; Johnston, 'Resisting a Genetic Identity: the Black Seminoles and Genetic Tests of Ancestry' (2003) 31 JLME 262, at 262–71.

[18] N. Rose, *The Politics of Life Itself* (2007), at 109.

[19] Gatter, 'Genetic Information and the Importance of Context: Implications for the Social Meaning of Genetic Information and Individual Identity' (2003) 47 St L ULJ 423, at 423–62.

[20] See eg Granet-Lambrechts, 'Le droit à l'identité', in F. Sudre (ed), *Le droit au respect de la vie privée au sens de la Convention européenne des droits de l'homme* (2005) 193, at 193–213; Benar, 'Vers les droits de l'homme de la quatrième dimension, essai de classification et de hiérarchisation des droits de l'homme', in K. Vasak, *Les droits de l'homme à l'aube du XXIe siècle, Karel Vasak amicorum liber* (1999) 109.

[21] M.-P. Peis-Hitier, *La protection de l'identité de l'être humain: contribution à l'étude des biodroits de l'homme* (2004), at 2. Peis-Hitier opposes identity-*idem* to identity-*ipse* in reference, respectively, to the permanent identity of the individual and identity as moving self-perception of the individual. See also Rose, *supra* note 18, at 133: 'the languages and aspirations of citizenship have shaped the way in which individuals understand themselves and relate to themselves and to others.'

'biological citizenship'[22] and constitute a major element in criminal law and paternity tests.[23]

By contrast, from the human rights perspective, the right to identity embraces the existential meaning of identity, ie the inner story people need to define the meaning of their lives.[24] International human rights law adopts a personality-identity framework to protect the psychological need for self-consciousness and the value of personhood.[25] The development of the human being constitutes the *raison d'être* of human rights. Accordingly, the child's right to identity is directed towards the development of the child's personality. The right to cultural identity or the right to one's own culture (the culture of the group into which one is born, in which one lives and with which one is identified[26]) also amount to human rights. Because identity is in constant transformation as a result of technological breakthroughs, it has become a subject of normative concern in international human rights law, and no more so than in the area of genetics. To put it crudely, health status shapes personhood. Individuals are increasingly obliged to formulate life strategies to seek to maximize their life chances, despite the fact that the penetrance[27] of any particular gene may be unknown. As such, the right to genetic identity informs the right to genetic privacy and the right to know about oneself.

2. *Genetic Identity of the Human Species and Right to Uniqueness*

Practices such as germ-line gene therapy, genetic screening, and pre-implantation genetic diagnosis[28] for selection of individuals before birth have the potential to have a very real impact on genetic diversity. Pluralism in genotypes ensures the

[22] See Rose, *supra* note 18, at 132 noting that 'biological citizenship' is used to 'encompass all those citizenship projects that have linked their conceptions of citizens to beliefs about the biological existence of human beings, as individuals, as men and women, as families and lineages, as communities, as populations and races, and as species.'

[23] Deutsch, 'Rights and Obligations of the Persons Concerning their Genetic Data' (2005) 22 Law and the Human Genome Rev 75, at 75–84. In the interests of international criminal cooperation, the 2006 Treaty of Prüm between EU Member States provides for the communication of lists of genetic profiles.

[24] See eg Bernstein, 'Accommodating Technological Innovation: Identity, Genetic Testing and the Internet' (2004) 57 VULS 963, at 963–1037; Marks, 'Tying Prometheus Down: the International Law of Human Genetic Manipulation' (2002) 3 Chicago J Intl L 119, at 123.

[25] Smith, 'Human Rights and Bioethics: Formulating a Universal Right to Health, Healthcare, or Health Protection?' (2005) 38 Vanderbilt J of Transnational Law 1295, at 1308–9.

[26] Stavenhagen, 'Cultural Rights: a Social Science Perspective', in A. Eide *et al* (eds), *Economic, Social and Cultural Rights* (2001) 85.

[27] *Ibid*, at 107. Penetrance means the likelihood a given gene will result in disease: eg if half (50%) of the people with the neurofibromatosis (NF) gene have the disease NF, the penetrance of the NF gene is 0.5.

[28] Pre-implantation genetic diagnosis (PGD) refers to diagnostic procedures that are performed on embryos prior to implantation, sometimes even on oocytes prior to fertilization. PGD is considered an alternative to prenatal diagnosis. Its main advantage is that it avoids selective pregnancy termination as the method makes it highly likely that the foetus will be free of the disease under consideration.

plasticity of the human genome and the capacity of evolution of the human species: hence, practices that reduce genetic diversity may impact on the future of the species. This explains the reason why the 'uniqueness and diversity' of individuals (Article 2) is protected as a guarantee of the 'inherent diversity' of the human family (Article 1) in the UDHGHR.[29] This is also the case in the Oviedo Convention.[30]

In sum, then, with the protection of the genetic identity of the individual and of the human species, international human rights law is resorted to protect the individual *per se*, on the one hand, and as the depositary of the genome of the human species, on the other.[31]

B. Protection of the Individual *Per Se*: The Right to Genetic Privacy

Genetic privacy has become 'a *terminus technicus* to refer to the specificity of legal problems related to the question of how to maintain control over the genetic information of the individual';[32] it has also been depicted as the 'traditional route for the protection of genetic information.'[33] Privacy protects the right of the individual to control information that is intrinsically linked to his/her identity. Specifically, it encompasses two rights, both of which are usually presented as autonomous: the first of these is the right to know (Article 5 UDHGHR), and the second is the right to confidentiality (Article 7 UDHGHR). Both are outlined in more details in what follows.

1. The Right to Know: The Individual as Information-Recipient

In the context of genetic testing, access to genetic health information has become a right, and wrong information has become a cause for action.[34] The right to know, and its corollary the right not to know in medical practice, and the right to feedback in research settings (Article 5c UDHGHR), equip the individual with information to appraise choices with respect to lifestyle, volunteering for research, and undertaking medical treatment. Presented as a natural extension of

[29] Article 2b lays down '[t]hat dignity makes it imperative not to reduce individuals to their genetic characteristics and to respect their uniqueness and diversity.' See Andorno, 'Seeking Common Ground on Genetic Issues: the UNESCO Declaration on the Human Genome', in J. Sandor (ed), *Society and Genetic Information: Codes and Laws in the Genetic Era* (2003).

[30] Article 1. [31] Peis-Hitier, *supra* note 21, at 8.

[32] Sandor, 'Genetic Information: Science, Society and Legal Norms', in Sandor (ed), *supra* note 29, at 31.

[33] Townend, 'Who Owns Genetic Information', in Sandor (ed), *ibid*, at 125. Spaak, 'The Idea of a Right to Genetic Privacy', SSRN Paper (2006) (accessed 1 October 2008).

[34] D. Dibre-Krajcman, *Information génétique et fonctions médicales: essai sur l'ambivalence de la condition juridique des médecins* (2004). See below at 264.

individual autonomy,[35] the right to know and the right not to know in genetic testing recognized in Article 5c UDHGHR are considered to be essential guarantors of the individual's ability to choose his/her lifestyle. The right to know one's genetic origin has been recognized by the European Court of Human Rights (ECtHR) as part of the right to privacy protected by Article 8 of the European Convention on Human Rights (ECHR). The Court has described this right as the 'right to identity and personal development, and the right to establish and develop relationships with other human beings and the outside world. The preservation of mental stability is in that context an indispensable precondition to effective enjoyment of the right to respect for private life.'[36] The reference to 'mental stability' and privacy echoes the link made above between identity and a person's development. *A contrario*, 'a state or regime that bars access to information effectively controls the destiny of its people.'[37]

Clearly, the recognition of the right to know is less arguable than its circumvention. Indeed, as outlined below, the fact that it is impossible to foresee how an individual will react to the result of a genetic test makes it particularly difficult not just to place limits on the right to know, but also to know how best to recognize, and place limits on, the right not to know.

(a) Limits on the right to know in one's own interest

The extent to which individuals have the right to find out information about themselves and what role, if any, the state should have in limiting access to such information[38] must be answered in accordance with the impact that genetic test results can have on the individual. However, this impact is difficult to evaluate for two reasons. First, apart from a very small number of cases where prediction is certain,[39] the penetrance of the information remains low; that is to say, there is no certainty that a disease predisposition will actually express itself. This means that results from genetic testing constitute a 'zone of foreseeability' and genetic information is merely an indicator of that individual's likely future health.[40] Secondly, the impact on individuals of misinterpreted or erroneous predictive health information which overstates the role of genetics in developing common

[35] Chizuka, 'Human Genome Analysis and Human Rights', in S. Btesh (ed), *Human Genome Research and Society, Proceedings of the Second International Bioethics Seminar in Fukui, 20–21 March 1992* (1992) 86, at 86.

[36] Case No 42326/98 *Odievre v France* [2003] ECHR 86 (13 February 2003).

[37] Hoffman, 'In Search of an International Human Right to Receive Information' (2003) 65 Loyola Los Angeles Intl and Comparative L Rev 165.

[38] Human Genetics Commission, 'Genes Direct: Ensuring the Effective Oversight of Genetic Tests Supplied Directly to the Public' (Department of Health London, March 2003), 29. There might be a role for state involvement if direct genetic testing resulted in increased burdens on NHS resources, such as increased calls on GPs' or clinical genetics professionals' time and on laboratory resources. Some drew parallels with private IVF treatment that may have led indirectly to an increased demand for neonatal specialist care for premature and multiple births.

[39] Huntington's disease is the main mono-genetic disease.

[40] Sandor (ed), *supra* note 29, at 22.

diseases makes it difficult for individuals to appraise the risks of genetic testing. This may result in delays in seeking proper medical advice or in a search for unnecessary medical treatment or expensive and unproven dietary or lifestyle changes.[41]

Because of these uncertainties, it remains the case that there is little guidance as to the disclosure of the positive information of a presymptomatic diagnosis. In a context of uncertainty, it seems appropriate that it should be the individual who evaluates the risks and decides whether or not to undertake genetic testing. This means, in turn, that genetic counselling is the main form of protection of the individual: it is recommended before and after genetic testing.[42] It has however been suggested that disclosure should be seen as essential in cases where a patient is to embark on having a family and where knowledge of the parental state may affect their decision.[43] It has also been suggested that the psycho-social harm that genetic testing can create would justify the prohibition of genetic testing of children for late-onset diseases, such as Alzheimer's or Huntington's disease.[44]

(b) The recognition of, and limits on, the right not to know in one's own interest

The right not to know has been recognized by the Council of Europe in the case of 'unexpected findings', such as information about a person and his/her genetic lineage concerning not only their health but also, for instance, paternity.[45] In this instance, the refusal of an individual to be informed of the result of a test is legitimate, notwithstanding the fact that this may impede the opportunity for relatives to know that they are, or are not, at risk.

At present, the only limit on the right not to know (ie the only circumstance in which there is a duty to inform the patient) appears to be when the patient seems to be at risk of a serious disease which it is possible to prevent or cure. Article 10 of the Oviedo Convention, for example, provides for the possibility of overriding the right not to know on the basis that 'it may be of vital importance for patients to know certain facts about their health, even though they have expressed the wish not to know them'. This limitation to the right not to know is repeated in Article 9 IDHGD and in the 2002 Guidelines on

[41] S. Hogarth, *The Regulation of Commercial Genetic Testing Services in the UK: A Briefing for the Human Genetics Commission* (2005), at 7.

[42] Recommendation 92(3) of the Council of Europe on Genetic Testing, Principle 3.

[43] Roberts, 'Medical Genetics, Social Issues and the Genome Programme', in Btesh (ed), *supra* note 35, at 102.

[44] Keller, 'Late-onset Disease and Children Genetic Testing', SSRN Paper, 2002 (accessed 1 October 2008).

[45] Recommendation 92(3) of the Council of Europe on Genetic Testing, Principle 11 'Unexpected Findings'; Explanatory Memorandum to Recommendation (92) 3 of the Committee of Ministers to Member States on Genetic Testing and Screening for Health Care Purposes, para 46.

epidemiological research of the Council for International Organizations of Medical Science (CIOMS).

2. *The Right to Genetic Confidentiality: The Individual as Information-Source*

The individual is the *information-source*, from which genealogies, medical records, and genetic sequences are derived.[46] Privacy (Article 5b UDHGHR) and confidentiality (Article 7 UDHGHR) are thus the core issues in genetic research which analyses participants' genetic characteristics. The principle of non-discrimination (Article 6 UDHGHR), which implies that genetic information should not be used for medical or non-medical health purposes to discriminate against the individual, is also central.

The right to know protects the intrinsic value of genetic information and the need to access information about one's life as perceived as a human right. As the other side of the same coin, confidentiality comes into play in reaction to the need of third parties to access information: it reconciles the intrinsic and the instrumental value of the information.[47] Confidentiality in relation to health information responds to 'an individual's claim to control the circumstances in which personal health information is collected, used, stored and transmitted'.[48] Accordingly, for the 'genetic data associated with an identifiable person' to be disclosed, this person must give an informed consent as established in Article 7 UDHGHR. However, because researchers are rarely in contact with participants in genetic research, re-contacting them can be a very cumbersome process that may impede the scientific enterprise. It might be thought that anonymization, that is, the dissociation of genetic data from identifier, could be used to avoid the problem in the first place: once 'traces of subjectivity' are removed, there is no residual interest for the tissue-source. However, anonymization[49] is particularly problematic in the case of genetic research for three reasons. First, although anonymization can protect confidentiality, because it ends an individual's rights to control his/her own information, it also neutralizes the right to know which presupposes conserving the link with the research participant.[50] Secondly, some

[46] Palsson and Rabinow, 'The Icelandic Genome Debate' (2001) 19 Trends in Biotechnology 166, at 166–71.

[47] Molinari, 'Vers un nouveau paradigme de protection de l'information biomédicale?', in B. M. Knoppers and C. Herve (eds), *Biological Material and Electronic Data: Much Ado About Nothing?* (2006) 77, at 77–8.

[48] L.O. Gostin, *Public Health Law: Power, Duty and Restraint* (2000), at 128.

[49] Tallachini, 'Rhetoric of Anonymity and Property Rights in Human Biological Materials (HBMs)' (2005) 22 Law and the Human Genome Rev 153.

[50] See Menzies, 'Information Gathering and Confidentiality: Data Bases, Monopolies of Knowledge and the Right to be Informed', in K.E. Mahoney and P. Mahoney (eds), *Human Rights in the Twenty-First Century: A Global Challenge* (1993) 793, at 793–800; Early, 'Science, Technology and Human Rights: the Role of Data Protection', *ibid*, at 801.

genetic research cannot be carried out in the absence of identifiers. Finally, as a technical form of protection, it is not completely reliable.

To summarize, then, this section of the chapter has outlined the refinement of two components of the right to health: namely, the right to know (Article 5c UDHGHR) and the right to confidentiality (Article 7 UDHGHR) in genetic practices. As regards the right to know, it was suggested that its expansion to genetic information is in line with the evolving interpretation of the right to identity. The judicial assessment of this right by the ECtHR ensures a strong legal basis for this refinement of the right to privacy. At the same time, however, there is no clear limit on either the right to know or the right not to know. The difficulty of managing the risks associated with individual reactions to genetic tests (ie the 'subjective risks') goes some way towards explaining why achievement of the right equilibrium between individual autonomy and protection of the individual has proved elusive. In the absence of legal limits, efforts can be made to awaken a sense of responsibility towards genetic relatives: individuals have a moral obligation to decide whether to know or not to know, taking into account the consequences for the family.

It can also be noted that protecting both the right to know and the right to confidentiality in genetics is a particularly difficult task. The protection of the right to confidentiality rests on the problematic process of anonymization of data used for research; at the same time, it short-circuits the claim of the individual over genetic information (eg benefit-sharing and right to know).[51] As to the right to know or not to know, although genetic counselling realizes the partnership between scientists/physicians and research participants/patients, it also transfers the task of risk assessment from the former to the latter.

C. Protection of Genetic Diversity: Collectivization of the Human Genome

The protection of the individual as the *depositary of the genetic code* is remarkable insofar as it ensures the protection of collective interests through individual empowerments. In this instance, therefore, human rights can be said to be at the interface between the protection of the future individual and the protection of the human species. The paragraphs that follow outline the four different legal constructions that have emerged to date in an attempt to protect the genetic identity of the human species: namely, the prohibition of practices contrary to human dignity, and in particular of reproductive human cloning (Article 11 UDHGHR); the right to genetic integrity; the right to uniqueness of the embryo; and the principle of non-discrimination.

[51] Tallachini, *supra* note 49.

1. The Prohibition of Practices Contrary to Human Dignity

The prohibition of practices contrary to human dignity (Article 11 UDHGHR) aims to prevent genetic manipulations that would endanger the genetic integrity of the human species.[52] This prohibition recalls the need to maintain and perpetuate humankind in the different expressions of its identity laid down in the 1994 La Laguna Declaration, which affirmed the right to genetic identity in reaction to direct intervention which would:

> undermine in any way whatsoever the human form of life, particularly through acts which compromise irreversibly and definitively the preservation of the human species, the human genome and humankind's genetic heritage, or acts which are designed to partly or completely destroy a national, ethnic, racial or religious group.

2. The Right to Genetic Integrity and the Right to a Non-Modified Genetic Heritage

In gene therapy, the right to genetic integrity impedes the modification of the genetic code of the individual. It was originally referred to as the 'right to a non-modified genetic heritage' in the 1982 Recommendation of the Parliamentary Assembly of the Council of Europe on Genetic Engineering.[53] It is linked in that instrument to the right to life and human dignity.[54] Article 2 of the Oviedo Convention and Article 3 of the IDHGD refer explicitly to 'genetic integrity' and protect it as a building block of individual personality.

3. The Right to a Unique Identity

The right to genetic identity means a right to a *unique* identity, to 'uniqueness' or to an 'open future'[55] when it accrues to the embryo. The embryo is therefore recognized as a 'prenatal identity'[56] that is protected as a right. This justifies the prohibition of reproductive human cloning which is 'an intervention seeking to create a human being genetically identical to another human being.'[57] The right of a child to his/her genetic identity can also be inferred from Principles 3 and 4

[52] Peis-Hitier, *supra* note 21, at 4.

[53] Recommendation 934 (1982) of the Parliamentary Assembly of the Council of Europe on Genetic Engineering, 26 January 1982, Point 7a.

[54] R.M. Lozano, *La protection européenne des droits de l'homme dans le domaine de la biomédecine* (2001), at 213.

[55] Marks, *supra* note 24, at 123.

[56] D. De Grazia, *Human Identity and Bioethics* (2005), at 244–94.

[57] See the Additional Protocol to the Convention for the Protection of Human Rights and Dignity of the Human Being with regard to the Application of Biology and Medicine, on the Prohibition of Cloning Human Beings, Paris, 12 January 1998, Art 1 in which 'genetically identical' means that two human beings share the same nuclear gene set: the nuclei of the cells contain identical genetic material.

of the Declaration of the Rights of the Child of 1959, which recognize the right to a name and nationality, and the right to grow and develop in health.

4. The Principle of Non-Discrimination

The principle of non-discrimination on the ground of genetic characteristics is embodied in Article 6 UDHGHR. Generally defined as discrimination, prejudice, and stigma arising from the knowledge or study of genetic information,[58] genetic discrimination differs from discrimination based on gender, race, ethnicity, religion, or physical characteristics in that it can have no physical manifestations, it affects family units and reproductive options, and it usually elicits stereotypic responses including genetic determinism and inevitability.[59]

The use of genetic information for non-health purposes has been a major issue since the Human Genome Project[60] and was the top priority of the 2006 report by the Special Rapporteur on the Human Genome.[61] At the national level, it is in the United States (which is the major producer of genetic information[62]) that genetic discrimination first appeared, finding its way into access to health care and employment, as well as social entitlements such as insurance, education, governmental services, and adoption. In all of these different fields, genetic tests are operated that impact on the advantages the individual can claim as a citizen.

The principle of non-discrimination could be used to prohibit eugenic practices in reproductive technologies.[63] Departing from the traditional model of international human rights law, this new interpretation would protect future generations, by impeding disadvantages arising *in the future*.[64] In this way, the prohibition of genetic discrimination could be seen, along with the prohibition of practices against human dignity and the right to a unique genetic identity, as a legal technique to protect future individuals.

To sum up: the need to protect the genetic identity of the individual, on the one hand, and of the human species, on the other, have informed the right to genetic privacy, the right to a non-modified genetic heritage of the adult, the right to the uniqueness of the embryo, and the freedom from discrimination

[58] See, also, the chapter by Iulia Voina Motoc in this volume.

[59] Billings, 'International Aspects of Genetic Discrimination', in Btesh (ed), *supra* note 35, 114, at 115.

[60] Rothstein, 'Legal and Ethical Issues by the Human Genome Project', in *Proceedings of the Conference held in Houston, Texas, March 1991*; Makino, 'The New Age of Genetic Testing is Coming', *ibid*, at 90–2.

[61] Motoc, Special Rapporteur on Human Rights and the Human Genome, Interim Report, E/CN.4/Sub.2/2005/38, 14 July 2005. See also her Preliminary Report, E/CN.4/Sub.2/2004/38, 23 July 2004.

[62] Billings, *supra* note 59, at 116.

[63] Campiglio, 'Human Genetics, Reproductive Technologies and Fundamental Rights' (2004) 14 Italian Yearbook of Intl Law 83, at 113–14.

[64] Sandor (ed), *supra* note 29, at 21.

among future individuals.[65] As pointed out earlier, in 'individualizing moments',[66] when the individual is protected in him/herself and artificially separated from the rest of the genetic 'pool', the right to confidentiality, and the right to know and not to know, can be seen as components of the universally recognized right to privacy. The biological sense of identification makes certain kinds of ethical demands possible: human rights encapsulate the individual need for self-perception. These rights can therefore be seen as interpretations of old rights and have arguably already crystallized. This is not the case, however, with respect to the rights attached to the protection of the individual for the sake of human genetic diversity. Here, the individual is protected as a depositary of the genetic heritage of the species and the right to genetic identity protects the interests of future individuals and of the human species over time. Human rights are thus associated to 'collectivizing' moments: the individual is protected as part of the whole. And, as outlined above, this collective dimension of the human genome can be seen as manifest in the protection of genetic identity associated with the right to genetic integrity, the right to uniqueness, the principle of non-discrimination, and the prohibition of practices contrary to human dignity. These provisions have a trans-temporal dimension, protecting present and future individuals, as well as limiting scientific practices. None of these rights, however, has crystallized into positive rights.

2. Legal Constructions to Protect the Interests of Genetic Relatives

As mentioned in the introduction to this chapter, genetic information is collective, that is, shared to a greater or lesser extent by the family, the community, society, and the human species. As a result, genetic research constitutes a powerful group-making process and threatens inter-individual relationships. Indeed, it has been said that, in the same way that social categories generate identities by clearly identifying their members, genetic research threatens 'to drive scientific wedges into the social cracks that already divide people from their neighbours.'[67] Four paradigmatic cases where the collective dimension of genetic information disrupts inter-individual relationships are identified below.

First, at the level of fundamental research, both epidemiological and population-based research target *groups* (ie 'genetic pools') in order to localize genes and understand their functions. In a few cases, groups of people volunteer to be research subjects or participants: these are the cases when people affected by a

[65] Lozano, *supra* note 54, at 209–16.

[66] The expressions 'individualizing' and 'collectivizing' moments are borrowed from Rose, *supra* note 18, at 131–54.

[67] Juengst, 'Population-based Genetic Research and Screening: Conceptual and Ethical Issues', in B. Steinbock (ed), *The Oxford Handbook of Bioethics* (2007) 471, at 474.

common disease mobilize themselves to foster research and to access clinical trials.[68] For the most part, however, it remains the case that groups do not *volunteer* but are asked to participate in research. In these cases, as soon as some individuals agree to participate in the genetic research, they may expose the whole group to the risk of discrimination or stigmatization. Problems can also arise when the decision to participate is taken by the representative of the group: his/her decision may be contrary to the religious beliefs of one of the members of the group.[69] There is also the fact that most human biological materials have economic value only in the context of a larger group;[70] hence, it is the whole set of samples which is evaluated for compensation and benefit-sharing rather than the individual contribution, which means that negotiations with the whole group have to take place.

Secondly, at the level of drug and diagnostic tools development, pharmacogenomics raises serious risks of discrimination. This branch of science targets 'genetic groups' so as to produce 'smart' drugs, ie drugs individualized so as to avoid resistance. However, the fact that only specific groups are targeted, following a market-oriented approach, means that there is a risk of 'orphan genotypes', ie genotypes for which no specific drugs are developed.

Thirdly, in case of genetic tests, family members and the 'proband'[71] may have comparable interests in obtaining the information. Indeed it has been argued that '[w]hen an illness or pathology is thought of as genetic, it is no longer an individual matter. It has become familial, a matter both of family histories and potential family futures.'[72] There is therefore a need to authorize the group of persons at risk to access this information. The ensuing social risk then is that of familial dilemmas and conflicts 'in which medical issues seem especially vexing.'[73]

Fourth, and finally, when genetic testing accompanies reproductive technologies (that is, in the case of pre-implantation genetic diagnosis or prenatal diagnosis), the risk expands to the level of the human species. The primary fear associated with these practices is that of 'private' eugenics, whereby individuals empowered with predictive knowledge decide on the future and the features of their embryo. Hence, the use of the terms 'designer babies' and 'designing parents'.[74]

[68] See eg the groups of women claiming that research be undertaken in reproductive health in the USA who also invoked the right to benefit from scientific development to obtain participation of pregnant women in research. See R.J. Cook *et al, Santé de la reproduction et droits humains* (2005), at 204–5. The same phenomenon took place under the pressure of HIV activists.

[69] Rohter, 'In the Amazon, Giving Blood but Getting Nothing', 260 *New York Times* (20 June 2007).

[70] Resnik, 'Are DNA Patents Bad for Medicine?' (2003) Health Policy 181, at 191: 'In most genetic research, researchers need hundreds or thousands of samples to acquire the data they need about DNA sequences, disease epidemiology, variations, protein coding, and so on. Most people who donate tissue will not be giving away a "gold mine". Nevertheless, the tissue still has some economic value and patients may feel that they deserve some type of compensation.'

[71] The 'proband' means the individual or member of a family being studied in a genetic investigation and through whom a family's medical history comes to light.

[72] Juengst, *supra* note 67, at 108.

[73] Strouse, *supra* note 13, at 140.

[74] M.J. Sandel, *The Case Against Perfection* (2007).

Against the backdrop of these tensions between collective and individual interests, the challenge for the norm-maker is to ensure that the rights accruing to the individual—namely, the right not to know and the right to genetic confidentiality presented in the first part of this chapter—do not impede genetic relatives from having access to information or from being consulted when genetic research would generate a risk of discrimination for them. The aim in short is to temper, in research participation and genetic testing, the process of individuation and subjectivization in human rights law. By making the individual responsible for choosing on behalf of the group, the human-rights based approach may neglect the interest of the group in being represented separately from the interest of society as a whole and from the interest of the individual. It is therefore regrettable that the UDHGHR is silent on the issue of the interests of the genetic line. The only protection offered to collective entities under the UDHGHR is that the interests of science cannot prevail over the human rights, fundamental freedoms, and human dignity of groups of peoples.[75] In addition, the Declaration repeats the possibility of limiting individual human rights 'for compelling reasons within the bounds of public international law and the international law of human rights.'[76]

In what follows, the different attempts made at the domestic level to limit individual rights in order to protect family members are sketched out and classified under a series of sub-headings. Thereafter, the final section of the chapter goes on to outline another trend at the domestic level: namely, empowering groups by means of collective entitlements.

A. Disclosure of the Results of Genetic Tests: Protection of Family Members

The limitation of confidentiality of health data in the interests of relatives is widely acknowledged by national courts and legislations. The question is whether the limitation should apply equally to genetic health data. Recommendation (92) 3 of the Council of Europe on Genetic Testing,[77] the work of the EU Working Group on Personal Data Protection,[78] and the 25 Recommendations of the European Union[79] all authorize that in the case of serious risk for other members

[75] Article 10.

[76] Article 9. In the IDHGD, Article 4(ii) merely notes that human genetic data 'may impact on the family, including offspring, extending over generations, and in some instances on the whole group to which the person belongs.'

[77] Principle 9: '[i]n the case of a severe genetic risk for other family members, consideration should be given, in accordance with national legislation and professional rules of conduct, to informing family members about matters relevant to their health or that of their future children.'

[78] Renegar *et al*, 'Returning Genetic Research Results to Individuals: Points-to-Consider' (2006) 20 Bioethics 24, at 24–36.

[79] The 25 Recommendations on the Ethical, Legal and Social Implications of Genetic Testing, European Commission, 2004, Recommendation 10 on Data Protection: Confidentiality, Privacy and Autonomy notes that 'the relevance for other family members' has to be addressed.

of the family, confidentiality is not/should not be absolute. In addition, in Recommendation (97)5 on health data, the Council of Europe has indicated that members of the 'genetic line'[80] should be afforded 'hybrid legal protection':[81] that is, they should not be treated as an 'ordinary' third party, rather they should have a right to access data protected by confidentiality where it is in the interests of their health.[82] The point of divergence among these different texts lies in the way the results of the genetic tests are communicated to the relatives of the 'proband'.

As outlined below, four different approaches can be identified in domestic law. In the first, the national data protection authority authorizes disclosure of the information in the interest of 'others', which includes family members. In the second system, shared medical data is considered as 'personal' data with respect to each family member.[83] The third system makes it the duty of the physician to inform relatives, and the fourth places a duty on the individual to initiate the process of giving information to his/her relatives, either directly and personally or through an intermediary body. As discussed below, the requirement of informed consent of the proband does not have the same relevance across these different systems.

1. Disclosure in the Interest of 'Others' by the National Data Protection Authority

National data protection authorities play a key role in the protection of research participants and their role has been strengthened since the emergence of biobanks and the massive storage of genetic data. In that context, a remarkable example of the limitation of confidentiality for the interest of the relatives was given by the Italian national data protection authority. In Italy, the first case of disclosure of health data for a genetic relative took place in 1999. The case concerned a

[80] Recommendation (97)5 of the Committee of Ministers of the Council of Europe on the Protection of Medical Data, Art 1: 'The genetic line is the line constituted by genetic similarities resulting from procreation and shared by two or more individuals.'

[81] Explanatory Memorandum to Recommendation (97)5 of the Committee of Ministers of the Council of Europe on the Protection of Medical Data, 13 February 1997, para 58.

[82] Recommendation (97)5, para 50 provides that 'family data comprise information about a person's parents, uncles, aunts, grandparents, brothers, sisters, children, as well as more distant relatives. Family data are genetic data only to the extent that the disease or trait in a given person is known to be genetically determined or genetically influenced, or if the occurrence of a trait or disease is such that in the given family(ies) it appears to be inherited or influenced by genes, even if it had previously not been suspected that the trait or disease could be of a genetic nature'. Paragraph 51 provides that 'family data also comprise information about marriages between related persons and information about numbers of offspring, stillborn children and abortions. Family data are essential for genetic analyses of normal traits as well as diseases. Records of family data are kept in physicians' files, in medical registers for use in the future in connection with genetic counselling or diagnosis, even in coming generations, or in research registers.'

[83] Taylor, 'Data Protection, Shared (Genetic) Data and Genetic Discrimination' (2006) 8 Medical Law Intl 51, at 51–77.

woman who was affected by a genetic disease called 'glaucoma bilaterale' and needed to have access to the medical data of her father in order to evaluate the penetrance of the gene and the risk of a future child being affected. The national data protection authority ('*Garante per la protezione dei dati personali*') authorized access to this personal data, reasoning that third parties must not be harmed by the confidentiality rule.[84] It can be argued that here the protection of relatives builds on a principle-exception approach, in which the interests of the relatives are appreciated objectively and *in concreto*.

2. *The Right to Access Medical Data*

To date, among the conceptual frameworks presented here, the one adopted by the Icelandic Supreme Court[85] in the *Ragnhildur* case is the most protective of the interests of genetic relatives. The case is remarkable as it considers that a right over genetic information accrues to each member of the family.[86] The claimant in the case asked the Court to overturn a refusal by a Medical Director of Health to accede to her request that health information in the medical records pertaining to her deceased father should not be entered into the Iceland Health Sector Database. The development and operation of this national database belongs to the licensee, deCODE Genetics, Inc.,[87] which is a for-profit organization. The Supreme Court concluded that the claimant could not exercise the right to prohibit the transfer acting as a substitute of her deceased father. However, it recognized that, on the basis of her right to protection of privacy, she might have an interest in preventing the transfer of health data concerning her father into the database as information could be inferred from such data relating to the hereditary characteristics of her father which might also apply to herself.

3. *The Duty of the Physician to Warn and the Right to Reparation for Relatives*

To the extent that substantive norms do not provide any clear answers, one may turn to substantive norms in conjunction with procedural norms. This correlation between substantive and procedural norms is characteristic of the common law: where there is a remedy, there is a right, ie individual rights can be derived from available and existing remedies. To illustrate this approach, we can look to

[84] Decision 22/5/99. See A. Santosuosso, *Corpo e libertà* (2001), at 259–60.

[85] Supreme Court, *Ragnhildur Guðmundsdóttir (Ragnar Aðalsteinsson, Attorney at law)* v *The State of Iceland (Skarphéðinn Þórisson, Attorney at law)*, 27 November 2003, No. 151/2003. Reproduced in (2004) 21 Law and the Human Genome Review 127–38.

[86] Gertz, 'Is It "Me" or "We"? Genetic Relations and the Meaning of "Personal Data" under the Data Protection Directive' (2004) 11 Eur J of Health Law 231, at 231–44; Taylor, *supra* note 83, at 51–77.

[87] The Ministry of Health and Social Security oversees contractual agreements between itself and deCODE Genetics, Inc., for the development and operation of the Health Sector Database.

US jurisprudence; specifically, to the case law wherein the courts have recognized a duty to warn relatives in the case of deceased persons where there was no explicit opposition before death.

Take, for example, *Safer v Pack,*[88] wherein the Court held that a physician had a duty to inform the children of a deceased patient that they were at risk of colon cancer. Academic commentary on this case has underlined the fact that, especially in terms of foreseeability, there is no essential difference between the type of genetic threat at issue here and the menace of infection, contagion, or physical harm.[89] In particular, the individual or group at risk is easily identified, and substantial future harm may be averted or minimized by a timely and effective warning. Although an overly broad and general application of the physician's duty to warn might lead to confusion, conflict, or unfairness in many types of circumstances, it is arguable that the duty to warn of avertable risk from genetic causes, by definition a matter of familial concern, is sufficiently narrow to serve the interests of justice. Further, as emphasized by Dolgin and Shepherd, it is appropriate that the duty be seen as owed not only to the patient himself but that it also 'extends beyond the interests of a patient to members of the immediate family of the patient who may be adversely affected by a breach of that duty.'[90]

4. *The Duty of the Individual to Inform*

The French system is unique insofar as it places the duty to initiate the family-information process directly on the individual.[91] Genetic information is treated as family information and the individual has the choice to inform relatives in person or to entrust the communication of the information to the Biomedicine Agency. The Agency is the body charged with accrediting centres of pre-genetic testing and authorizing practices of organ transplantation. Once asked by the proband, the Agency will inform his/her genetic relatives that medical information related to them is available, while maintaining the anonymity of the proband. This kind of resort to a health agency, the normal competence of which is the field of public health, highlights the growing trend for state intervention in the medical field. This is justified by the need to maintain inter-family relationships in the case of genetic testing.

5. *The Informed Consent of the 'Proband'*

The final point to be noted under this topic is that an injunction to warn genetic relatives about the result of a genetic test does not say anything on whether or not the informed consent of the 'proband' is required or can be derogated. In the case

[88] *Safer v Pack* 677 A.2d 1188 (Super Ct NJ 1996), discussed in J.L. Dolgin and L.L. Shepherd (eds), *Bioethics and the Law* (2005), at 100 ff.

[89] Dolgin and Shepherd (eds), *ibid*, at 100.

[90] *Ibid.*

[91] Article 1131 Code of Public Health.

of the French system outlined immediately above, the question is not relevant insofar as the individual him/herself initiates the process of information. Equally, in the *Ragnhildur* case decided by the Icelandic Supreme Court, the informed consent of the 'proband' did not come into play as the right to privacy accrued directly to the genetic relative. Conversely, when a physician informs a relative or a data protection authority authorizes disclosure, the question of the need for informed consent of the 'proband' is relevant and highly significant.

At the present time, there is no consistent approach to this issue in international documents. Some instruments require the patient's consent so that it is not possible to override his/her refusal to warn relatives. This is the case with the Guidelines of the CIOMS, which also lay down that the cultural context where relatives 'would usually expect to be informed of such results' must be taken into account: in this respect, precautions should be taken in setting up any research agreement and explanations of the research project and eventual future uses of research findings should be provided during the process of obtaining informed consent.[92]

Other instruments recommend that physicians communicate directly with the patient's relatives under special circumstances. This is the position adopted by the World Medical Assembly (WMA) in its 2005 Statement on Genetics and Medicine, which allows the principle of confidentiality to be derogated in the case of any direct or imminent threat to the life or health of an individual.[93] The Advisory Committee on Health Research of the World Health Organization (WHO), in its 2002 report, stressed the link between, on the one hand, the protection of confidentiality and, on the other, the degree of increase in risk and the existence of an intervention that could eliminate or decrease that risk. The nature of the interests at stake appears to be decisive: so, for example, if the result impacts on the reproductive choices of family members or health, disclosure may be justified and confidentiality need not be promised to patients in these circumstances.[94]

B. The Collectivization of Entitlements in Epidemiological Research: Protection of the Group

Precedents created by indigenous peoples suggest at least three forms of entitlements that could be used to protect group interests in genetic research: first,

[92] CIOMS, International Ethical Guidelines for Biomedical Research Involving Human Subjects, Geneva, 2002, Commentary on Guideline 18.

[93] WMA, Statement on Genetics and Medicine, Santiago 2005, para 12: 'In the case of a positive test result that may have implications for third parties such as close relatives, the individual tested should be encouraged to discuss the results of the test with such third parties. In cases where not disclosing the results involves a direct and imminent threat to the life or health of an individual, the physician may reveal the results to such third parties, but should usually discuss this with the patient first. If the physician has access to an ethics committee, it is preferable to consult such a committee prior to revealing results to third parties.'

[94] WHO, 'Genomics and World Health: Report of the Advisory Committee on Health Research', 2002, 156.

human rights and non-discrimination claims; second, self-determination claims; and third, claims as genetic groups (*sui generis*).[95] At the same time, however, as explained in Section 1 below, the recognition of group rights remains highly problematic in general international human rights law. Section 2 outlines how for indigenous peoples the question of group rights is bypassed by the principle of self-determination, which is recognized as a collective human right *par excellence* and which is currently being expanded to include human genetic resources. However, as Section 3 explores, this solution is not universally shared and, for indigenous and other groups, the most likely solution will be the recognition of *sui generis* entitlements as different from rights.

1. Group Rights

Group rights are intended to preserve the distinctiveness of an identity group and to achieve its public recognition as a permanent feature that will imprint a mark into the very basis of the state. They refer to those rights accorded to particular identity groups, by contrast to rights recognized for the benefit of a community (eg the right to environment) or rights exercised collectively (eg freedom of association).[96] Group rights have so far been accorded to national, ethnic, religious, and linguistic minorities in the contexts of cultural property, education, religion, and genocide.[97]

Among group rights, indigenous rights[98] seem to raise the least objections.[99] In general, however, group rights encounter strong ideological opposition on the grounds that they threaten the fragile equilibrium that links the individual to the

[95] Kingsbury, 'Reconciling Five Conceptual Structures of Indigenous Peoples' Claim in International and Comparative Law' (2001) 34 NYU J Intl L and Politics, at 289 (republished in P. Alston (ed) *Peoples' Rights* (2001), at 69–110. In the case of indigenous people, the 'historic sovereignty claim' constitutes a fifth form of claim which does not fit the case of genetic groups.

[96] Rivero, 'Le problème des nouveaux droits de l'homme', in *Droits de l'homme. Droits collectifs ou droits individuels Actes du Colloque de Strasbourg du 13 et 14 Mars 1979* (1980), at 17–18.

[97] Underkuffler, 'Human Genetics Studies: the Case for Group Rights', in *Duke Law School Working Paper* (2006), at 8–9.

[98] Kingsbury, *supra* note 95. This results from the definition given by the World Bank Operational Directive on Indigenous Peoples: 'the terms "indigenous peoples", "indigenous ethnic minorities", "tribal groups", and "scheduled tribes" describe social groups with a social and cultural identity distinct from the dominant society that makes them vulnerable to being disadvantaged in the development process.' See also T. Meron, *The Humanization of International Law* (2006), at 365–9. By contrast to ILO Convention No 107 (1957), ILO Convention No 109 concerning indigenous and tribal peoples in independent countries (1989) illustrates the more recent trend to grant ethnically distinct groups a degree of political and economic autonomy within existing state boundaries.

[99] Proposed American Declaration on the Rights of Indigenous Peoples, doc. OEA/Ser/L/V/. II.95, doc 6, Art 1(2) (1997). The proposed Declaration, drafted by the Inter-American Commission on Human Rights, provides that 'Self-determination as indigenous shall be regarded as a fundamental criterion for determining peoples to which the provisions of this declaration apply.'

state in the social contract.[100] In the case of genetics, there is an additional, structural reason that impedes the use of the model of rights: the vagueness of the definition of the community or the multi-generational dimension of the entities.[101] International law requires the identification of a specific legal entity as rights-holder and the definition of the 'genetic group' by the close genetic ties between its members does not meet the requirement. This makes the question of group rights a delicate one: it requires us to rationalize the interests that the family, the tribe, and the cohort of people involved in epidemiological studies, or in clinical trials, could have in the control, access, commercialization, or benefit-sharing related to the patentability of the genetic information.

Notwithstanding the conceptual and political difficulty in recognizing group rights, it is the solution retained by the WHO Advisory Committee on Health Research. Using a collective-rights based approach, its 2002 Report on Genomics and World Health stressed that respect for cultural practice should lead to 'collective consent' (eg obtaining the informed consent of the person representing the whole community).[102] The Report deduced the need for the dual requirements of strong confidentiality, and some limits to it in family and reproductive contexts, from the fact that genetic testing is predictive of one's future health and that biological relatives are at a similar risk.[103] Along the same line of thought, the WHO 2003 Draft Report on Indigenous Peoples noted that, depending on the structure of the indigenous community concerned, it may be necessary for the settlement of the research agreement to obtain three levels of consent: from the individual, from the community, and from an umbrella organization.[104] With simpler structures, the consent of individuals and community leadership may suffice.[105] Respect for cultural practice also influences the form of informed consent (ie written or oral) and the information that needs to be provided to those from whom consent is sought.

[100] N.T. Casals, *Group Rights as Human Rights: A Liberal Approach to Multiculturalism* (2006), at 8. In light of liberal theories of rights, the issue of group rights calls into question some paramount political ideals: liberalism postulates that the state is neutral with regard to the different backgrounds of its citizens, relegating aspects such as culture, ethnicity, or religion to the private sphere, and also that assigning the same rights to all citizens is sufficient to guarantee the legitimate forms of diversity in a democratic society.

[101] See Varella, 'Access to Genetic Resources and Benefit-Sharing: Main Aspects of some Legal Frameworks', in *Presentation in the Open Meeting of the Global Environmental Change Research Community, Montreal, 16–18 October*, at 11–12. Underkuffler, *supra* note 97, at 8; Pullman and Latus, 'Clinical Trials, Genetic Add-Ons, and the Question of the Benefit-Sharing', (2003) 362 The Lancet 242, at 243; Juengst, *supra* note 67, at 471–90.

[102] *Supra* note 94. [103] *Ibid*, at 147–8.

[104] A working group within the Economic Council of the United Nations, led by E.I.A. Daes, is drafting a code of conduct for institutions working on human, natural, and animal genetic resources linked to indigenous peoples. It provides for informed consent, which could impede research if contrary to the vision of indigenous groups. See Hermitte, 'L'encadrement juridique de la recherche scientifique', in M.-A. Hermitte (ed), *La liberté de la recherche et ses limites* (2001).

[105] *Draft Indigenous Peoples and Participatory Health Research, Planning & Management: Preparing Research Agreements*, 2003, 8.

2. *The Right to Self-Determination*

Conceptually, the right to self-determination has been interpreted by scholars as the general right to control genetic resources and genetic information and as the legal basis to claim a share in the benefits derived from research on genetic resources and information.[106] This interpretation is innovative insofar as it provides the right to self-determination with a new content and new rationale which merge the principle of individual autonomy as understood in bioethics and the collective right to self-determination of international human rights law.

The right to self-determination in international law,[107] as recognized in the first article of the two international human rights covenants of 1966, is conceived either as a collective right[108] or, in the words of the UN Human Rights Committee, as an 'essential condition for the effective guarantee and observance of individual human rights.'[109] If the substantive elements of the international human right to self-determination apply to groups and individuals, remedies based on the right are limited to groups.[110] In its content, the right to self-determination protects and enables groups to prosper and transmit their culture as well as to participate fully, as a group, in political, economic, and social processes. The identification of the group within the state constitutes 'external'

[106] See in particular, Lenzerini, 'Biogenetic Resources and Indigenous Peoples' Rights', in F. Francioni (ed), *Biotechnologies and International Human Rights* (2007), 191, at 201–7.

[107] Article 1(1) of the International Covenant on Civil and Political Rights (ICCPR) and Art 1(1) ICESCR. The right to self-determination is also recognized in the African Charter and the Helsinki Act. On the origins of the concept of self-determination and its emergence at national and international levels, see eg R. McCorquodale (ed), *Self-Determination in International Law* (2000); J. Castellino, *International Law and Self-Determination* (2000), at 11–43; A. Cassese, *Self-Determination of Peoples: A Legal Reappraisal* (1995), at 11–65; McCorquodale, 'Human Rights and Self-Determination', in M. Sellers (ed), *The New World Order: Sovereignty, Human Rights and Self-Determination of Peoples* (1996) 9, at 9–34. It has been suggested that the rights-based approach was the appropriate structure in which to consider self-determination. The rights-based approach would afford flexibility in finding solutions for protecting that right and resolving conflicts without dictating one solution. In other words, it would acknowledge the various ways in which the right of self-determination may be exercised. This is not the only conception of the rights-based approach. It is contended here that the rights-based approach cannot be deformed so as to integrate the plurality of situations. See J.S. Anaya, *Indigenous Peoples in International Law* (2004); M.W. Reisman and B.H. Weston (eds), *Towards World Order and Human Dignity* (1976), at 206–7.

[108] The ICJ seemingly recognized peoples as holders of the legal right to self-determination: Legal Consequences for States of the Continued Presence of South Africa in Namibia (South West Africa) notwithstanding Security Council Resolution 276 (1970), 1971 ICJ Rep 16 (Adv Op 21 June). On the issue, see Cassese, *ibid*, at 141 ff. On the customary international law of indigenous peoples, see Anaya, *ibid*, at 61–94.

[109] Committee on Civil and Political Rights, General Comment No 12, The Right to Self-Determination (Article 1), para 2, Twenty-first session, 1984.

[110] The Human Rights Committee has so far always refused to deal with the right to self-determination under the complaints of breaches of Art 1 ICCPR, on the basis that only individuals can bring complaints against a state party before the Committee. See also Ching, 'Indigenous Self-Determination in an Age of Genetic Patenting: Recognizing an Emerging Human Rights Norm' (1997) 66 Fordham L Rev, at 695.

self-determination, while protection of the identity of the group is 'internal' self-determination.[111] In its 'external dimension',[112] self-determination involves major structural and institutional changes in a state and affects groups and individuals in, and even beyond, the state. By contrast, 'internal' self-determination refers to a people's ability to control all aspects of their lives without challenging the territorial integrity of the nation in which the group is located. In its dual dimension, the right to self-determination puts on states an obligation to free peoples or individuals from exploitation.

In the management of genetic resources, the right to self-determination came into play in reaction to biopiracy, ie the fact that corporations from the developed countries claim ownership of, free-ride on, or otherwise take advantage of the genetic resources and indigenous knowledge and technologies of developing countries. With the Human Genome Diversity Project (HGDP),[113] the right of indigenous people to decide whether, and to what extent, to participate in genetic research was inferred from the right to self-determination.[114]

There is no doubt that the concept of the right to self-determination could travel better if separated from some of its underlying justification. The line between the legal and ethical understandings of self-determination has become blurred in the genetic-control context and the right appears here as a 'mixed' (individual and collective) human right which borrows the meaning of the bioethical principle of autonomy. In other words, the issue of biopiracy while addressing the content of the principle (control over genetic identity) generated its conversion into a right and even a mixed (individual and collective) right. It is noteworthy that, at the same time, any research participant can invoke the right to self-determination, in the same way as indigenous peoples participating in research do.

3. Sui Generis *Entitlement*

The proposal for collective informed consent forwarded by the WHO, mentioned above, is difficult to achieve and may not be conceptually appropriate for large populations. It may instead be appropriate to proceed to a collective debate

111 Cassese, *supra* note 108, at 67–133.

112 Rosas, 'The Right to Self-Determination', in Eide *et al* (eds), *supra* note 26, at 111–18, 111–12.

113 Unlike the Human Genome Project, which has attempted to map the entire human genome, the HGDP has attempted to map the DNA that varies between humans, which is a less than 1% difference. It was started by Stanford University's Morrison Institute and led by L. Cavalli-Sforza. Partially funded by UNESCO, these funds were suspended after being decried by representatives of indigenous peoples.

114 Indigenous peoples have objected strenuously to the HGDP. Several indigenous representatives have called on the international community for an end to the HGDP. See for example Discrimination Against Indigenous Peoples: Report of the Working Group on Indigenous Populations on its Fourteenth Session, Commission on Human Rights, Sub-Commission on Prevention of Discrimination and Protection of Minorities, 48th Sess, Agenda Item 14, 27-28, UN Doc E/CN.4/Sub.2/1996/21 (16 August 1996). For a complete list of the instruments adopted on the subject, see Underkuffler, *supra* note 97.

before the start of a project and before individual consents are pursued: in this way, a person can take a decision that does not involve only him/herself as an individual, but will also take into account the dimensions underlined through this collective debate. It is, in this sense, that the 25 EU Recommendations foresee a 'consultation' of the community. Along the same lines, the 2005 Universal Declaration on Bioethics and Human Rights (UDBHR) states that '[i]n appropriate cases of research carried out on a group of persons or a community, additional agreement of the legal representatives of the group or community concerned may be sought.'[115]

To summarize: in this section, the protection of genetic group interests was considered from two different perspectives. The first was the possibility for relatives to access the result of genetic tests, given that they may have a comparable interest to the 'proband' in obtaining it. Some decisions can be found at the national level, which limited individual rights or permitted genetic relatives to obtain communication of these results, from the individual or a third party. The second perspective considered the protection of group interests through a form of collective empowerment. This requires a much more substantial conceptual change, which may help to explain why the measures adopted to date seem akin to 'favours', such as consultation of the community, rather than the recognition of collective rights.

3. Conclusion

This chapter has provided an overview of the main developments occurring in international human rights law in reaction to the growth of genetic health practices. In so doing, it has emphasized that the process of 'geneticization' of rights (ie the reference to the 'right to genetic identity', the 'right to genetic privacy', and the 'right to genetic confidentiality') that is taking place in legislation and literature does not refer to any new right but highlights the new context in which the 'old rights' (ie rights enshrined in human rights treaty or customary international law) are to be protected. It has also been argued that although international human rights law is widely seen as a system of individual entitlements to oppose state power, the debate has moved on with genetics. Informed consent and human rights likely to embody the requirement of individual entitlements do not exhaust the protection of all interests. In fact, family, communities, and the human species arguably have interests distinct from that of society or of the individual, thereby providing a justification for the partial refinement of the rationale of international human rights law and the search for additional models of protection.

[115] Article 6(3) UDBHR: 'In no case should a collective community agreement or the consent to a community leader or other authority substitute for an individual's informed consent.'

Particular attention was given to two points. First, the fact that the rationale of human rights allocation has changed and the individual is empowered not only as such but also as a depositary of the genetic heritage of the human species. As a result, the process of individualization usually associated with human rights is not at odds with 'collectivizing moments' whereby the individual is protected as part of the whole. In 'individualizing moments', the rationale of human rights has not changed and the emphasis is placed on the form of protection of the rights, namely genetic counselling for the protection of the right to know and anonymization of data for the protection of the right to confidentiality. Because of the highly predictive power given to genetic information, these issues have attracted tremendous attention. In 'collectivizing moments', however, human rights are recognized with a trans-temporal approach which departs from the traditional model whereunder acts are captured only in their past and present consequences. Legal anticipation is realized through 'old' human rights tools (namely, the principle of non-discrimination and the prohibition of practices contrary to human dignity) and it is also aligned with the more and more often claimed right to the uniqueness of the embryo and the right to genetic integrity. All these constructions protect interests that cross over time and space.

Secondly, the human rights approach is abandoned when it appears insufficient to protect interests of genetic relatives from the desiderata of the 'proband'. However, to date, the recognition of *sui generis* entitlements is nothing more than a 'favour' and in no way a great achievement. Of more significance is the intervention of the state in the control of genetic information. Interposing an independent regulatory body, like in France, in the communication process of the result of a genetic test to relatives is a remarkable illustration of the 'publicization' of the private sphere. It is a strong reminder of the role of the state in preserving social order and preventing the inter-familial tensions that genetic testing can create.

Overall, the chapter has depicted how international human rights law has unarguably expanded to the new branch of biomedicine that constitutes genetic practices. An account was given of the extent to which this extension has created problems for human rights thinking and practice. In fact, international human rights law was not originally designed for the protection of the common good and still hardly copes with future and group interests. On the one hand, the new context offered by genetic practices exacerbates old issues. On the other hand, the interplay between ethical and legal instruments in this specific field offers a unique regulatory context in which the new normative constructions that were presented can be allowed to develop.

Index

Abortion
citizen participation 219
reproductive choice 197–200
Access to information
see also **Data protection; Privacy**
Aarhus Convention
fundamental principles 185–186
scope 184–185
special status of GMOs 186–187
protection of genetic relatives 263
risk analysis 179–184
Accountability
reproductive choice 202–203
threat of new technologies 190–191
Assisted reproductive technologies
choice
abortion argument 197–200
bad argument 196–197
child welfare 209–210
conflict with success rates 210–213
gender issues 204–209
linkage with genetic technologies 200–202
overview 195–196
PGD compared 216–218
for whom? 202–204
citizen participation 219–220
defining the new technologies 16

Bioethics *see* **Ethics**
Biotechnologies
bioethical triangle
dignitarian alliance 25–26
human rights approach 23–25
introduction 2, 20
UNESCO's adoption of Universal Declaration 21–22
utilitarianism 22–23
conclusions 83–84
consent 35–37
explicit regulatory standards 61–71
harm principle 33–35
implicit regulatory standards 71–76
internal regulatory standards 60–61
key approaches 39
PGD and PTT
UKHL judgment 55–60
overview 50–51, 60
Science and Technology Committee response 51–55
precautionary principle 37–39
Cartagena Protocol 169–174
regulation of genetics
bioethical triangle 103–115
conclusions 126–127
legislative challenge 94–103
Dutch Minister's opinion reviewed 116–126
need for critical assessment 85–88
PGD and neo-eugenics 88–94
regulatory legitimacy
overview 40–41
PGD and PTT 50–60
therapeutic cloning 41–50
rights revolution 19–20, 76–83
role of human dignity
constraint 28–29
empowerment 26–27

Cartagena Protocol
genetically modified organisms 181–182
risk criteria for precautionary principle 169–174
Child welfare 209–210
Choice
genetic privacy
best interests limitation 253–254
individual as information-recipient 252–253
individual as information-source 255–256
right not to know 254–255
reproductive technologies
abortion argument 197–200
bad argument 196–197
child welfare 209–210
conclusions 220–221
conflict with success rates 210–213
ethnographic studies 213–220
gender issues 204–209
linkage with genetic technologies 200–202
overview 195–196

Choice *(cont)*
speakability 218–220
for whom? 202–204
sex selection 124–125
Citizen participation
constitutional patriotism
basis of united Europe 128–131
conclusions 159–160
constitutional culture of privacy 143–155
meaning and scope 131–134
right to privacy 134–143
tolerant and intolerant constitutional patriotism 159
genetic technologies 17–18
indigenous peoples 266–267
introduction 6–7
PGD 93–94
precautionary principle 16
democratic governance 190–191
general principle 178–179
genetically modified organisms 184–187
interaction between science and politics 179–184
litigation practice 187–190
reproductive choice 218–220
reproductive technologies 16–17
vulnerable groups 241–244
Cloning
fear of cloning 3–4
human dignity 111
protection of diversity 256
regulatory legitimacy
moral status of embryo 45–48
overview 41–42
permissive approach 42–43
preferred regulatory positions 43–45
Select Committee response to objections 48–50
Codex Alimentarius 183–184
Collateral effects 162
Collective rights
genetic technologies
candidate human rights 248–249
conclusions 270–271
epidemiological research 265–270
genetic privacy 252–256
identity protection 249–252
overview 246–248
protection of diversity 256–259
protection of family members 261–265
indigenous peoples 266–267
Commercialization 29
Commodification 29
Confidentiality *see* **Privacy**
Consent
bioethical requirement 249
genetic screening 232
international law of genetic discrimination 226
protection of genetic relatives 264–265
utilitarianism 35–36
Constitutional patriotism
basis of united Europe 128–131
conclusions 159–160
constitutional culture of privacy 143–155
meaning and scope 131–134
right to privacy 134–143
tolerant and intolerant constitutional patriotism 159
Courts *see* **Review bodies**
Criminal investigation
DNA forensics 237–240
proteomic data and samples 241
'Cultural lives'
introduction 2–3

Damage criteria 174
Data protection
see also **Privacy**
additional EU legislation 142–143
Art 8 136–139
conflict with freedom of expression 144–151
DNA databases 77–78
fundamental rights 139
genetic discrimination 228–229
protection of genetic relatives
access to medical data 263
disclosure in interests of others 262–263
overview 261–262
Deliberative democracy *see* **Citizen participation**
Democracy
demanding form of rule 132–133
threat of new technologies 190–191
Different capacities principle 178
Digital technologies 7–8
Dignitarian alliance
bioethical triangle 25–26
introduction 4
key approaches 39
rights revolution 14

Dignity *see* **Human dignity**
Discrimination
  international law of genetic discrimination
    challenge of privacy 231–233
    conclusions 244–245
    criminal investigation 237–241
    genetic discrimination defined 224–225
    insurance and employment 233–237
    international legal instruments 225–231
    overview 222–224
    vulnerable groups 241–244
  protection of genetic diversity
    non-modified heritage 257
    practices contrary to human dignity 257
    principle of non-discrimination 258–259
    unique identity 257–258
  reproductive choice
    abortion argument 197–200
    bad argument 196–197
    child welfare 209–210
    conclusions 220–221
    conflict with success rates 210–213
    ethnographic studies 213–220
    gender issues 204–209
    linkage with genetic technologies 200–202
    overview 195–196
    speakability 218–220
    for whom? 202–204
DNA databases
  consent 248
  data protection 77–78
  genetic discrimination
    conclusions 245
    criminal investigation 237–240
    employment screening 237
  international law 224

'East Coast' model 81
Embryology
  legitimacy of stem cell research
    moral status of embryo 45–48
    overview 41–42
    permissive approach 42–43
    preferred regulatory positions 43–45
    Select Committee response to objections 48–50
  new meanings 10–11
  PGD and PTT
    UKHL judgment 55–60
    overview 60
    Science and Technology Committee response 51–55
Employment
  genetic discrimination 236–237
Empowerment 26–27
Environmental protection
  genetically modified organisms 95
  human dignity 31–32
  precautionary principle
    fundamental principles 185–186
    litigation practice 188
    scope 184–185
    special status of GMOs 186–187
Epidemiological research 265–270
Ethics
  biotechnologies
    conclusions 83–84
    consent 35–37
    explicit regulatory standards 61–71
    harm principle 33–35
    implicit regulatory standards 71–76
    internal regulatory standards 60–61
    PGD and PTT 50–60
    precautionary principle 37–39
    rights revolution 76–83
    three-way plurality 20
    two-dimensional approach 32–33
  consent
    bioethical requirement 249
    genetic screening 232
    international law of genetic discrimination 226
    utilitarianism 35–36
  genetic technologies
    human dignity 106–108
    human rights approach 104–105
    utilitarianism 103–104
  reproductive choice
    abortion argument 197–200
    bad argument 196–197
    child welfare 209–210
    conclusions 220–221
    conflict with success rates 210–213
    ethnographic studies 213–220
    gender issues 204–209
    linkage with genetic technologies 200–202
    overview 195–196
    speakability 218–220
    for whom? 202–204
Ethnography
  importance 220
  PGD and other ARTs compared 216–218
  reproductive choice 'in action' 213–216

Eugenics
association with PGD 91–93
international law of genetic discrimination
challenge of privacy 231–233
conclusions 244–245
criminal investigation 237–241
genetic discrimination defined 224–225
insurance and employment 233–237
international legal instruments 225–231
overview 222–224
vulnerable groups 241–244
introduction 17
need for critical assessment 85–88
Exceptionalism
conflict between privacy and freedom of expression 144–151
human dignity as constraint 28–29
introduction 4–5
morality exclusion
market access 65–71
patentability 63–65
proportionality 152
public interest 152
Sunday trading 148
Exclusionary technologies
'East coast' model 81
'West coast' model 81–83
Exclusivity principle
bioethical triangle 63–65
market access 65–71
patentability 61–63

Fear
eugenics 87
genetically modified organisms 100
introduction 3–4
Fiction science 3–4
Fundamental rights
conflict between privacy and freedom of expression 147
data protection 139

Genetic technologies
Aarhus Convention
fundamental principles 185–186
scope 184–185
special status of GMOs 186–187
citizen participation 17–18
collective and individual interests compared
candidate human rights 248–249
conclusions 270–271
epidemiological research 265–270
genetic privacy 252–256
identity protection 249–252
overview 246–248
protection of diversity 256–259
protection of family members 261–265
international law of genetic discrimination
challenge of privacy 231–233
conclusions 244–245
criminal investigation 237–241
genetic discrimination defined 224–225
insurance and employment 233–237
international legal instruments 225–231
overview 222–224
vulnerable groups 241–244
introduction 6
PGD and PTT
UKHL judgment 55–60
overview 60
Science and Technology Committee response 51–55
precautionary principle
interaction between science and politics 182
regulation
bioethical triangle 103–115
conclusions 126–127
legislative challenge 94–103
Dutch Minister's opinion reviewed 116–126
need for critical assessment 85–88
PGD and neo-eugenics 88–94
repetition 8
Genetically modified organisms
Aarhus Convention
fundamental principles 185–186
scope 184–185
special status of GMOs 186–187
Cartagena Protocol 181–182
defining the new technologies 7–8
environmental protection 95
extra-territorial effects 109
fear 87
importance of transparency 116
precautionary principle 38, 112, 182
rejection of safety measures 99–100
risk regulation 17–18
Genome *see* **Human genome**

Harm principle
human rights approach 34–35
reproductive choice 195–196

UNESCO 23
utilitarianism 33–34
HLA-typing 91, 122–124
Hope 2–3
Human dignity
bioethical triangle
dignitarian alliance 25–26
human rights approach 23–25
introduction 20
UNESCO's adoption of Universal Declaration 21–22
utilitarianism 22–23
bioethics
consent 37
harm principle 35
key approaches 39
precautionary principle 38–39
cloning 111
environmental protection 31–32
exclusivity principle
market access 65–71
patentability 63–65
free movement of services 148–149
genome regulation 228
introduction 4, 17
pre-implantation genetic diagnosis 106–108, 110–114
protection of genetic diversity 257
reference points
alternative approaches 29–30
tension between different approaches 30–31
regulatory legitimacy of stem cell research
overview 41–42
permissive approach 42–43
preferred regulatory positions 43–45
role in biotechnologies
constraint 28–29
empowerment 26–27
stem cell research
moral status of embryo 45–48
Select Committee response to objections 48–50
Human genome
collective and individual interests in genetics
introduction 246–248
first regulatory instruments 228
genetic privacy
best interests limitation 253–254
individual as information-recipient 252–253
individual as information-source 255–256
right not to know 254–255
growing cause for concern 226–227
international law 223–224
protection of genetic diversity
non-modified heritage 257
practices contrary to human dignity 257
principle of non-discrimination 258–259
unique identity 257–258
Human rights
bioethics
bioethical triangle of bioethics 23–25
consent 36–37
exclusivity principle 63–65
harm principle 34–35
legitimacy of stem cell research 41–50
pre-implantation genetic diagnosis 104–105, 114–115
precautionary principle 38–39
two-dimensional approach 32–33
genetic technologies
candidate rights 248–249
conclusions 270–271
epidemiological research 265–270
genetic privacy 252–256
identity protection 249–252
overview 246–248
protection of diversity 256–259
protection of family members 261–265
need for reworking 18
repetition 7–8
'what works' 15–16
Hype *see* **fear**

Identity protection
individual and human right 250–251
right to uniqueness 251–252
Indigenous peoples
group rights 266–267
overview 265–266
self-determination 268–269
sui generis entitlement 269–270
Individual rights
candidate human rights 248–249
conclusions 270–271
epidemiological research 265–270
genetic privacy 252–256
identity protection 249–252
protection of diversity 256–259
protection of family members 261–265
Insurance
genetic discrimination 233–236

International law *see* **Regulation**
Intracytoplasmic sperm injection 205
IVF
abortion argument 198–200
additional procedure 119
linkage with genetic technologies 200–202
prenatal tests for PGD 89
repetition 8
reproductive revolution 6
risk 90
success rates 210–213

Judicial deference 157
Juridicization
new modes of action 167–168
new modes of thought 166–167

Law *see* **Regulation**
Legitimacy
difficult questions 9
embryology
moral status of embryo 45–48
overview 41–42
permissive approach 42–43
preferred regulatory positions 43–45
Select Committee response to objections 48–50
genetic technologies 95–102
introduction 20
Life, right to
ethical plurality 71–76
Linkage
citizen participation and precautionary principle 179
introduction 1–2

Margin of appreciation
abandonment of rights 156
ECJ approach 149–150
ECtHR approach 151–152
universal moral core 157–158
Market access 65–71

Nanotechnologies
importance 20
introduction 8
Neurotechnologies 20
New technologies
collateral effects 161
defined 161
repetition 8
risk 162–163

Panopticon technologies 79–81
Patentability 61–63
Pre-implantation genetic diagnosis
applications 88–89
association with eugenics 91–93
carriers 119–120
citizen participation 93–94
defining the new technologies 7–8
global regulation 87–88
UKHL judgment 55–60
human rights as metanorm 108–115
lessons for regulating 102–103
linkage with genetic technologies 201
Dutch Minister's opinion reviewed
availability 119–125
critical analysis 125–126
ethical plurality 117–119
history 116
importance of transparency 116–117
overview 60
reproductive choice
ARTs compared 216–218
choice in action 213–216
risk regulation 17–18, 89–91
Science and Technology Committee response 51–55
success rates 210–213
techniques 89
Pre-implantation genetic screening
availability 119–125
critical analysis 125–126
ethical plurality 117–119
history 116
importance of transparency 116–117
Pre-implantation tissue-typing
UKHL judgment 55–60
overview 60
Science and Technology Committee response 51–55
Precautionary principle
advantages in international law
anticipatory approach 166
emphasis on interdependence 165
new approach 163–164
regulating uncertainty 164–165
approach to ethics 32–33
bioethics
human rights approach 38–39
overview 37–38
utilitarianism 38
citizen participation 16
democratic governance 190–191

general principle 178–179
genetically modified organisms 184–187
interaction between science and politics 179–184
litigation practice 187–190
fundamental criteria
damage criteria 174
different capacities principle 178
risk criteria 168–174
scientific uncertainty principle 174–178
PGD 95–102
regulation of new concerns
new modes of action 167–168
new thought methods 166–167
rights revolution 12–14
societal dimension 191–193
World Trade Organization 115
Privacy
see also **Access to information; Data protection**
Art 8 ECHR 135–136
comparisons of ECtHR and ECJ
constitutional culture 143
ECJ approach 144–151
ECtHR approach 151–155
conclusions 159–160
data protection
additional EU legislation 142–143
Art 8 136–139
fundamental rights 139
development 130
genetic privacy
best interests limitation 253–254
individual as information-recipient 252–253
individual as information-source 255–256
right not to know 254–255
national diversity of approach 129
protection against genetic discrimination 231–233
reproductive choice 207–208
tolerant and intolerant constitutional patriotism compared 155–159
vehicle for exploring tensions 134–135
Proportionality 152
Public interest 152
Public participation *see* **Citizen participation**

Regulation
bioethical triangle
dignitarian alliance 25–26
human rights approach 23–25
introduction 2, 20
UNESCO's adoption of Universal Declaration 21–22
utilitarianism 22–23
bioethics
conclusions 83–84
explicit regulatory standards 61–71
implicit regulatory standards 71–76
internal regulatory standards 60–61
overview 40–41
PGD and PTT 50–60
rights revolution 76–83
therapeutic cloning 41–50
genetic technologies
bioethical triangle 103–115
conclusions 126–127
legislative challenge 94–103
Dutch Minister's opinion reviewed 116–126
need for critical assessment 85–88
PGD and neo-eugenics 88–94
international advantages of precautionary principle
anticipatory approach 166
emphasis on interdependence 165
new approach 163–164
regulating uncertainty 164–165
international law of genetic discrimination
challenge of privacy 231–233
conclusions 244–245
criminal investigation 237–241
genetic discrimination defined 224–225
insurance and employment 233–237
international legal instruments 225–231
overview 222–224
vulnerable groups 241–244
legitimacy of stem cell research
moral status of embryo 45–48
overview 41–42
permissive approach 42–43
preferred regulatory positions 43–45
Select Committee response to objections 48–50
linkage 1–2
new concerns of precautionary principle
new modes of action 167–168
new thought methods 166–167
protection of genetic relatives
disclosure of tests 261–265
epidemiological research 265–270
overview 259–261
repetition 8–9

Regulation *(cont)*
rights revolution 14–15
role of human dignity 28–29
Repetition
courts 9–10
human rights definitions 7–8
new technologies definitions 8
regulation 8–9
Reproductive technologies
ART
choice 195–218
citizen participation 219–220
defining the new technologies 16
choice
abortion argument 197–200
bad argument 196–197
child welfare 209–210
conclusions 220–221
conflict with success rates 210–213
ethnographic studies 213–220
gender issues 204–209
linkage with genetic technologies 200–202
overview 195–196
speakability 218–220
for whom? 202–204
citizen participation 16–17
fear of cloning 3–4
introduction 5–6
new meanings 10–11
PGD
applications 88–89
association with eugenics 91–93
carriers 119–120
citizen participation 93–94
defining the new technologies 7–8
global regulation 87–88
UKHL judgment 55–60
human rights as metanorm 108–115
lessons for regulating 102–103
linkage with genetic technologies 201
Dutch Minister's opinion reviewed 116–126
overview 60
reproductive choice 213–8
risk regulation 17–18, 89–91
Science and Technology Committee response 51–55
success rates 210–213
techniques 89
PGS
availability 119–125
critical analysis 125–126
ethical plurality 117–119
history 116
importance of transparency 116–117
PTT
UKHL judgment 55–60
overview 60
Science and Technology Committee response 51–55
repetition 8
rights revolution
rogue candidates 11–12
Research and development
bioethical triangle 2
collective and individual interests in genetics
candidate human rights 248–249
conclusions 270–271
epidemiological research 265–270
genetic privacy 252–256
identity protection 249–252
overview 246–248
protection of diversity 256–259
protection of family members 261–265
indigenous peoples
group rights 266–267
overview 265–266
self-determination 268–269
sui generis entitlement 269–270
legitimacy of stem cell research
moral status of embryo 45–48
overview 41–42
permissive approach 42–43
preferred regulatory positions 43–45
Select Committee response to objections 48–50
Review bodies
comparisons of ECtHR and ECJ
basis of united Europe 128–131
conclusions 159–160
constitutional culture of privacy 143–155
constitutional patriotism 131–134
right to privacy 134–143
tolerant and intolerant constitutional patriotism 159
difficult questions of legitimacy 9–10
freedom of expression
ECJ approach 143–151
ECtHR approach 151–155
PGD and PTT 55–60
precautionary principle
damage criteria 174
different capacities principle 178
litigation practice 187–190

risk criteria 169–174
scientific uncertainty principle 175–178
regulation of genetics
immense challenge 94–95
international legitimacy 95–102
repetition 9–10
reproductive choice 206–209
risk criteria for precautionary principle 169–174
Rights revolution
biotechnologies 19–20
dignitarian alliance 14
importance 10
introduction
citizen participation 6–7
exceptionalism 4
genetic technologies 6
reproductive technologies 5–6
precautionary principle 12–14
regulation 14–15
reproductive technologies
new meanings 10–11
rogue candidates 11–12
Risk
criteria for precautionary principle
components 169–174
damage criteria 174
meaning and content 168–169
introduction 4
IVF 90
new technologies 162–163
PGD 89–91
scientific uncertainty principle compared 175–178

Scientific uncertainty principle
content and meaning 174–175
risk analysis compared 175–178
Sex selection 124–125
Speakability 218–220
Stem cell research
regulatory legitimacy
moral status of embryo 45–48
overview 41–42
permissive approach 42–43
preferred regulatory positions 43–45
Select Committee response to objections 48–50
reproductive revolution 6
Stratified reproduction 203

Therapeutic cloning
moral status of embryo 45–48
overview 41–42
permissive approach 42–43
preferred regulatory positions 43–45
Select Committee response to objections 48–50

Uncertainty
international law 164–165
precautionary principle 179
scientific uncertainty principle
content and meaning 174–175
risk analysis compared 175–178
threat to modern society 190–191
Unequal choice *see* **Choice**
UNESCO
bioethical triangle 21–22
consent in biotechnologies 35
genetic discrimination
genome regulation 229
vulnerable groups 242
harm principle 23
introduction 2
lack of unanimity among regulators 114
lead in bioethics 105
utilitarianism 22–23
Utilitarianism
consent 35–36
exclusivity principle
market access 65–71
patentability 63–65
harm principle 33–34
limited approach 32–33
meaning and scope 22–23
pre-implantation genetic diagnosis 103–104, 109–110
precautionary principle 38
regulatory legitimacy of stem cell research
moral status of embryo 45–48
overview 41–42
permissive approach 42–43
preferred regulatory positions 43–45
stem cell research 48–50

Vulnerable groups
genetic discrimination 241–244

Warnock Committee 48–50
'West Coast' model 81–83
'What works' 15–16

World Trade Organization
access to medicines 11
bioethics 71
cosmopolitan minimal standards 93–94
exceptionalism 66
importance of transparency 116
legitimacy 95–102
precautionary principle 115
regulation of eugenics 87–88
risk analysis 175–178
risk regulation 17–18

Lightning Source UK Ltd.
Milton Keynes UK
UKOW05n1928230414

230469UK00001B/37/P

9 780199 562572